MW00604382

Bayesian Networks and Probabilistic Inference in Forensic Science

Statistics in Practice

Statistics in Practice is an important international series of texts, which provide detailed coverage of statistical concepts, methods and worked case studies in specific fields of investigation and study.

With sound motivation and many worked practical examples, the books show in down-to-earth terms how to select and use an appropriate range of statistical techniques in a particular practical field within each title's special topic area.

The books provide statistical support for professionals and research workers across a range of employment fields and research environments. The subject areas covered include medicine and pharmaceutics; industry, finance and commerce; public services; the earth and environmental sciences, and so on.

The books also provide support to students studying statistical courses applied to the above areas. The demand for graduates to be equipped for the work environment has led to such courses becoming increasingly prevalent at universities and colleges.

It is our aim to present judiciously chosen and well-written workbooks to meet everyday practical needs. The feedback of views from readers will be most valuable to monitor the success of this aim.

A complete list of titles in this series appears at the end of the volume.

Bayesian Networks and Probabilistic Inference in Forensic Science

Franco Taroni
University of Lausanne, Switzerland

Colin Aitken
University of Edinburgh, UK

Paolo Garbolino
IUAV University, Venice, Italy

Alex Biedermann
University of Lausanne and Federal Office of Police, Berne, Switzerland

John Wiley & Sons, Ltd

Telephone (+44) 1243 779777

Email (for orders and customer service enquiries): cs-books@wiley.co.uk
Visit our Home Page on www.wiley.com

Reprinted November 2006

Other Wiley Editorial Offices

John Wiley & Sons Inc., 111 River Street, Hoboken, NJ 07030, USA

Jossey-Bass, 989 Market Street, San Francisco, CA 94103-1741, USA

Wiley-VCH Verlag GmbH, Boschstr. 12, D-69469 Weinheim, Germany

John Wiley & Sons Australia Ltd, 42 McDougall Street, Milton, Queensland 4064, Australia

John Wiley & Sons (Asia) Pte Ltd, 2 Clementi Loop #02-01, Jin Xing Distripark, Singapore 129809

John Wiley & Sons Canada Ltd, 22 Worcester Road, Etobicoke, Ontario, Canada M9W 1L1

Wiley also publishes its books in a variety of electronic formats. Some content that appears in print may not be available in electronic books.

Library of Congress Cataloging-in-Publication Data

Bayesian networks and probabilistic inference in forensic science / Franco Taroni ... [et al.].
p. cm. – (Wiley series on statistics in practice)
Includes bibliographical references and indexes.
ISBN 0-470-09173-8 (alk. paper)
1. Bayesian statistical decision theory – Graphic methods. 2. Uncertainty (Information theory) – Graphic methods. 3. Forensic sciences – Graphic methods. I. Taroni, Franco. II. Statistics in practice.

QA279.5.B3887 2006
363.2501′519542 – dc22

2005057711

British Library Cataloguing in Publication Data

A catalogue record for this book is available from the British Library

ISBN-13: 978-0-470-09173-9 (HB)

Typeset in 10/12pt Times by Laserwords Private Limited, Chennai, India
Printed and bound in Great Britain by Antony Rowe Ltd, Chippenham, Wiltshire
This book is printed on acid-free paper responsibly manufactured from sustainable forestry in which at least two trees are planted for each one used for paper production.

To Our Families

Contents

Preface

Forensic scientists may have hardly ever been able to gather and offer as much information, analytical or otherwise, as is possible today. Owing to advances made in science and technology, today's forensic scientists can choose amongst a broad scope of methods and techniques, applicable to various kinds of evidence even in their remotest quantities. However, despite the potentially abundant amount of available information, there now is an increased awareness amongst a significant part of the forensic community – including legal scholars – that there are risks associated with there is a sense of some sort of overconfidence, for example.

Scientific evidence as encountered in the real world is always incomplete to some degree, thus uncertainty is a prevalent element and one with which forensic scientists have to deal. Evidence does not say anything in itself; its significance needs to be elucidated in the light of competing propositions and background knowledge about the case at hand. There is a great practical necessity for forensic scientists to advise their clients, be they lawyers, prosecutors, jurors or decision makers at large, of the significance of their findings. Forensic scientists are required to qualify and, where possible, quantify their states of knowledge and to be consultants in the assessment of uncertainties associated with the inferences that may be drawn from forensic evidence.

For this task, forensic scientists should consider probability theory as the fundamental concept to govern their reasoning. The aim of this book will be to show that the practical application of probabilistic reasoning in forensic science can be assisted and its rationale substantially clarified if it is conducted in a graphical environment; *i.e.*, conducted through the use of a formalism known as Bayesian networks.

Thus, the idea for a book on *Bayesian networks and probabilistic inference in forensic science* is guided by a series of questions currently asked by forensic scientists and other participants in the criminal justice system. The aim is to offer theoretical and practical elements to help solve the following questions.

- What are the relationships among a set of (usually unobservable) causes and a set of (observable) scientific evidence?

- What are the structural relationships among arguments based on different kinds of evidence in an inference to one or more propositions of interest?

- How can we construct coherent, credible and defensible arguments in reasoning about the evidential value of scientific evidence?

- Given that the known set of evidence is not sufficient to determine the cause(s) of its origins with certain degrees of certainty; *i.e.*, given that inductive inferences are risky, what additional information should be obtained?

- What is the value of each of these additional pieces of information?

- Can we build expert systems to guide forensic scientists and other actors of the criminal justice system in their decision making about forensic evidence?

- How can one collect, organise, store, update and retrieve forensic information (hard data or linked with expert judgement) in expert systems?

The current state of the art in forensic science, notably in scientific evidence evaluation, does not allow scientists to cope adequately with the problems caused by the complexity of evidence (typically the combination of evidence, especially if contradictory) even if such complexity occurs routinely in practice.

Methods of formal reasoning have been proposed to assist the forensic scientist to understand all of the dependencies which may exist between different aspects of evidence and to deal with the formal analysis of decision making. Notably, graphical methods, such as Bayesian networks, have been found to provide valuable assistance for the representation of the relationships amongst characteristics of interest in situations of uncertainty, unpredictability or imprecision. Recently, several researchers (mainly statisticians) have begun to converge on a common set of issues surrounding the representation of problems which are structured with Bayesian networks. Bayesian networks are a widely applicable formalism for a concise representation of uncertain relationships among parameters in a domain (in this case, forensic science). The task of developing and specifying relevant equations can be made invisible to the user and the arithmetic can be almost completely automated. Most importantly, the intellectually difficult task of organising and arraying complex sets of evidence to exhibit their dependencies and independencies can be made visual and intuitive. Bayesian networks are a method for discovering valid, novel and potentially useful patterns in data where uncertainty is handled in a mathematically rigorous, but simple and logical, way. A network can be taken as a concise graphical representation of an evolution of all possible stories related to a scenario. However, the majority of scientists are not familiar with such new tools for handling uncertainty and complexity.

Thus, attention is concentrated here on Bayesian networks essentially because they are relatively easy to develop and – from a practical point of view – they also allow their user to deduce the related formulae (expressed through likelihood ratios) for the assessment of scientific evidence. Examples of the use of Bayesian networks in forensic science have already been presented in several papers. In summary, the use of Bayesian networks has some key advantages that could be described as follows:

- the ability to structure inferential processes, permitting the consideration of problems in a logical and sequential fashion;

- the requirement to evaluate all possible narratives;

- the possibility to calculate the effect of knowing the truth of one proposition or piece of evidence on the plausibility of others;

- the communication of the processes involved in the inferential problems to others in a succinct manner, illustrating the assumptions made;
- the ability to focus the discussion on probability and underlying assumptions.

A complete mastery of these aspects is fundamental to the work of modern forensic scientists.

The level of the book is the same as that of Aitken and Taroni (2004), namely those with a modest mathematical background. Undergraduate lawyers and aspiring forensic scientists attending a course in evidence evaluation should be able to cope with it though it would be better appreciated by professionals in law or forensic science with some experience of evidence evaluation.

The aim of the authors is to present a well-balanced book which introduces new knowledge and challenges for all individuals interested in the evaluation and interpretation of evidence and, more generally, the fundamental principles of the logic of scientific reasoning. These principles are set forth in Chapter 1. Chapter 2 shows how they can be operated within a graphical environment – Bayesian networks – with the reward of being applicable to problems of increased complexity. The discussion of the logic of uncertainty is then continued in the particular context of forensic science (Chapter 3) with studies of Bayesian networks for dealing with general issues affecting the evaluation of scientific evidence (Chapter 4). Later chapters will focus on more specific kinds of forensic evidence, such as DNA (Chapter 5) and transfer evidence (Chapter 6). Bayesian network models studied so far will then be used for the analysis of aspects associated with the joint evaluation of scientific evidence (Chapter 7). In Chapter 8 the discussion will focus on case-preassessment, where the role of the forensic scientists consists of assessing the value of expected results *prior* to laboratory examination. Here, Bayesian networks will be constructed to evaluate the probability of possible outcomes in various cases together with their respective weight. Chapter 9 will emphasise the importance of the structural dependencies among the basic constituents of an argument. It will be shown that qualitative judgements may suffice to agree with the rules of probability calculus and that reasonable ideas can be gained about a model's properties through sensitivity analyses. The book concludes with a discussion of the use of continuous variables (Chapter 10) and further applications including offender profiling and Bayesian decision analysis (Chapter 11).

An important message of the present book is that the Bayesian network formalism should primarily be considered as an aid to structure and guide one's inferences under uncertainty, rather than a way to reach 'precise numerical assessments'. Moreover, none of the proposed models is claimed to be, in some sense, 'right'; a network is a direct translation of one's subjective viewpoint towards an inference problem, which may be structured differently according to one's extent of background information and knowledge about domain properties. It is here that a valuable property of Bayesian networks comes into play: they are flexible enough to accommodate readily structural changes whenever these are felt to be necessary.

The authors believe that their differing backgrounds (*i.e.*, forensic science, statistics and philosophy of science) have common features and interactions amongst them that enable the production of results that none of the disciplines could produce separately. A book on this topic aims to offer insight not only for forensic scientists but also for all persons faced with uncertainty in data analysis.

We are very grateful to Glenn Shafer for his permission to use his own words as the heading of Section 1.1.1, to Michael Dennis for help with Chapter 10 and to Silvia Bozza for commenting on Section 11.2. We thank Hugin Expert A/S who provided the authors with discounted copies of the software to enable them to develop the networks described throughout the book. Other software used throughout the book are R, a statistical package freely available at `www.r-project.org`, and XFIG, a drawing freeware running under the X Window System and available at `www.xfig.org`.

One of us, Paolo Garbolino, has been supported by research grant COFIN 2003 2003114293, "Probabilistic judgments and decision processes" of the Italian Ministry of Education, University and Research.

We also express our appreciation to the Universities of Lausanne and Edinburgh and the IUAV University of Venice for support during the writing of the book.

F. Taroni, C.G.G. Aitken, P. Garbolino, A. Biedermann
Lausanne, Edinburgh, Venezia, Lausanne

Foreword

In the past we have been far more adept at gathering, transmitting, storing and retrieving information than we have been at making sense out of a mass of information and drawing defensible and persuasive conclusions from it. However, significant progress has been made in recent years in efforts to close this important methodological gap. The authors of this book have a long and distinguished record of success in their attempts to close this gap in the forensic sciences. Forensic scientists, in common with investigators in so many other areas, have the task of establishing the relevance, credibility and inferential or probative force of the various kinds of information they will take as evidence in probabilistic inferences of concern to them. The establishment of these three important credentials of evidence rests on arguments constructed to show how probabilistic variables and their evidential bases are linked in certain ways. These arguments can indeed be complex and are commonly called inference networks.

This book concerns a class of methods for the generation and probabilistic analysis of complex inference networks. The first basis for these methods consists of the generation of graphical representations, in the form of directed acyclic graphs, showing what the persons generating the inference network believe to be patterns of probabilistic linkages amongst the variables or propositions of interest in the situation being represented. As the authors note, the generation of these graphical representations rests upon the knowledge and imagination of the persons generating the inference network. Critical reasoning is also involved as the persons generating the inference network attempt to avoid illogical connections or *non sequiturs* in the network under construction. As the authors acknowledge, and for which they supply numerous examples, an inference network can often be constructed in different ways to emphasise various different distinctions that might be important in the problem at hand. Particularly good examples are provided in their discussion of DNA evidence in Chapter 5. So, one way to view a generated inference network is that it represents an argument showing how the elements of some complex process are linked in their influence upon one or more variables that are of basic interest.

The method for probabilistic analysis of inference networks discussed in this work involves applications of Bayes' rule. Consequently, these networks are commonly called *Bayesian networks*. In most of their discussions of applications of Bayesian networks in the analysis of a wide assortment of different forms of trace evidence of interest in the forensic sciences, the authors wisely focus on likelihoods and their ratios, the crucial ingredients of Bayes' rule that concern the inference or probative force of evidence. In their discussions of the force of evidence they acknowledge that some of the probabilistic ingredients of the force of evidence may have a basis in statistical relative frequencies, but other probabilistic ingredients will rest upon epistemic, judgmental or subjective estimates. Chapters 1 through

3 of this book concern basic tutorial discussions of the graphical and probabilistic elements of Bayes' nets. In these introductory discussions the authors use some examples from Sherlock Holmes stories to begin their explanation of the variety of concepts that are part and parcel of Bayesian inference networks. The inferential issues here are often subtle or complex. Consequently, the terminology that has arisen in the analysis of Bayesian networks is complex, particularly with reference to the graphical structures that can emerge in such analyses. Careful attention to these tutorial comments in Chapters 1 through 3 will be well rewarded as you proceed.

In Chapters 4 through 7 the authors provide an array of examples concerning the construction and analysis of Bayes' nets for a variety of forms of trace evidence including footwear marks, stain evidence, fibre traces, DNA evidence related to various crimes or disputed parentage, handwriting and fingermarks evidence, and ballistics evidence. The authors' discussion of these applications demonstrates their considerable expertise in understanding the complexity of the inferential issues surrounding these forms of trace evidence. But their discussions of these matters bring to my mind one of the most important features of this book. Most prior works on Bayes's nets leave out what I regard as a major virtue of performing the kinds of analyses the authors advocate in this work. What is so often overlooked is the heuristic value of such analyses in the enhancement of investigative or discovery-related processes in the forensic sciences and in so many other contexts of which I am aware. This kind of analysis, including their probabilistic elements, can prompt you to ask questions you might not have thought of asking if you did not perform this detailed analysis. In short, Bayesian analysis can help you open up new lines of inquiry and new potentially valuable evidence. These matters are well documented in Chapter 7, regarding what the authors call Pre-assessment.

Those of us who have employed Bayesian network analyses for various purposes come to recognize their virtues in allowing us to tell different stories about some complex process depending upon what probabilistic ingredients we will include in them. The authors discuss these matters in Chapter 9 concerning sensitivity analyses. The Bayesian probabilistic underpinnings of such analyses allow us to tell how each of these alternative stories will end. In our analysis of the ballistics and other evidence in the case of Nicola Sacco and Bartolomeo Vanzetti, Professor Jay Kadane and I were able to tell different stories, on behalf of the prosecution and on behalf of the defence (Kadane and Schum 1996). We would have experienced considerable difficulties in constructing these stories and saying what their endings should be if we had not used this form of analysis.

I have been pleased and honoured to be asked to provide you with these introductory comments on a truly excellent work of scholarship that will be most helpful to persons in the forensic sciences and to the many other persons to whom I will recommend this book.

David A. Schum

1

The logic of uncertainty

1.1 Uncertainty and probability

1.1.1 Probability is not about numbers

The U.S. Federal Rule of Evidence 401 says that

> 'Relevant evidence' means evidence having any tendency to make the existence of any fact that is of consequence to the determination of the action more probable or less probable than it would be without the evidence. (Mueller and Kirkpatrick 1988, p. 33)

The term *probable* here means the degree of belief the fact finder entertains that a certain fact occurred. If it is not known whether the fact occurred, only a degree of belief less than certainty may be assigned to the occurrence of the fact, and there can then be discussion about the strength of this degree. Sometimes we are satisfied with speaking loosely of 'strong' or 'weak' beliefs; sometimes we would prefer to be more precise because we are dealing with important matters. A way to be more precise is to assign numerical values to our degrees of beliefs, and to use well defined rules for combining them together.

People are usually not very willing to assign numbers to beliefs, especially if they are not actuaries or professional gamblers. In this book we shall ask our readers to assign numbers, but these numbers are not important by themselves: what really matters is the fact that numbers allow us to use powerful rules of reasoning which can be implemented by computer programs. What is really important is not whether numbers are 'precise', whatever the meaning of 'precision' may be in reference to subjective degrees of belief based upon personal knowledge. What is really important is that we are able to use sound rules of reasoning to check the logical consequences of our propositions, that we are able to answer questions like: 'What are the consequences with respect to the degree of belief in A of assuming that the degree of belief in B is high, let us say x, or between x and y?'; 'how the degree of belief in A does change, if we lower the degree of belief in B by, let us say, z?'. If we are willing to take seriously the task of making up our mind to quantify, as best we can, our degrees of belief, then the reward will be the possibility of using the laws

Bayesian Networks and Probabilistic Inference in Forensic Science F. Taroni, C. Aitken, P. Garbolino and A. Biedermann

of *probability calculus* to answer questions like those formulated above. As a distinguished scholar of the logic of uncertainty, Glenn Shafer once said: 'Probability is not really about numbers; it is about the structure of reasoning'.

1.1.2 The first two laws of probability

In order to be able to 'measure' our degree of belief that a certain fact occurred, it is necessary to be precise about what 'a degree of belief' is. In this book it will be defined as a *personal degree of belief* that a proposition of a natural language, describing that fact, is true. 'Evidence' bearing on that proposition is expressed by means of other propositions. Therefore, it shall be said, on first approximation, that a proposition B is *relevant*, according to an opinion, for another proposition A if, and only if, knowing the truth (or the falsity) of B would change the degree of belief in the truth of A.

Having defined what is to be measured, it is next necessary to choose a function that assigns numbers to propositions. There are several alternatives available and, according to the choice, there are different rules for combining degrees of belief. In this book a *probability function* has been chosen, denoted by the symbol $Pr()$ where the () contain the event or proposition, the probability of which is of interest. Numerical degrees of belief must satisfy, for any propositions A and B, the laws of the mathematical theory of probability.

The first two laws can be formulated as follows:

- Degrees of belief are real numbers between zero and one: $0 \leq Pr(A) \leq 1$.
- If A and B are mutually exclusive propositions, *i.e.*, they cannot be both true at the same time, then the degree of belief that one of them is true is given by the sum of their degrees of belief, taken separately: $Pr(A \text{ or } B) = Pr(A) + Pr(B)$ (*The addition law*).

The addition law can be extended to any number n of exclusive propositions. If $A_1, A_2, \ldots, A_n$ cannot be true at the same time, then

$$Pr(A_1 \text{ or } A_2 \text{ or } \ldots A_n) = Pr(A_1) + Pr(A_2) + \cdots + Pr(A_n).$$

Satisfying the first law means that, if it is known that proposition A is true, then the degree of belief should take the maximum numerical value, *i.e.*, $Pr(A) = 1$. The degrees of belief in propositions not known to be true, like, for example, the proposition A, 'this coin lands heads after it is tossed', are somewhere between the certainty that the proposition is true and the certainty that it is false. A straightforward consequence of the probability laws is that, when the degree of belief for heads is fixed, then it is necessary to assign to the proposition B, 'this coin lands tails after it is tossed', the degree of belief $Pr(B) = 1 - Pr(A)$, assuming that pathological results such as the coin balancing on its edge do not occur.

This is the simplest example of how probability calculus works as a *logic for reasoning under uncertainty*. The logic places constraints on the ways in which numerical degrees of belief may be combined. Notice that the laws of probability require the degrees of belief in A and in B to be such that they are non-negative and their sum is equal to one. Within these constraints, there is not an obligation for A to take any particular value. Any value between the minimum (0) and the maximum (1) is allowed by the laws.

This result holds, in general, for any two propositions A and B which are said to be mutually exclusive and exhaustive: one and only one of them can be true at any one time and together they include all possible outcomes.

- Given that a proposition and its logical negation are mutually exclusive and exhaustive, the degree of belief in the logical negation of any proposition A is $Pr(not - A) = Pr(\bar{A}) = 1 - Pr(A)$

where $\bar{A}$ is read as *A-bar*. The logical negation of an event or proposition is also known as the *complement*.

1.1.3 Relevance and independence

A proposition B is said to be *relevant* for another proposition A if and only if the answer to the following question is positive: if it is supposed that B is true, does that supposition change the degree of belief in the truth of A? A judgement of relevance is an exercise in hypothetical reasoning. There is a search for a certain kind of evidence because it is known in advance that it is relevant; if someone submits certain findings maintaining that they constitute relevant evidence, a hypothetical judgement has to be made as to whether or not to accept the claim. In doing that, a distinction has to be drawn, not only between the *hypothesis* A and *evidence* B for the hypothesis A, but also between that particular evidence B and whatever else is known.

When a proposition's degree of belief is evaluated, there is always exploitation of available *background information*, even though it is not explicit. An assessment of the degree of belief in the proposition 'this coin lands heads after it is tossed' is made on the basis of some background information that has been taken for granted: if the coin looks like a common coin from the mint, and there is no reason for doubting that, then it is usually assumed that it is well balanced. Should it be realised, after inspection, that the coin is not a fair coin, this additional information is 'evidence' that changes the degree of belief about that coin, even though it is still believed that coins from the mint are well balanced. A *relevant proposition* is taken to mean a proposition which is not included in the background information. The distinction between 'evidence' and 'background information' is important because sometimes it has to be decided that certain propositions are to be considered as evidence, while others are to be considered as part of the background information.

For example, suppose a DNA test has been evaluated. Assume that all scientific theories which support the methodology of the analysis are true, that the analysis has been done correctly, and that the chain of custody has not been broken. These assumptions all form part of the background information. Relevant evidence is only those propositions which describe the result of the test, *plus* some other propositions reporting statistical data about the reliability of the evidence. Alternatively, propositions concerning how the analysis has been done, and/or the chain of custody, can also be taken to be part of the evidence while scientific theories are still left in the background. Therefore, it is useful to make a clear distinction between what is considered in a particular context to be 'evidence', and what is considered to be 'background'. For this reason, background information is introduced explicitly from time to time in the notation.

Let $Pr(A \mid I)$ denote 'the degree of belief that proposition A is true, given background information I', and let $Pr(A \mid B, I)$ denote 'the degree of belief that proposition A is true, given that proposition B is assumed to be true, *and* given background information I'.

- B is relevant for A if and only if the degree of belief that A is true, given that B is assumed to be true, is different from (greater or smaller than) the degree of belief that A is true, given background information I only, thus

$$Pr(A \mid B, I) \neq Pr(A \mid I). \tag{1.1}$$

It is assumed that, for the evaluation of forensic evidence, the distinction between the background knowledge of the Court or of the expert, and the findings submitted for judgement, is clear from the context and the distinction between 'background information' and 'relevant information' shall not be dwelt on further.

One more remark is pertinent here, however. The definition of *epistemic relevance*, and, implicitly, *irrelevance*, has been given in terms of *probabilistic dependence*, and *probabilistic independence*, respectively. 'Probabilistic independence' is a subtle concept that must be handled with care, always making up one's mind about what counts, personally, as background information. For example, imagine that a coin is to be tossed twice: is the outcome of the first toss relevant for the belief in the outcome of the second toss, *i.e.*, does the degree of belief in the outcome of the second toss change, knowing the outcome of the first one? The answer is 'It depends'. If it is known that the coin is well balanced, that it is a coin from the mint, then the answer is *no*. But, if it is not known if the coin is well balanced, then the answer is *yes*. Surely, one toss is not enough to influence beliefs above a significant psychological threshold, but if a sequence of many tosses is considered, the difference between the two situations can be seen. If it is not known whether the coin is fair, a sequence of tosses is just an experiment to test the hypothesis that the coin is fair, and the outcomes of the tosses are not independent.

It is not necessary to go into the subtleties of the concept of probabilistic independence. The purpose here is only to emphasise the point that *probability is always conditional on knowledge*. It is obvious that personal beliefs depend upon the particular knowledge one has. If the choice is made to represent degrees of belief by means of probabilities, then it must be kept in mind that it will always be the case that probabilities are relative to the knowledge available *and* the assumptions made. Statisticians say that probability is always relative to a *model*. The assumption that the coin is well balanced is a model. An alternative assumption that, possibly, the coin is not well balanced is another model or, better, is equivalent to the postulate of a cluster of models. For this example, a continuous probability distribution may be used to represent this set of models. Continuous probability distributions will be introduced in Chapter 10.

1.1.4 The third law of probability

Another constraint on one's degrees of belief is given by the so-called *multiplication law of probability*.

- For any propositions A and B, the degree of belief that A and B are both true, given background information I, is equal to the degree of belief that A is true, given background information I, times the degree of belief that B is true, given that one assumes that A is true: $Pr(A, B \mid I) = Pr(A \mid I) \times Pr(B \mid A, I)$.

In what follows, $Pr(A, B \mid I)$ will denote the degree of belief that propositions A and B are both true. The mathematical law is commutative, so one's numerical degrees of belief must satisfy commutativity as well. One can calculate the degree of belief in A and B in two different ways, but the final result must be the same:

$$Pr(A, B \mid I) = Pr(A \mid I) \times Pr(B \mid A, I) = Pr(B \mid I) \times Pr(A \mid B, I). \qquad (1.2)$$

It can be checked that probabilistic dependence is a symmetrical relationship. If B is relevant for A, then A is also relevant for B: if $Pr(A \mid I)$ is different from $Pr(A \mid B, I)$, then $Pr(B \mid I)$ must also be different from $Pr(B \mid A, I)$, for (1.2) is to be satisfied.

Consider again the example of coin tossing, with B and A denoting, respectively, the propositions 'the outcome of the first toss is heads' and 'the outcome of the second toss is heads'. If one knows that the coin is fair, then B is probabilistically independent from A: the degree of belief in the outcome of the second toss would not change, if one were able to know in advance the outcome of the first toss. Therefore, in this case, the multiplication law reads:

$$Pr(A, B \mid I) = Pr(B \mid I) \times Pr(A \mid I).$$

Just as for the addition law, the multiplication law can be extended to any number of propositions $A_1, A_2, \ldots, A_n$:

$$Pr(A_1, A_2, \ldots, A_n \mid I) = Pr(A_1 \mid I) \times Pr(A_2 \mid A_1, I) \\ \times \cdots \times Pr(A_n \mid A_1, A_2, \ldots, A_{n-1}, I)\ .$$

Let $A_1, A_2, \ldots, A_n$, be the propositions 'the outcome of the first toss is heads', 'the outcome of the second toss is heads', and so on. If one knows that the coin is fair, then the outcomes of the tosses 1 to $(n-1)$ are not relevant for the degree of belief about the outcome of the toss n:

$$Pr(A_1, A_2, \ldots, A_n \mid I) = Pr(A_1 \mid I) \times Pr(A_2 \mid I) \times \cdots \times Pr(A_n \mid I).$$

1.1.5 Extension of the conversation

A quite straightforward theorem of probability calculus, that follows immediately from the second law plus propositional calculus, puts another constraint on one's degrees of belief and is an all-important rule of the logic of reasoning under uncertainty. Dennis Lindley has underlined its role by dubbing it the *extension of the conversation rule* (Lindley 1985, p. 39):

- For any propositions A and B, the degree of belief that A is true, given background information I, is given by the sum of the degree of belief that A and B are both true, given background information I, and the degree of belief that A is true and B is false, given background information I: $Pr(A \mid I) = Pr(A, B \mid I) + Pr(A, \bar{B} \mid I)$.

Its importance comes from the fact that it is easier to assign a degree of belief to a proposition if it is thought about it in a context, extending the conversation to include other propositions that are judged to be relevant. Applying the multiplication law, the formula reads:

$$Pr(A \mid I) = [Pr(A \mid B, I) \times Pr(B \mid I)] + [Pr(A \mid \bar{B}, I) \times Pr(\bar{B} \mid I)]. \qquad (1.3)$$

For example, let A be the proposition 'the outcome of the first toss is heads'. Assume that there are two, and only two hypotheses about the state of the coin: it is either fair (proposition B) or it has two heads (proposition $\bar{B}$). Then one's overall degree of belief in A can easily be calculated via the formula. We can extend the conversation to any number of propositions, provided they are exclusive and exhaustive. In the example of coin tossing, suppose one knows that only three propositions B_1, B_2, B_3 are possible, respectively, the hypotheses that 'the coin is fair', 'the coin has two tails', 'the coin has two heads'. Then the overall degree of belief in A is given by:

$$Pr(A \mid I) = [Pr(A \mid B_1, I) \times Pr(B_1 \mid I)] + [Pr(A \mid B_2, I) \times Pr(B_2 \mid I)] + [Pr(A \mid B_3, I) \times Pr(B_3 \mid I)].$$

It is convenient to have a more compact notation in order to deal with more than two alternative propositions, and, following mathematical usage, the symbol $\sum_{i=1}^{n}$, that reads 'add the results for each value of i from 1 to n', shall be used. If $B_1, B_2, \ldots, B_n$ are mutually exclusive and exhaustive propositions, then, for any proposition A:

$$Pr(A \mid I) = \sum_{i=1}^{n} Pr(A \mid B_i, I) \times Pr(B_i \mid I).$$

Finally, notice that the multiplication law, under the form (1.3), implies that definition (1.1) of relevance is equivalent to saying that:

- B is relevant for A if and only if

$$Pr(A \mid B, I) \neq Pr(A \mid \bar{B}, I). \tag{1.4}$$

Indeed, suppose that B is not relevant for A, that is, $Pr(A \mid B, I) = Pr(A \mid \bar{B}, I)$. Then, given that $Pr(\bar{B} \mid I) = 1 - Pr(B \mid I)$, this implies that the right-hand side of (1.3) is identically equal to the left-hand side, independent of the value of $Pr(B \mid I)$.

1.1.6 Bayes' theorem

So far it has been said that a relevant proposition B, with respect to another proposition A, is a proposition such that if it were true, then such information would change our degree of belief in A. The rule according to which the degree of belief in A must change has not yet been provided. This rule is given by an elementary manipulation of (1.2):

- For any propositions A and B, the degree of belief that A is true, given that one assumes that B is true, is equal to the degree of belief that A and B are both true, given background information I, divided by the degree of belief that B is true, given background information I, provided that $Pr(B \mid I) > 0$:

$$Pr(A \mid B, I) = \frac{Pr(A, B \mid I)}{Pr(B \mid I)}. \tag{1.5}$$

(1.5) is called *Bayes' theorem* because it is the algebraic version of the formula first proved in the second half of the eighteenth century by the Reverend Thomas Bayes, a member of the English Presbyterian Clergy, and Fellow of the Royal Society (Bayes 1763). The importance of Bayes' theorem is due to the fact that it is a rule for *updating degrees of belief on receiving new evidence.*

The process of evidence acquisition may be modelled as a two-step process in time. At time t_0, it is planned to look for evidence B because B is believed to be relevant for hypothesis A. At time t_0, the change in degree of belief in A, if it were to be discovered that B were true, may be calculated by use of (1.5). Denote the degrees of belief at time t_0 by $Pr_0()$. Then, by use of the multiplication rule and the extension of the conversation rule, (1.5) may be rewritten as

$$Pr_0(A \mid B, I) = \frac{Pr_0(B \mid A, I) Pr_0(A \mid I)}{[Pr_0(B \mid A, I) Pr_0(A \mid I)] + [Pr_0(B \mid \bar{A}, I) Pr_0(\bar{A} \mid I)]}. \tag{1.6}$$

The probability $Pr_0(A \mid B, I)$ is also called the probability of A, *conditional on* B (at time t_0). $Pr_0(A \mid I)$ and $Pr_0(\bar{A} \mid I)$ are also called *prior*, or *initial*, probabilities. $Pr_0(B \mid A, I)$ is called the *likelihood* of A given B (at time t_0). Analogously, $Pr_0(B \mid \bar{A}, I)$ is the *likelihood* of $\bar{A}$ given B (at time t_0). This use of a likelihood is another exercise in hypothetical reasoning. Assessment of the likelihood requires consideration of the question: Supposing that hypothesis A is true (or false), what is the degree of belief that B is true?

At time t_1, it is discovered that B is true. Denote the degree of belief at time t_1 by $Pr_1()$: what is the degree of belief in the truth of A at time t_1, *i.e.*, $Pr_1(A \mid I)$? A reasonable answer seems to be that, if it has been learned at time t_1 that B is true, *and nothing else*, then knowledge that B is true has became part of the background knowledge at time t_1; therefore, the overall degree of belief in A at time t_1 is equal to the degree of belief in A, conditional on B, at time t_0:

$$Pr_1(A \mid I) = Pr_0(A \mid B, I). \tag{1.7}$$

The situation is the same as when there is information at time t_1 that a proposition B that had not been thought of before time t_1, is true. Assessment of the relevance of B for A, and the effect of this relevance on the degrees of belief, require thought about likelihoods and initial probabilities as if one were at time t_0: what would have been the degree of belief that B were true, given the background knowledge less the information that B is true, and supposing that hypothesis A is true, or false? What would have been the prior probability for A, given the background knowledge less the information that B is true?

1.1.7 Another look at probability updating

It may be helpful in clarifying the general mechanism underlying probabilistic updating to consider (1.7) in another way. Consider all the possible scenarios that can be derived from the combination of two logically compatible propositions A and B at time t_0. These scenarios can be represented graphically by means of a *probability tree* (*see* Figure 1.1).

A probability tree is a type of graphical model which consists of a series of branches stemming from nodes, usually called *random nodes,* which represent uncertain events. At every random node, there are as many branches starting as the number of the possible outcomes of the uncertain event. In this context, outcomes of uncertain events are described

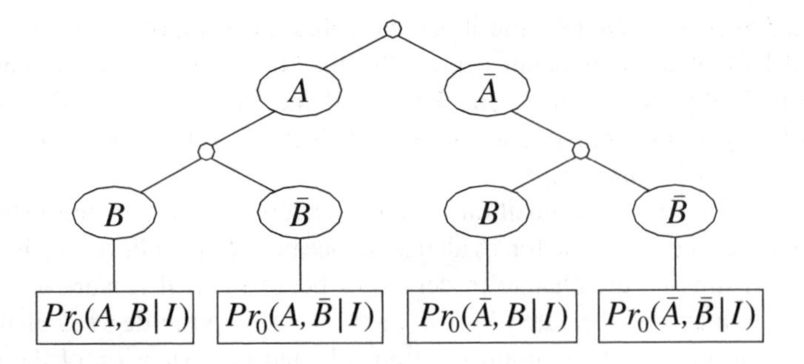

Figure 1.1 The probability tree for the uncertain propositions A and B at time t_0, given background information I.

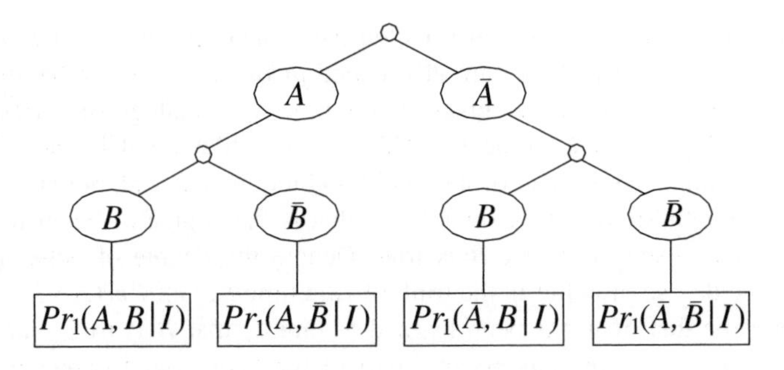

Figure 1.2 The probability tree for the uncertain propositions A and B at time t_1, given background information I.

by propositions, and branches containing more than one node correspond to the logical conjunction of as many propositions as the number of nodes. Branches have associated with them probabilities of the corresponding conjunctions of propositions calculated *via* the multiplication rule, and the probability of each proposition is given by the sum of the probabilities of the branches containing it (extension of the conversation). The sum of the probabilities of all the branches must add to one (addition rule).

Then, at time t_1, proposition B is known to be true. The probability tree is shown in Figure 1.2. As B is known to be true, only two scenarios are possible in the new state of knowledge and thus two of the branches in Figure 1.2 have probability zero.

Again, the sum of the probabilities of the branches in Figure 1.2 must add to one. The original probabilities of the branches must be amended in such a way that their sum turns out to be one. This can be done by multiplying them by the constant factor $1/Pr_0(B \mid I)$, and hence

$$
\begin{aligned}
Pr_1(A, B \mid I) &= Pr_0(A, B \mid I) \times \frac{1}{Pr_0(B \mid I)} = \frac{Pr_0(A, B \mid I)}{Pr_0(B \mid I)};\\
Pr_1(\bar{A}, B \mid I) &= Pr_0(\bar{A}, B \mid I) \times \frac{1}{Pr_0(B \mid I)} = \frac{Pr_0(\bar{A}, B \mid I)}{Pr_0(B \mid I)}.
\end{aligned} \tag{1.8}
$$

Indeed, addition of the terms in (1.8) gives the desired result.

$$\begin{aligned} Pr_1(A, B \mid I) + Pr_1(\bar{A}, B \mid I) &= \frac{Pr_0(A, B \mid I)}{Pr_0(B \mid I)} + \frac{Pr_0(\bar{A}, B \mid I)}{Pr_0(B \mid I)} \\ &= \frac{Pr_0(B \mid I)}{Pr_0(B \mid I)} = 1. \end{aligned}$$

The new probability of A is now identically equal to the probability of the branch containing both A and B. Because $Pr_1(A, \bar{B} \mid I) = 0$,

$$Pr_1(A \mid I) = Pr_1(A, B \mid I).$$

Therefore, substitution of this result in (1.8) gives (1.7).

$$\begin{aligned} Pr_1(A \mid I) &= \frac{Pr_0(A, B \mid I)}{Pr_0(B \mid I)}; \\ Pr_1(\bar{A} \mid I) &= \frac{Pr_0(\bar{A}, B \mid I)}{Pr_0(B \mid I)}. \end{aligned} \tag{1.9}$$

Result (1.8) demonstrates that the use of Bayes' theorem for probability updating is equivalent to the redistribution of probabilities among possible scenarios in a symmetric way; given that the state of information has changed on learning only that B is true, and nothing else, there is no reason to make a change biased for or against certain particular scenarios. This important point will be mentioned again in Section 1.2.6. In what follows, time subscripts will be omitted unless strictly necessary.

1.1.8 Likelihood and probability

In this book, likelihoods will play an important role. In some less technical talks, the expressions 'it is likely' and 'it is probable' are used interchangeably; so it is worthwhile to emphasise the difference between the *likelihood* of A, given B, and the *probability* of A, conditional on B.

As an illustration of the importance of the difference between likelihoods and probabilities, consider the following example. An eyewitness has seen Mr Jones running away from the house where a crime has been committed, at approximately the time of the crime. Let B be the proposition 'Mr Jones was running away from the scene of the crime at the time when it was committed', and A the proposition 'Mr Jones committed the crime'. It is reasonable to believe that the likelihood $Pr(B \mid A, I)$ is high, but not necessarily $P(A \mid B, I)$ is high as well. If Mr Jones actually committed the crime, it is expected that he would want to hasten away from the scene of the crime; the hypothesis of culpability is a good explanation of the evidence. However, the fact that he was running away from the house does not, by itself, make it very probable that he committed the crime; there are many other possible explanations of that fact.

The confusion of the likelihood of a hypothesis A, given evidence B, with the probability of the same hypothesis, conditional on the same evidence, is known as the *fallacy of the transposed conditional* (Diaconis and Freedman 1981). The fallacy occurs when, from the fact that if A has occurred, then B occurs with a high probability, it is erroneously concluded that if B has occurred, then A occurs with high probability. The best antidote against the fallacy is to remember Bayes' theorem; inspection of (1.6) shows that the two numbers are not necessarily close. In the second term of the sum in the denominator, the likelihood of $\bar{A}$, given B, could be high as well: there can exist another good explanation of evidence B even in the case that A is false. Moreover, the degree of belief of A, given the background knowledge only, has also to be taken properly into account.

The conversation can obviously be extended to n exclusive and exhaustive propositions (A_i, $i = 1, \ldots, n$), which represent possible alternative explanations of B, and in the general case, for any A_i, Bayes' theorem reads:

$$Pr(A_i \mid B, I) = \frac{Pr(B \mid A_i, I)Pr(A_i \mid I)}{\sum_{i=1}^{n} Pr(B \mid A_i, I)Pr(A_i \mid I)}. \tag{1.10}$$

1.1.9 The calculus of (probable) truths

A meaningful proposition can be true or false. Twentieth-century logicians have introduced the 'trick' of associating to any proposition, say A, a function f, called its *truth-function*, that assigns to A either the value $f(A) = 1$ if A is true, or the value $f(A) = 0$ if A is false. In this way a one-to-one correspondence can be established between the truth-function of a proposition A and a *random variable*, say X_A, whose value is uncertain. Therefore,

$$Pr(X_A = 1 \mid I)$$

may be written instead of $Pr(f(A) = 1 \mid I)$ to denote the probability that A is true given information I, *i.e.*,

$$Pr(A \mid I) = Pr(X_A = 1 \mid I).$$

In order to simplify notation, explicit mention of the background information I will be omitted when using random variables.

The concept of a random variable is especially helpful in dealing with continuous quantities. Continuous random variables will be considered in Chapter 10. At present only discrete quantities, like 'truth-values' that can take only two values, 0 and 1, are considered. In general, a *discrete random variable* X is an uncertain quantity that can take a discrete number of mutually exclusive and exhaustive values x with probabilities $Pr(X = x)$, such that

$$0 \le Pr(X = x) \le 1 \text{ and } \sum_x Pr(X = x) = 1.$$

Note the use of the abbreviated notation $\sum_x$, that reads 'add the results for each value x of the variable X'. The *probability distribution* of a random variable X is the set of probabilities $Pr(X = x)$ for all possible values x of X. In the special case of the random variable associated to a proposition A, this set contains only two values, $\{Pr(X_A = 1), Pr(X_A = 0)\}$, and:

$$Pr(X_A = 1) + Pr(X_A = 0) = 1.$$

Let $\{X_1, X_2, \ldots, X_n\}$ be a set of random variables. The *joint distribution* of these variables is the set of probabilities:

$$Pr((X_1 = x_1), (X_2 = x_2), \ldots, (X_n = x_n))$$

for all possible values x_i of X_i ($i = 1, 2, \ldots, n$).

For instance, for any pair of propositions A and B, the elements of the *joint probability distribution* of the random variables X_A and Y_B are the probability that A and B are both true, the probability that A is true and B is false, and so on:

$$\begin{aligned}
&Pr((X_A = 1), (Y_B = 1));\\
&Pr((X_A = 1), (Y_B = 0));\\
&Pr((X_A = 0), (Y_B = 1));\\
&Pr((X_A = 0), (Y_B = 0)).
\end{aligned}$$

Through the multiplication law, the joint distribution of two random variables X and Y can be expressed as a product of two other distributions, one being the distribution of the single variable X and the other the *conditional distribution* of the single variable Y, conditional on X:

$$\begin{aligned}
Pr((X = x), (Y = y)) &= Pr(X = x) \times Pr(Y = y \mid X = x)\\
&= Pr(Y = y) \times Pr(X = x \mid Y = y).
\end{aligned}$$

In this book, conditional distributions for propositions are tabulated by means of conditional probability tables like Table 1.1 where, by abuse of language, the symbols denoting propositions are used instead of the corresponding random variables (t stands for 'true' and f for 'false').

In this context, the probability distribution of a variable X_A can be obtained as the sum over y of its joint distribution with Y_B:

$$Pr(X = x) = \sum_y Pr(X = x, Y = y).$$

This operation is called *marginalisation*, and it can be seen that it is the same operation as 'extending the conversation'. The probability $Pr(X = x)$ is referred to as the *marginal distribution* of X and in Chapter 2 shorthand notation of the following type:

$$\begin{aligned}
Pr(A) &= \sum_B Pr(A, B) = Pr(A, B) + Pr(A, \bar{B})\\
Pr(A) &= \sum_{BC} Pr(A, B, C)\\
&= Pr(A, B, C) + Pr(A, B, \bar{C}) + Pr(A, \bar{B}, C) + Pr(A, \bar{B}, \bar{C})\\
Pr(A, B) &= \sum_C Pr(A, B, C) = Pr(A, B, C) + Pr(A, B, \bar{C}),
\end{aligned}$$

shall be used for the marginalisation of probabilities of propositions.

Table 1.1 Illustrative conditional distribution of B given A.

	A:	t	f
B:	$Pr(B = t \mid A)$	0.4	0.1
	$Pr(B = f \mid A)$	0.6	0.9

Table 1.2 Illustrative joint and marginal distributions of A and B.

	A:	t	f	Marginal
B:	t	0.32	0.02	0.34
	f	0.48	0.18	0.66
	Marginal	0.80	0.20	1.00

Table 1.3 Conditional distribution of A given B derived from Table 1.2

	B:	t	f
A:	$Pr(A = t \mid B)$	0.94	0.73
	$Pr(A = f \mid B)$	0.06	0.27

A way of representing a joint distribution is to draw the *table of marginals*. Assume that the distribution of variable A is $Pr(A) = (0.8, 0.2)$ and the conditional distribution of B given A is as given in Table 1.1. Then the table of marginals for the joint distribution of A and B is Table 1.2, where the bottom row with the column totals provides the marginal distribution of A, and the rightmost column, with the row totals, provides the marginal distribution of B. For example:

$$Pr(A = t, B = t) = Pr(A = t) \times Pr(B = t \mid A = t) = 0.8 \times 0.4 = 0.32.$$

Calculations of conditional probabilities via marginal probability tables are straightforward. In Table 1.3, the (rounded off) entries are calculated from Table 1.2. For example,

$$Pr(A = t \mid B = t) = \frac{Pr(A = t, B = t)}{Pr(B = t)} = \frac{0.32}{0.34} = 0.94.$$

All these definitions and operations generalise to three or more random variables. The joint distribution of several random variables may be decomposed via the multiplication law into a product of distributions for each of the variables individually.

1.2 Reasoning under uncertainty

1.2.1 *The Hound of the Baskervilles*

The following is a passage from the *Devon County Chronicle* of May 14th, 189...:

> The recent sudden death of Sir Charles Baskerville, whose name has been mentioned as the probable Liberal candidate for Mid-Devon at the next election, has cast a gloom over the county. (····) The circumstances connected with the death of Sir Charles cannot be said to have been entirely cleared up by

> the inquest, but at least enough has been done to dispose of those rumours to which local superstition has given rise. There is no reason whatever to suspect foul play, or to imagine that death could be from any but natural causes. (····) In spite of his considerable wealth he was simple in his personal tastes, and his indoor servants at Baskerville Hall consisted of a married couple named Barrymore, the husband acting as butler and the wife as housekeeper. Their evidence, corroborated by that of several friends, tends to show that Sir Charles's health has for some time been impaired, and points especially to some affection of the heart (····) Dr James Mortimer, the friend and medical attendant of the deceased, has given evidence to the same effect.
>
> The facts of the case are simple. Sir Charles Baskerville was in the habit every night before going to bed of walking down the famous yew alley of Baskerville Hall. The evidence of the Barrymores shows that this had been his custom. On the fourth of May (····) he went out as usual for his nocturnal walk (····) He never returned. At twelve o'clock, Barrymore, finding the hall door still open, became alarmed, and, lighting a lantern, went in search of his master. The day had been wet, and Sir Charles's footmarks were easily traced down the alley. Halfway down this walk there is a gate which leads out on to the moor. There were indications that Sir Charles had stood for some little time here. He then proceeded down the alley, and it was at the far end of it that his body was discovered. One fact which has not been explained is the statement of Barrymore that his master's footprints altered their character from the time he passed the moor-gate, and that he appeared from thence onward to have been walking upon his toes. (····) No signs of violence were to be discovered upon Sir Charles's person, and though the doctor's evidence pointed to an almost incredible facial distortion (····) it was explained that is a symptom which is not unusual in cases of dyspnoea and death from cardiac exhaustion. This explanation was borne out by the post-mortem examination, which showed long-standing organic disease, and coroner's jury returned a verdict in accordance with the medical evidence.

Many readers will have recognised a passage from Conan Doyle's novel *The Hound of the Baskervilles* (Conan Doyle 1953, pp. 676–77). The newspaper report is read by Dr Mortimer himself to Sherlock Holmes and his friend John Watson, whose counsel he is asking for because certain circumstances he knows make the case less simple than the public facts let suppose to be. So he relates to Holmes the strangest of those circumstances (*The Hound of the Baskervilles* (Conan Doyle 1953, p. 679)):

> I checked and corroborated all the facts which were mentioned at the inquest. I followed the footsteps down the yew alley, I saw the spot at the moor-gate where he seemed to have waited, I remarked the change in the shape of the prints after that point, I noted that there were no other footsteps save those of Barrymore on the soft gravel, and finally I carefully examined the body, which has not been touched until my arrival. (····) There was certainly no physical injury of any kind. But one false statement was made by Barrymore at the inquest. He said that there were no traces upon the ground round the body. He did not observe any. But I did - some little distance off, but fresh and clear.

'Footprints?'

'Footprints.'

'A man's or a woman?'

(····) 'Mr. Holmes, they were the footprints of a gigantic hound!'

The readers will remember that at the core of the plot there was the legend of the devilish hound which punished for his sins a wicked ancestor of the Baskervilles, information which is part of Holmes' and Watson's background knowledge, given that they have been told by Dr Mortimer. The beginning of this famous story is used as an example to show how the logic of probability works and how inferences licensed by probabilistic reasoning do agree with common sense inferences, as far as the problem is simple enough to be tackled by intuitive reasoning. But Conan Doyle's story will also offer a good example of what was claimed before, namely, that probability is always relative to a model, and that seemingly 'simple' problems might be less simple than they seem.

Readers' common sense, besides that of Holmes and Watson, will conclude, from the public and private facts given by Dr Mortimer, that the official inquest verdict is highly credible, but that there is some ground, although very thin, to arouse curiosity about the circumstances of Sir Charles Baskerville's death. Dr Mortimer will relate some more facts which will make Holmes cry that 'it is evidently a case of extraordinary interest, and one which presented immense opportunities to the scientific expert' (Conan Doyle 1953, p. 680), but consideration here is limited to two pieces of evidence as mentioned earlier, namely, medical evidence and the footprints.

The goal is now to show how well old Dr John Watson, who is not naturally endowed with the genius and the prodigious insight of Sherlock Holmes, but is learned about Bayes' theorem, can build up a probabilistic argument on the basis of these two pieces of evidence. First, reasoning is done step by step, taking into account the coroner's report, and then the footprints. Afterwards, reasoning is done by taking into account all the evidence at once.

1.2.2 Combination of background information and evidence

Given Watson's background information *only*, that is, given what he knew before considering the coroner's report, three mutually exclusive and exhaustive hypotheses were possible:

H_1 : Sir Charles Baskerville died by natural causes.

H_2 : Sir Charles Baskerville died by a criminal act.

H_3 : Sir Charles Baskerville committed suicide.

It is assumed that a nineteenth-century positivist held to be impossible, given his background information, the hypothesis that Baskerville's death was caused by a ghost. It would have not been considered admissible by a Court in Sherlock Holmes' times, or in our times either. This only reflects the fact that background information is historically given: some centuries ago, the fourth hypothesis might have been taken seriously into account by a Court. It is not the task of logic alone to decide what counts as background information.

Suppose Watson has to formulate his opinion knowing only the coroner's report. This report is such that Watson is warranted to believe that the following proposition is true:

R : The proximate cause of Sir Charles Baskerville's death was heart attack.

What is the effect of evidence R on his hypotheses? Proposition R, of course, is evidence *for* the hypothesis of accidental death, and *against* the other two hypotheses, according to common sense. We know from Section 1.1.6 that the question may be reformulated in probabilistic terms to read: what are the probabilities of the hypotheses, conditional on R?

Watson can make some reasonable judgements about prior probabilities. First of all, the three hypotheses cover all the logical possibilities and hence the sum of their probabilities is one. Moreover, he could think that the probability of H_1 is greater than both the probabilities of H_2 and H_3, and also greater than their logical disjunction, given only background knowledge about frequencies of causes of death:

$$
\begin{aligned}
&Pr(H_1 \mid I) > Pr(H_2 \mid I);\\
&Pr(H_1 \mid I) > Pr(H_3 \mid I);\\
&Pr(H_1 \mid I) > Pr(H_2 \text{ or } H_3 \mid I) = Pr(H_2 \mid I) + Pr(H_3 \mid I).
\end{aligned} \tag{1.11}
$$

Notice that Watson has expressed only comparative probability judgements: so far, he doesn't need to assign precise numerical probabilities.

As regards the likelihoods, he considers first the suicide hypothesis: it is *a priori* possible, but its likelihood, given that the proximate death cause was heart attack, can be safely assumed to be practically equal to zero.

He does not have reasons to exclude the crime hypothesis *a priori*. According to his background knowledge, a person, especially an aged person with bad health, can be scared to death. Surely, that would be quite a complicated way to kill a person, but this means that his prior probability for such an hypothesis will be very, very low, but not equal to zero. Of course, it is quite impossible to give an estimate of the likelihood of the hypothesis that a criminal act has been committed, given that the proximate cause of death is heart attack, but, again, it is not necessary to provide a precise estimate.

The likelihood of the accident hypothesis is, of course, much higher than the crime hypothesis, and Watson could even try to estimate it using statistical data, if he has any, about the death rate by heart attack in the class of people in Baskerville's range of age and health conditions. Therefore, the following judgements seem reasonable:

$$
\begin{aligned}
&Pr(R \mid H_1, I) > Pr(R \mid H_2, I);\\
&Pr(R \mid H_3, I) = 0.
\end{aligned} \tag{1.12}
$$

It is true that, with comparative probability judgements like (1.11) and (1.12), it is not possible to calculate the numerical value of the denominator in formula (1.6), but it is not necessary to do that to check that Bayes' theorem does yield a result in agreement with common sense. It is sufficient to notice that, by (1.6) and (1.12):

$$Pr(H_3 \mid R, I) = 0 \tag{1.13}$$

and to rewrite Bayes' formula (1.5) for hypotheses H_1 and H_2 as follows:

$$Pr(H_1 \mid R, I) \times Pr(R \mid I) = Pr(R \mid H_1, I) \times Pr(H_1 \mid I)$$
$$Pr(H_2 \mid R, I) \times Pr(R \mid I) = Pr(R \mid H_2, I) \times Pr(H_2 \mid I). \quad (1.14)$$

By (1.11) and (1.12), it follows that the probability of the accident hypothesis, that for Watson was already greater than the probability of the crime hypothesis, given background knowledge only, is even higher knowing the coroner's report:

$$Pr(H_1 \mid R, I) > Pr(H_2 \mid R, I). \quad (1.15)$$

It is worthwhile to notice that: (i) exact numerical probabilities are not always needed to draw probabilistic inferences; (ii) conclusion (1.15) is a *deductive inference*, being a necessary consequence of premisses (1.11) and (1.12) according to Bayes' theorem. Moreover, in those cases where the truth of a particular hypothesis is not logically compatible with some particular evidence, then observation of that evidence falsifies the hypothesis. In this example, proposition R would have been false if proposition H_3 had been true (premiss (1.12)); therefore, the truth of R implies the falsity of H_3 according to Bayes' theorem (conclusion (1.13)).

1.2.3 The odds form of Bayes' theorem

The combination of the two formulae that appear in (1.14) gives the so-called *odds form of Bayes' theorem*:

$$\frac{Pr(H_1 \mid R, I)}{Pr(H_2 \mid R, I)} = \frac{Pr(R \mid H_1, I)}{Pr(R \mid H_2, I)} \times \frac{Pr(H_1 \mid I)}{Pr(H_2 \mid I)}. \quad (1.16)$$

The left-hand side is the odds in favour of H_1, conditional on R, called the *posterior odds* in favour of H_1. In the right-hand side, the first term is the *likelihood ratio*, also called *Bayes' factor*, and the second term is the *prior odds* in favour of H_1. The effect of evidence on hypotheses can be calculated multiplying the likelihood ratio by the prior odds.

The likelihood ratio, say V, has been suggested as a measure of the probative value of the evidence with respect to two alternative hypotheses (Aitken and Taroni 2004; Robertson and Vignaux 1995; Schum 1994). If $V > 1$ in (1.16), we shall say that the probative value of evidence R is in favour of H_1, if $V < 1$ we shall say that it is in favour of H_2, and if $V = 1$, we shall say that R is not relevant for the hypotheses in question, or that the evidence is 'neutral' with respect to them.

It is necessary, again, to emphasise that the *probative value* of a piece of evidence with respect to two hypotheses is to be accurately distinguished by *conditional degrees of belief* for the same hypotheses: as (1.16) clearly shows, the probability of hypotheses conditional on that particular piece of evidence depends not only on the likelihood ratio, but on the prior odds as well, *i.e.*, on the background knowledge and *all other relevant evidence that is known.*

1.2.4 Combination of evidence

Now consider Dr Mortimer's evidence. For the sake of simplicity, it shall be taken that Watson can believe the following proposition to be true:

F : A gigantic hound was running after Sir Charles Baskerville.

Now, Watson has to evaluate the probability:

$$Pr(H_1 \mid F, R, I) = \frac{Pr(F \mid H_1, R, I)Pr(H_1 \mid R, I)}{Pr(F \mid R, I)}.$$

He can argue that likelihoods are against the accident hypothesis, and in favour of the crime hypothesis; the probability that a big dog was in that place of the moor at that time is smaller than the probability that this event would have happened if an intentional scheme was at work:

$$Pr(F \mid H_1, R, I) < Pr(F \mid H_2, R, I). \quad (1.17)$$

At this step, the initial probabilities are (1.15), and they are in favour of H_1, so: what is the overall effect of evidence F? Consider the odds form of Bayes' rule.

$$\frac{Pr(H_1 \mid F, R, I)}{Pr(H_2 \mid F, R, I)} = \frac{Pr(F \mid H_1, R, I)}{Pr(F \mid H_2, R, I)} \times \frac{Pr(H_1 \mid R, I)}{Pr(H_2 \mid R, I)}.$$

From this form, it can be seen immediately that, given the initial odds ratio at this point is in favour of H_1, the posterior odds can be reversed only if the likelihood ratio in favour of H_2 is greater than the initial odds ratio.

$$Pr(H_2 \mid F, R, I) > Pr(H_1 \mid F, R, I)$$

if and only if

$$\frac{Pr(F \mid H_2, R, I)}{Pr(F \mid H_1, R, I)} > \frac{Pr(H_1 \mid R, I)}{Pr(H_2 \mid R, I)}. \quad (1.18)$$

Even if Watson is not able to quantify precisely the relevant likelihoods, he can surely say that the left-hand ratio in (1.18) cannot be greater than the right-hand one. Therefore, Watson concludes that the probability of natural death is higher by far, even though the probability of a criminal scheme has been raised by the evidence of the footprints evidence, and this conclusion is reached by comparative probability judgements only, which Watson is able to make.

1.2.5 Reasoning with total evidence

Would the conclusion change if Watson does apply Bayes' rule to both pieces of evidence at the same time, instead of building up the argument in two steps? The probability Watson must evaluate now is the following

$$Pr(H_1 \mid F, R, I) = \frac{Pr(H_1, F, R, I)}{Pr(F, R, I)}.$$

Given that the multiplication law is commutative, Watson knows that the denominator can be factorised both ways, and that the final result must be the same.

$$Pr(H_1 \mid I) \times Pr(R \mid H_1, I) \times Pr(F \mid H_1, R, I)$$
$$= Pr(H_1 \mid I) \times Pr(F \mid H_1, I) \times Pr(R \mid H_1, F, I).$$

According to which factorisation he chooses, the overall effect of evidence will be calculated either using the likelihood ratio:

$$\frac{Pr(R \mid H_1, I)}{Pr(R \mid H_2, I)} \times \frac{Pr(F \mid H_1, R, I)}{Pr(F \mid H_2, R, I)} \tag{1.19}$$

or the likelihood ratio:

$$\frac{Pr(F \mid H_1, I)}{Pr(F \mid H_2, I)} \times \frac{Pr(R \mid H_1, F, I)}{Pr(R \mid H_2, F, I)}. \tag{1.20}$$

Watson is free to choose the formula which is more convenient for him, in the sense that he might be better able to estimate probabilities taking one factorisation rather than the other. We can notice that path (1.19) follows the temporal order by which evidence has been acquired, whereas path (1.20) follows the temporal order by which the hypothetical scenarios occurred.

If Watson relies upon his judgements (1.12) and (1.17), then formula (1.19) looks like the easiest way for him to follow. But suppose Watson wishes to give a comparative evaluation of the likelihoods in (1.20). He could sensibly say that the likelihood of H_2, given F, is higher than the likelihood of H_1, given F:

$$Pr(F \mid H_2, I) > Pr(F \mid H_1, I). \tag{1.21}$$

Then he might argue that, given that a big dog was running after Sir Charles Baskerville, the probabilities he would die from a heart attack are the same, the reason for the presence of the dog does not matter. Therefore:

$$Pr(R \mid H_1, F, I) = Pr(R \mid H_2, F, I). \tag{1.22}$$

Following (1.20), with the seemingly reasonable premisses (1.21) and (1.22), the conclusion is reached that the total body of evidence is in favour of the crime hypothesis. Indeed, given (1.22), evidence R is 'neutral' with respect to the choice of H_1 versus H_2, whereas, given (1.21), evidence F is in favour of H_2:

$$\frac{Pr(F \mid H_2, I)}{Pr(F \mid H_1, I)} > \frac{Pr(R \mid H_1, F, I)}{Pr(R \mid H_2, F, I)} = 1.$$

This does not mean that *posterior odds* will be in favour of H_2, for the calculus of posterior odds includes the *prior odds*, and these are overwhelmingly in favour of H_1:

$$\frac{Pr(H_1 \mid F, R, I)}{Pr(H_2 \mid F, R, I)} = \frac{Pr(F \mid H_1, I)}{Pr(F \mid H_2, I)} \times \frac{Pr(R \mid H_1, F, I)}{Pr(R \mid H_2, F, I)} \times \frac{Pr(H_1 \mid I)}{Pr(H_2 \mid I)}.$$

A correct way for Watson to report his inference would be to say that, although the evidence is in favour of the crime hypothesis, the prior credibility of that hypothesis is so low that the accident hypothesis is still the most credible. To be more precise, Watson should try to assign numerical values to his opinions (1.11), (1.21) and (1.22).

It shall be seen in Section 2.1.8 the reason why it is easier to assign numerical probabilities following the temporal order by which the possible scenarios occurred rather than the temporal order of evidence acquisition.

1.2.6 Reasoning with uncertain evidence

Suppose that the coroner's report is such that there is still some uncertainty left about the proximate cause of Baskerville's death. Assume, for the sake of argument, that Watson's degree of belief in the truth of proposition R ('The proximate cause of Sir Charles Baskerville's death was heart attack') at time t_1, after having known the report, is higher than his initial degree of belief at time t_0, but it falls short of certainty:

$$1 > Pr_1(R \mid I) > Pr_0(R \mid I).$$

What is the effect of this *uncertain evidence* upon the hypotheses?

The problem is that Watson *cannot* take the probability of H_1 conditional on R as his new degree of belief, because *he does not know* R for certain. One might say that Watson could make explicit some proposition, say E, that he knows for sure and such that he can extend the conversation taking into account the probability of R conditional on E at time t_0. For the hypothesis H_1, for instance, this would require the calculation of

$$\begin{aligned} Pr_0(H_1 \mid E, I) &= Pr_0(H_1, R \mid E, I) + Pr_0(H_1, \bar{R} \mid E, I) \\ &= \{Pr_0(H_1 \mid R, E, I) \times Pr_0(R \mid E, I)\} \\ &\quad + \{Pr_0(H_1 \mid \bar{R}, E, I) \times Pr_0(\bar{R} \mid E, I)\}. \end{aligned}$$

This suggestion has two drawbacks. The first is that Watson would have to evaluate the likelihoods of H_1 given R and E, which might not turn out to be an easy task. The second, and more important, is that it might be difficult for Watson to formulate explicitly such a proposition E that he knows for certain, considering that this proposition would be the logical conjunction of many other propositions, each one of which should be known for certain. Indeed, if there is any uncertainty left in the report, the best that Watson can do is to assess directly the effect of information on his degrees of belief, and, indeed, this is all that Watson is reasonably asked to do. A probabilistic rule can be formulated that allows him to update directly on uncertain evidence.

Consider, for simplicity, the reduced probability tree of Watson's problem given in Figure 1.3, *i.e.*, a probability tree containing only propositions H_1 and R. The difference with the tree in Figure 1.2 is that the probabilities of all the four scenarios at time t_1 are now

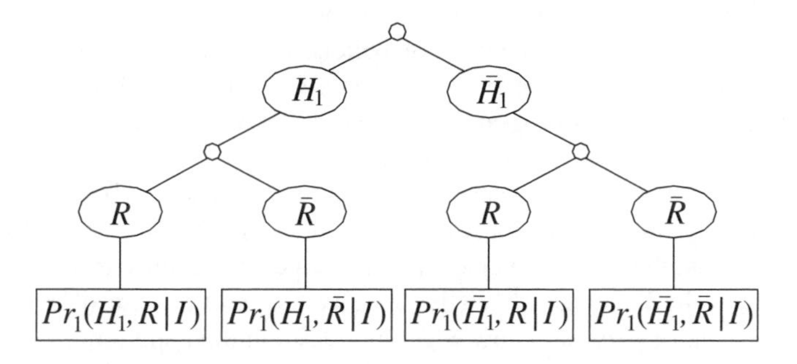

Figure 1.3 The probability tree for the propositions H_1 and R at time t_1.

greater than zero. The problem facing Watson is: *how can he redistribute the probabilities of scenarios in such a way that they add up to one?*

There are many ways of doing the redistribution but, considering that the only change in Watson's state of information that has occurred is a change in the probability of R, and no information has been given about the *probability ratio* of different scenarios, then the least biased way is to leave unchanged that ratio. Thus, a reasonable answer is that

- the probabilities are redistributed such that the ratio between the new and the old probabilities of the scenarios is the same as the ratio between the new and the old probabilities of R:

$$\frac{Pr_1(H_1, R \mid I)}{Pr_0(H_1, R \mid I)} = \frac{Pr_1(R \mid I)}{Pr_0(R \mid I)}. \tag{1.23}$$

From (1.23) the rule for calculating the new probabilities for any branch of the tree is then derived:

$$Pr_1(H_1, R \mid I) = Pr_0(H_1, R \mid I) \times \frac{Pr_1(R \mid I)}{Pr_0(R \mid I)}. \tag{1.24}$$

Formula (1.24) obeys the same principle as (1.8). The constant factor now is $Pr_1(R \mid I)/Pr_0(R \mid I)$ instead of $1/Pr_0(B \mid I)$, but the probabilities have been redistributed among the possible scenarios in a symmetric way as before: given that the state of information has changed on learning only the new probability $Pr_1(R \mid I)$, and nothing else, there are no reasons to make a change biased for or against certain particular scenarios. Formula (1.24) is a straightforward generalisation of Bayes' theorem, known in the philosophical literature under the name of *Jeffrey's rule*, because it was the philosopher of science Richard Jeffrey who first argued that it was a reasonable general updating rule (Jeffrey 1983). In the next Chapter (*see* Section 2.2.6) it is shown that (1.24) can be considered as the 'logical core' of updating algorithms that have been developed for Bayesian Networks, following Lauritzen and Spiegelhalter's path-breaking work (Lauritzen and Spiegelhalter 1988).

1.3 Frequencies and probabilities

1.3.1 The statistical syllogism

In this book, a normative point of view about the logic of uncertain reasoning is adopted. It is not assumed that forensic scientists, and people at large, satisfy the laws of probability calculus in their common sense reasoning. It is recommended that forensic scientists *should* use probabilities to express their degrees of beliefs, and the laws of probability to draw inferences, when they evaluate the probative force of evidence. This normative point of view is the core of a research programme known as 'Bayesian subjectivism', that ranges from statistical theory to artificial intelligence.

The evaluation of scientific evidence in Court is the expression of one's personal degree of conviction. This usually requires a combination of the knowledge of the frequencies for reference classes together with a personal knowledge of circumstances for a particular case. Probability is the language of relative frequencies, and the use of probability as the quantitative language of personal knowledge provides a unified and powerful framework to deal with the combination of 'objective' statistical data and 'subjective' evaluation of the facts of the case.

Subjective Bayesians consider 'objective' probability judgements to be judgements about whether it is possible to achieve an inter-subjective agreement. For example, a degree of belief in a particular case based upon the knowledge of relevant statistical data can be deemed more 'objective' than a degree of belief in a particular case not based upon the knowledge of statistical data, in the sense that reasonable people tend to agree that, if statistical data are known, personal degrees of belief should take into account those data. This *does not* amount to saying that a probability estimate based upon the knowledge of statistical data is 'objective' in the sense that there exists some kind of 'intrinsic' link between frequencies and probabilities which is independent of all non-frequency related information.

Any probability judgement in a particular case, even when the judgement is frequency-based, has a component based on personal knowledge. A singular probability judgement is subsumed under a statistical law by an argument which is sometimes called the *'statistical syllogism'*. The scheme of the argument is as follows, where (1)-(3) are the premisses and (4) is the conclusion:

Statistical Syllogism

1. The relative frequency of property Q in the population R is γ.
2. a_i is an individual in the population R ($i = 1, 2, \ldots, n$).
3. Nothing else is known about a_i which is relevant with respect to possession of property Q by a_i.
4. The probability that a_i has property Q is γ.

An example of a statistical syllogism is the following, where 'the probability that a particular individual has DNA profile Q' is to be understood as the *profile probability* as distinct from the so-called *random match probability* (Aitken and Taroni 2004, p. 408):

1. The relative frequency of DNA profile Q in the population R is γ.
2. The unknown criminal is an individual in the population R.
3. Nothing else which is relevant is known about the criminal with respect to possession of profile Q.
4. The probability that the criminal has profile Q is γ.

The third premiss shows how anyone who follows the argument takes into account his personal state of knowledge, besides knowledge of statistical data. The argument goes traditionally under the name of 'syllogism' but, as it is stated, it is not really a *deductive* argument. It cannot be said that the conclusion follows necessarily from the premisses, even though it is the conclusion that would be accepted by most people. Conversely, the subjective Bayesian version of the statistical syllogism *is* a deductive inference.

Bayesian Statistical Syllogism

1. The relative frequency of property Q in the population R is γ.
2. a_i is an individual in the population R ($i = 1, 2, \ldots, n$).

3. a_i has, for me, the same probability of possessing property Q as any other individual in the population R.
4. The probability, for me, that a_i has property Q is γ.

To see that the conclusion of this schema follows necessarily from the premisses, we shall introduce the concept of *expectation*.

1.3.2 Expectations and frequencies

Let X be a *discrete random variable* which can take the values x with probabilities $Pr(X = x)$. Even when we do not know the value of X we can calculate a representative value known as the *expectation* or *expected value* of X, denoted $E(X)$ and given by:

$$E(X) = \sum_x x Pr(X = x). \tag{1.25}$$

Now suppose that X_i is a random variable which takes the values $\{1, 0\}$ (respectively, 'true' and 'false'), associated with the proposition A_i: 'The individual a_i has property Q', where i belongs to a population R of n individuals. By (1.25) it immediately follows that *the expectation of this random variable is the probability that the proposition is true*:

$$E(X_i) = \{Pr(X_i = 1) \times 1\} + \{Pr(X_i = 0) \times 0\} = Pr(X_i = 1).$$

Given that expectation is additive, the sum of probabilities of propositions A_i is equal to the expected value of the sum of random variables X_i:

$$\begin{aligned} Pr(X_1 = 1) + \cdots + Pr(X_n = 1) &= E(X_1) + \cdots + E(X_n) \\ &= E(X_1 + \cdots + X_n). \end{aligned} \tag{1.26}$$

But the sum of random variables is the number of individuals who are Q, among the n individuals in the population R, *i.e.*, the *frequency* of Q among R.

Dividing each side of (1.26) by n, and applying the linearity of E, *i.e.*, $E(X_i)/n = E(X_i/n)$, we have that *average probability is equal to expected relative frequency*:

$$\frac{Pr(X_1 = 1) + \cdots + Pr(X_n = 1)}{n} = E\left(\frac{X_1 + \cdots + X_n}{n}\right). \tag{1.27}$$

Suppose the distribution $Pr(X_i = x_i)$ is one's subjective probability. Suppose, also, that one knows the relative frequency γ: then γ may be substituted on the right-hand side of (1.27), and, if one does not know which $n\gamma$ of R are Q, and one believes that all of them have the same probability of being Q, then it immediately follows that:

$$Pr(X_i = 1) = \gamma, \ \ (i = 1, \ldots, n).$$

Therefore, if probabilities are given a subjective interpretation, relative frequencies are known, and the probabilities of membership of Q for all members of R are taken to be equal, then relative frequencies determine the individual probabilities.

On the other hand, if one would have evidence that individual a_i belongs to a sub-population S of R, with different characteristics which are relevant with respect to Q, then the probability that a_i is Q would not be equal to the probability that an individual in the general population R is Q. One should change one's premisses to obtain a new, valid, argument of the same form, by substituting S for R and a new value, say γ', for γ, corresponding to the relative frequency of Qs in S, if it is known (*see* for example, Section 3.3.3.3).

1.3.3 Bookmakers in the Courtrooms?

The argument that has been usually used in the philosophical and statistical literature to justify the core of the Bayesian research programme makes appeal to a notion of pragmatic coherence. It has been known for centuries that degrees of belief can be numerically expressed by betting quotients, and the idea that numerical probabilities are fair betting quotients can be traced back to the founders of the theory, in the middle of the seventeenth century, Blaise Pascal and Christian Huygens. But it was only around 1930 that the mathematician Bruno de Finetti rigorously proved the so-called *Dutch Book theorem* (de Finetti 1937). The theorem shows that, if one's numerical degrees of belief do not satisfy the probability laws, then there exists a set of bets which would be *prima facie* acceptable to us, in the sense that one would consider the odds to be fair, and such that one shall be doomed to lose, no matter what the outcome is of the event on which the bets are made.

The 'Dutch Book' argument can be convincing or not, and often it is not for 'unbelievers', but the most forceful argument for accepting the Bayesian programme lies in the fact that the programme does work in practice. The best argument for the application of Bayesian theory in forensic science is to show that the theory agrees with personal intuitions, when inference problems are simple and intuitions are reliable, and that it helps to go beyond them, when problems become complicated and intuitions are not so reliable. The proof of the pudding is in the eating and readers will be able to judge by themselves whether the Bayesian programme is convincing.

1.4 Induction and probability

1.4.1 Probabilistic explanations

Kadane and Schum in their reference book *A Probabilistic Analysis of the Sacco and Vanzetti Evidence* synthesised as follows the differences between three basic forms of reasoning (Kadane and Schum 1996, p. 39):

> As far as concerns the process of discovery and the generation of hypotheses, evidence, and arguments linking them, there is an important issue being debated these days. The issue is: Can we generate new ideas by the processes of deduction and induction alone? The American philosopher Charles S. Peirce believed that there was a third form of reasoning according to which we generate new ideas in the form of hypotheses. He called this form of reasoning *abduction* or *retroduction* (· · · ·). Deduction shows that something is *necessarily* true, *induction* shows that something is *probably* true, but abduction shows that something is *possibly*, or *plausibly* true. Most human reasoning tasks, such as those encountered by the historian and criminal investigator, involve mixtures of these three forms of reasoning.

Abduction is classically considered the first step of an investigation when a particular event has been considered, and it is desired to *explain the occurrence of that event*. Therefore, in order to understand what 'abduction' is, it has to be understood what is meant by 'explanation'.

Deductive-Nomological (D-N) explanations are the simplest case of sound explanations, and their schema is as follows. Statements (1) and (2) constitute what is called the *explanans*, (3) constitutes what is called the *explanandum*, and the *explanandum* is assumed to be true. If premiss (2) is not actually true, but only hypothetically true, then we have a potential explanation.

Deductive-Nomological explanation

1. The statement of a scientific law saying that, if events of type B and $C_1, C_2, \ldots, C_n$, occur, then an event of type A occurs.
2. The statement that a particular event is of type B and $C_1, \ldots, C_n$.
3. The statement that a particular event is of type A.

A straightforward *D-N* explanation of the event that can be described by the proposition 'Mr Smith's blood sample and the blood stain from the crime scene share the same DNA profile' is:

1. If a stain of organic liquids comes from a person, and it has not been in contact with extraneous organic material, then the stain shares the DNA profile of that person.
2. The blood stain found on the crime scene comes from Mr Smith and it has not been in contact with extraneous organic material.
3. The blood stain found on the crime scene shares Mr Smith's DNA profile.

The *explanans* contains here a *common sense generalisation* instead of a scientific law or, if you wish, a shorthand for a set of scientific laws too large for the enumeration of all of them. Authors like Schum, Twining and Anderson have underlined the role of common sense generalisations in evidential reasoning, and we shall turn on this important issue in the next chapter (Anderson and Twining 1998; Schum 1994).

D-N explanations are only one among many kinds of 'explanations'. In social science, biological sciences, medicine and, last but not least, forensic science, the scientific laws available are, in general, statistical laws and whatever the philosophical view of the world is, it cannot make use of *probabilistic explanations*, namely explanations which subsume particular facts under statistical laws. The schema of the *Inductive-Statistical (I-S)* explanation originally proposed by Hempel (1965) was as follows:

Inductive-Statistical explanation

1. The statement of the probability γ of occurrence of events of type Q in the reference class of events of type R.
2. The statement that a particular event is of type R.
3. The statement that a particular event is of type Q with probability γ.

If it is not known whether premiss (2) is actually true, then it is a *potential I-S* explanation which is being used. In order to establish the first premiss, the argument presupposes the use of the statistical syllogism (*see* Section 1.3.1), as acknowledged by Hempel himself, who introduced the proviso that reference classes should be chosen on the basis of all relevant knowledge available prior to the *explanandum*. This requirement has been criticised for relativising statistical explanations to the knowledge of scientists at a given time. Subjective Bayesians acknowledge this fact not as a shortcoming of statistical explanations but as an unavoidable matter of fact.

Suppose it is desired to explain the particular event 'the blood stain from the crime scene shares DNA profile Q with the suspect' by means of an *I-S* explanation:

1. The probability that a stain of organic liquids coming from a member of the population R, different from the suspect, has DNA profile Q is γ.
2. The blood stain found on the crime scene comes from a member of the population R, different from the suspect.
3. The probability that the blood stain found on the crime scene has DNA profile Q is γ.

Here the probability γ is the *random match probability*, based not only upon knowledge of the relative frequency but also upon knowledge of the fact that another person, the suspect, has the profile: the random match probability γ can be different from the relative frequency (Aitken and Taroni 2004, p. 404); see also Sections 4.1, 5.1 and 5.12. This means that the third premiss of the statistical syllogism should be modified as follows: 'nothing else which is relevant is known about the member of the population with respect to the possession of profile Q, but the fact that another member of the same population shares the same profile'.

Is the above explanation a legitimate *I-S* explanation? The answer is negative, because Hempel's *I-S* explanations required a high probability γ for the *explanandum*, a probability at least greater than 50%. That means that in DNA evidence analysis we cannot use *I-S* explanation in conjunction with the 'random match' hypothesis, because the random match probability is usually very low. It is true that the 'random match' hypothesis can be considered only an explanation *by default*, so to speak, of the fact that the blood stain found on the scene of the crime shares the suspect's DNA profile but, nevertheless, it is a *potential alternative* explanation. Moreover, a high probability is neither a sufficient nor a necessary condition for explanation, as the following example can help to make clear (Salmon 1999, p. 27):

> Suppose that Bruce Brown has a troublesome neurotic symptom. He undergoes psychotherapy and his symptom disappears. Can we explain his recovery in terms of the treatment he has undergone? (· · · ·) If the rate of recovery for people who undergo psychotherapy is no larger than the spontaneous remission rate, no matter how large is this rate, it would be a mistake to consider treatment a legitimate explanation of recovery. (· · · ·) If, however, the recovery rate is not very large, but is greater than the spontaneous remission rate, the fact that the patient underwent psychotherapy has at least some degree of explanatory force.

The so-called *Statistical-Relevance (S-R)* model of explanation (Salmon et al. 1971) has been put forward to obviate this shortcoming of Hempel's classical analysis. All that is required by this model to provide a probabilistic explanation of a particular event is that there exists a partition of the relevant population into reference classes such that the probability of the *explanandum* is different for any member of the partition. What does matter are differences in probabilities, not their size. The basic idea is that there is an explanation or, at least, an approximation to a satisfactory explanation, whenever factors can be identified that make a difference for the probability of the *explanandum.*

A *partition* of a population R is a collection $\{S_1, \ldots, S_n\}$ of sub-populations of R such that the S_i are exclusive and exhaustive, *i.e.*, they do not have common members and they jointly contain all the members of R.

Statistical-Relevance explanation

1. The statement of a partition $\{S_1, \ldots, S_n\}$ of the relevant population R such that each element S_i of the partition is probabilistically relevant for Q.
2. The statement of probabilities γ_i of occurrence of events of type Q, in the reference class of events of type S_i.
3. The statement that a particular event is of type S_k.
4. The statement that a particular event is of type Q.

In Salmon's example, the reference classes are, in the population R of people who suffer the symptom, the class of people who undergo psychotherapy (S_1), and the class of people who undergo no therapies (S_2). An *S-R* explanation is not an argument; a conclusion does not follow from the premisses and, in order to explain why Bruce Brown recovered (Q), all that can be done is to point out the fact that he underwent therapy and that the probability of recovery is higher for members of that class than for members of the other class.

A *S-R* explanation for DNA analysis containing the 'random match' in the *explanans* hypothesis can be framed as follows:

1. The statement of a partition $\{S_1, \ldots, S_n\}$ of the *potential perpetrator population*, defined as the population of the possible perpetrators of the crime excluding the suspect, where the elements of the partition are all the sub-populations which are probabilistically relevant with respect to the property of having DNA profile Q.
2. The statement of random match probabilities (as distinct from profile probabilities) γ_i that a blood stain coming from a member of the sub-population S_i has profile Q.
3. The statement that the blood stain found on the crime scene comes from a member of the sub-population S_k.
4. The statement that a blood stain found on the crime scene has profile Q.

Sub-populations $S_1, \ldots, S_n$ are identified by a conjunction of the factors which provide information relevant to the allelic frequencies, like ethnic group membership and degrees of relatedness with the suspect. For example, a coarse partition (that could be refined) of the suspect population is:

{S_1 = Relatives, S_2 = Not-relatives and members of ethnic group X_1, S_3 = Not-relatives and members of ethnic group $X_2, \ldots, S_n$ = Not-relatives and members of ethnic group X_{n-1}}.

The sub-population S_k is identified by the hypothesis proposed by the defence: it is the *relevant population* as defined by (Aitken and Taroni 2004, p. 281). The magnitude of the probability γ_k does not matter for a *S-R* explanation: what does matter is that it is different from all the other probabilities.

1.4.2 Abduction and inference to the best explanation

It is now possible to give the general schemata of abductive and inductive arguments. For the sake of simplicity in what follows, the proposition describing the particular event of type Q that is to be explained is denoted by E, and the proposition describing the particular event of type R or S_k, mentioned in the *explanans* of *potential D-N* or *S-R* explanations is denoted by H. The particular event described by H is defined as an *explanatory fact*. Then the general schema of *abductive arguments* is:

Abduction

1. E is observed.
2. H is an explanatory fact in a potential explanation of E.
3. H is possibly true.

In the contemporary debate about the term 'abduction', it is often taken to mean what the philosopher Gilbert Harman has called the rule of *inference to the best explanation* (Harman 1965). In summary, the rule says that, given some evidence E and alternative explanatory facts H_1 and H_2, H_1 should be inferred rather than H_2 if, and only if, H_1 belongs to a better potential explanation of E than H_2. The two concepts must be carefully distinguished, as philosopher Paul Thagart pointed out (Thagart 1988, pp. 75–143):

> Mechanisms such as abduction and conceptual combination can lead to the formation of new theories (· · · ·). But we clearly do not want to *accept* a theory merely on the basis that there is something that it explains (· · · ·). Clearly, we want to accept a theory only if it provides the best explanation of the relevant evidence (· · · ·). Abduction only generates hypotheses, whereas inference to the best explanation evaluates them.

The abductive schema presented above serves to generate hypotheses, and not to evaluate them, because no alternative explanations are mentioned. As Holmes pointed out in the short story *The Adventure of Black Peter* (Conan Doyle 1953, p. 567):

> One should always look for a possible alternative, and provide against it. It is the first rule of criminal investigation.

In order to 'provide against' a potential alternative explanation, or against many of them, some philosophers have suggested we should reason by inference to the best explanation.

Inference to the Best Explanation

1. E is observed.
2. H_1 and H_2 are explanatory facts in two potential alternative explanations of E.
3. The explanation containing H_1 is overall better than the explanation containing H_2.
4. H_1 is provisionally accepted.

Hence, 'abduction' by itself is only a part of an inference to the best explanation: it is what introduces the second premiss of the argument. An important proviso must be added: the 'best explanation' must always be the best *overall* explanation of all available evidence. A critical point, and a long debated one, is the meaning of the 'best overall explanation'. The Bayesian answer is given in the next section, Section 1.4.3.

1.4.3 Induction the Bayesian way

In the Bayesian approach, generation and evaluation of the *prior* probabilities of scenarios, like those exemplified in Figure 1.1, correspond to the abductive step. In order to qualify as an explanatory fact, H must satisfy two necessary conditions.

- H is an explanatory fact for E only if: (i) the probability of H, given background knowledge only, is greater than zero and (ii) the likelihood of H, given E and background knowledge, is greater than zero:

  (i) $Pr(H \mid I) > 0$.

  (ii) $Pr(E \mid H, I) > 0$.

In order to define the best overall explanation in probabilistic terms it is first necessary to consider when an explanation is better than another.

- An explanation of E containing H_1 is better than an explanation containing H_2 if, and only if, the likelihood of H_1, given E, is greater than the likelihood of H_2, given E: $Pr(E \mid H_1, I) > Pr(E \mid H_2, I)$.

A rule can now be given to decide which one, between two alternative hypotheses, provides an explanation which is better overall for E.

- An explanation of E containing H_1 is overall better than an explanation containing H_2 if, and only if, the ratio of the likelihoods, given E, is greater than the reciprocal of the ratios of their probabilities, given background knowledge only:

$$\frac{Pr(E \mid H_1, I)}{Pr(E \mid H_2, I)} > \frac{Pr(H_2 \mid I)}{Pr(H_1 \mid I)}.$$

It follows immediately from the odds form of Bayes' theorem (1.16) that the following schema, to be called *Bayesian inference*, is valid.

Bayesian inference

1. E is observed.
2. H_1 and H_2 are explanatory facts in two potential alternative explanations of E.
3. The explanation containing H_1 is overall better than the explanation containing H_2.
4. H_1 is more probable than H_2, conditional on E.

It can be recognised that this scheme of inference has been applied in the discussion of the Baskerville's case: see (1.14) and (1.18). Note the following very important point. A hypothesis can be the best explanation of a given piece of evidence, given the competing hypotheses, without being the best *overall* explanation. It is not sufficient that the likelihood of H_1 is higher, given E, to conclude that it is also more credible, unless it is *also* assumed that H_1 and H_2 have the *same prior* probabilities. Such an assumption, for example, is made in the traditional Essen-Möller test for paternity, but it is not always true that the knowledge is such to justify this assumption (Taroni and Aitken 1998).

An important practical consequence to be stressed that follows from the Bayesian analysis is that the inductive step is to be made by the Court, and not by the expert, for it is only the Court which masters the total evidence of the case, and which has the viewpoint to evaluate which is the best *overall* explanation. As correctly pointed out by Henri Poincaré in the 'expert opinion' he gave during the second appeal of the notorious Dreyfus trial in 1904 (English translation of Darboux et al. (1908, p. 504), quoted from Taroni et al. (1998, p. 192)):

> Since it is absolutely impossible for us [the experts] to know the a priori probability, we cannot say: this coincidence proves that the ratio of the forgery's probability to the inverse probability has that particular value. We can only say: following the observation of this coincidence, this ratio becomes X times greater than before the observation.

It has already been noted that conclusions which are only probably true are obtained *deductively* by premisses which are probably true, via the laws and theorems of the mathematical theory of probability *plus* the updating rule (1.24). This new viewpoint on 'inductive logic' constitutes one of the major philosophical achievements of the twentieth century, insofar as it has made it possible to give a constructive answer to David Hume's sceptical challenge to induction, and it has provided solid grounds to the Artificial Intelligence quest for mechanising uncertain reasoning. Hume was right: the traditional idea that ampliative induction is possible, starting from a collection of bare facts, was an illusion. People always face observations carrying with them their background knowledge and prior beliefs. But if this fact is recognised, then there is no obligation, as Hume thought, to trust a mere innate psychological habit; a powerful logical tool is now available. As the philosopher and logician Frank P. Ramsey wrote in his seminal essay, written in 1926 (Ramsey 1931, pp. 182–189):

> (····) a precise account of the nature of partial beliefs reveals that the laws of probability are laws of consistency, an extension to partial beliefs of formal

> logic, the logic of consistency. (····) We do not regard it as belonging to formal logic to say what should be a man's expectation of drawing a white or black ball from an urn; his original expectations may within the limits of consistency be any he likes, all we have to point out is that if he has certain expectations, he is bound in consistency to have certain others. This is simply bringing probability into line with ordinary formal logic, which does not criticise premisses but merely declares that certain conclusions are the only ones consistent with them.

The same idea was expressed at the same time by de Finetti without knowing Ramsey's paper and from a different philosophical background, but sharing with him a common pragmatist attitude (English translation of de Finetti (1930a, p. 259) quoted from Aitken and Taroni (2004, p. 154)):

> Probability calculus is the logic of the probable. As logic teaches the deduction of the truth or falseness of certain consequences from the truth or falseness of certain assumptions, so probability calculus teaches the deduction of the major or minor likelihood, or probability, of certain consequences from the major or minor likelihood, or probability, of certain assumptions.

Therefore, according to this opinion, 'inductive logic' is just a matter of deducing complex, maybe non-intuitive, probabilistic beliefs from premisses containing simpler, and more intuitive, probabilistic beliefs, and this is the reason why, in the last talk he gave before retirement in 1976 at the University of Rome, Bruno de Finetti declared that the expression 'Bayesian induction' is as likely to be said to be redundant, as it would be to say 'Pythagoric arithmetic' is redundant (de Finetti 1989, p. 165).

1.5 Further readings

The 'subjectivist interpretation' of probability defended in this Chapter can be traced back to the origins of the modern mathematical theory of probability: see Daston (1988) and the classical study of Hacking (1975), in particular, his Chapter 11, on Christian Huygens' book *De Ratiociniis in Aleae Ludo*, published in 1657, and his concept of 'expectation'. Huygens' expectation is the same concept used here and that is explained in Section 1.3.2. Blaise Pascal, as well, defined probabilities in terms of expectations in his essay on the arithmetical triangle, written in 1654: see Edwards (1987). At the beginning of his essay, Thomas Bayes (1763) defined probability in terms of betting odds. Modern Bayesianism originates in the papers that the English logician and philosopher Frank P. Ramsey and the Italian mathematician Bruno de Finetti wrote independently of each other (de Finetti 1930a, 1937; Ramsey 1931). The approach began to be widely known with the work of Savage (1972), first published in 1954. Modern classics are Lindley (1965) and de Finetti (1975). Related, but different, recent viewpoints about the foundations of probability may also be mentioned: the game-theoretic approach of Shafer and Vovk (2001), the 'objective Bayesianism' of Jaynes (2003), and the 'prequential' approach of Dawid and Vovk (1999). An accessible introduction to the different interpretations of the concept of probability and relationship about probability and inductive logic is Galavotti (2005).

In the framework of the Bayesian programme, the tendency to base degrees of belief upon relative frequencies, as discussed in Section 1.3.1, is not only fully acknowledged as 'reasonable', but takes the form of a mathematical theorem, de Finetti's *representation theorem* (de Finetti 1930b, 1937). The theorem says that the convergence of one's personal probabilities towards the values of observed frequencies, as the number of observations increases, is a logical consequence of Bayes' theorem if a condition called *exchangeability* is satisfied by our degrees of belief, prior to observations.

A sequence of observations, for which non-Bayesian statisticians would say that it is governed by a 'true' but unknown probability distribution P (one in which observations are *conditionally independent and identically distributed*, given P), is a sequence of observations that subjective Bayesians would say is *exchangeable*.

De Finetti's theorem, and other so-called convergence theorems, that have been proved lately demonstrating a generalisation of de Finetti's exchangeability, enable it to be said that a probability estimate situated around the value of an empirical frequency is 'objective' in the sense that several persons, whose *a priori* probabilities were different, would converge towards the same *posterior* probabilities, were they to know the same data and share the same likelihoods. This usually happens in the conduct of statistical inference, where likelihoods are provided by the choice of appropriate probability distributions, *i.e.*, statistical models, so that they are the same for any observer who agrees on the choice of the model. For a survey of convergence theorems in Bayesian statistics, one can see Bernardo and Smith (1994) and Schervish (1995).

With regard to the likelihood ratio as a measure of the probative value of the evidence (*see* Section 1.2.3), the philosopher Charles Peirce, at the end of nineteenth century, was the first to suggest the use of the *logarithm* of the likelihood ratio as a measure of the information contained in the evidence, calling it the *weight of evidence*, a suggestion made anew by the mathematician Alan Turing, when he was working to break the *Enigma* code during the Second World War (Good 1985). The choice is motivated by the fact that the logarithm has many of the properties a measure of information should have. For example, if the value, $V = 1$, the logarithm of V is zero, and insofar as the evidence is inconclusive, *i.e.*, it does not falsify one of the alternative hypotheses, the logarithm is different from zero. Another desirable intuitive property of such a measure is that the overall information content of two pieces of evidence, which are probabilistically independent from each other, should be given by the sum of the information. With the logarithmic measure, if two pieces of evidence are probabilistically independent, then Bayes' factors multiply and weights of evidence add.

Jeffrey's rule (*see* Section 1.2.6) has been discussed at length in the philosophical literature on induction and probability, after it was put forward by Jeffrey (1983). As a matter of fact, the rule was proposed by the English astronomer William Donkin in the middle of nineteenth century (Donkin 1851). The rule is another form of Fisher's factorization theorem (Diaconis and Zabell 1982), and equivalent versions of it have been put forward, independently by Jeffrey, in the field of artificial intelligence (AI) by Lemmer and Barth (1982), and, with far reaching consequences, by Lauritzen and Spiegelhalter (1988). The formal equivalence between Jeffrey's rule and a *minimum principle* has been also demonstrated by many authors (Domotor et al. 1980; Williams 1980). The equivalence can be stated as follows: (1.24) operates the minimal change in probabilities which is needed to take into account the new information $Pr_1(R \mid I) \neq Pr_0(R \mid I)$ *and only this information*, in the sense that it minimises the 'distance' between old $Pr_0()$ and new $Pr_1()$ as measured by the *Kullback–Leibler measure of relative entropy* (Kullback 1959).

The Kullback–Leibler measure $H(Pr_0, Pr_1)$ of the distance between $Pr_0()$ and $Pr_1()$ is defined as

$$H(Pr_0, Pr_1) = \sum_e \log \left[\frac{Pr_1(e)}{Pr_0(e)} \right]$$

where all the events e together form a partition of the space of events, *e.g.*, the sample space; the set $\{R, \bar{R}\}$ in (1.24) is such a partition.

Relative entropy $H(Pr_0, Pr_1)$ is only one among several possible measures of 'nearness' for probability distributions, and a *symmetry* argument has been provided by the philosopher of science van Fraassen (1986, 1989) for supporting this particular choice. He has shown that (1.24) is the only updating rule for probabilities which satisfies a very general invariance requirement that can be simply formulated using Edwin Jaynes' words: 'In two problems where we have the same state of knowledge, we should assign the same probabilities' (Jaynes 1983, p. 284). Tikochinsky et al. (1984) have put forward another argument, which is less general than van Fraassen's, in that it makes use of the concept of independent repetitions of the same experiment.

The Bayesian approach to induction outlined briefly in Section 1.4.3 has been defended by many philosophers as a valid unifying 'rational reconstruction' of scientific reasoning (Horwich 1982; Howson and Urbach 1993; Jaynes 2003; Jeffrey 1992, 2004; Salmon 1990). Without dwelling on the merits (or demerits) of the Bayesian approach in general, it is worth noting that some of the criticisms raised against the claim that Bayesian inference is a good 'rational reconstruction' of scientific inference do not apply in the context of forensic inference. First, it has been observed that, in the case where H_1 is a high-level scientific theory, it is very often impossible to specify an alternative theory H_2 in such a way that the likelihood $Pr(E \mid H_2)$ would turn out to be something different from 'anybody's guess' (Earman 1992, p. 164). In the case of forensic inference, however, the competing 'theories' are well defined: they are the prosecution hypothesis and the defence hypothesis.

Second, Bayesian inference cannot give any rational guidance when the body of evidence for H_1 is different from the body of evidence for H_2 for, obviously, the odds form of Bayes' theorem can be applied only when the evidence, E, is the same for both theories. However, in the case of forensic evidence, the body of evidence accepted by the Court is the same for prosecution and defence.

Third, when both hypotheses are deterministic and entail the evidence E, then the ratio of their probabilities conditional on E reduces to the ratio of their prior probabilities. In the case of forensic evidence, rarely is this the case.

In summary, it has been shown that Bayesian inference is a good 'rational reconstruction' of forensic inference as a particular instance of scientific reasoning.

2

The logic of Bayesian networks

2.1 Reasoning with graphical models

2.1.1 Beyond detective stories

Sherlock Holmes was a lucky guy. He had Conan Doyle at his disposal to arrange the plot in such a way that the mechanism of eliminative induction worked neatly, the very mechanism simple-minded Watson was apparently unable to appreciate, as Holmes remarked in the novel *The Sign of Four* (Conan Doyle 1953, p. 111):

> "You will not apply my precept", he said, shaking his head. "How often have I said to you that when you have eliminated the impossible, whatever remains, *however improbable*, must be the truth?"

Holmes' friend did not really deserve such a rebuke: the 'precept' is not that easy to apply. The philosopher of science John Earman has observed, with respect to Holmes' *dictum*, that (Earman 1992, p. 163)

> (· · · ·) the presupposition of the question can be given a respectable Bayesian gloss, namely, no matter how small the prior probability of the hypothesis, the posterior probability of the hypothesis goes to unity if all of the competing hypotheses are eliminated. This gloss fails to work if the Bayesian agent has been so unfortunate as to assign the true hypothesis a zero prior.

In different words, the gloss fails to work if the agent has been so unfortunate as to choose the wrong model. This does not usually happen in literary fiction (Earman 1992, pp. 163–64):

> The classic English detective story is the paradigm of eliminative induction at work. The suspects in the Colonel's murder are limited to the guests and servants at a country estate. The butler is eliminated because two unimpeachable witnesses saw him in the orangery at the time the Colonel was shot in the library. Lady Smyth-Simpson is eliminated because at the crucial moment she was dallying with the chauffeur in the Daimler, *etc*. Only the guilty party, the Colonel's nephew, is left.

Bayesian Networks and Probabilistic Inference in Forensic Science F. Taroni, C. Aitken, P. Garbolino and A. Biedermann

In real life, it can happen that the true hypothesis is not considered. This is neither the only problem in applying the precept of the down-to-earth Holmes, nor the biggest. If the model does not include the true hypothesis, this is not a shortcoming of probability theory. Probability rules merely deduce probable consequences from probable premisses: if the premisses are wrong, it is not the fault of the theory. The real problem is different; it is the very application of Bayesian machinery to real-life problems that seems to be practically impossible. Besides background knowledge, the application of Bayes' theorem to a new piece of evidence requires the totality of the evidence already considered, with the omission of nothing, for the logic of probability is *non-monotonic*: the probability of B, conditional on A, can be different from the probability of B, conditional on A and C. For example, the probability that footprints are left in the meadow, given that a certain person walked through it, is different, according to whether the soil was wet or not.

Denote by E the totality of the evidence less B. Evaluation of the probability of a proposition H, $Pr(H|B, E, I)$ requires account to be taken of the fact that E is a logical conjunction of many components. As the number of those components increases, the task will rapidly grow to an intractable level of complexity. At the dawn of the twentieth century, Watson and Holmes might have known Bayes' theorem, but they would have stopped before the wall of that complexity. At the dawn of the twenty-first century, Bayesian networks have come to the rescue.

2.1.2 What Bayesian networks are and what they can do

A Bayesian network is a type of *graphical model* whose elements are *nodes*, *arrows* between nodes, and *probability assignments*. A finite set of nodes together with a set of arrows (*directed links*) between nodes forms a mathematical structure called a *directed graph*.

If there is an arrow pointing from node X to node Y, it is said that X is a *parent* of Y and Y is a *child* of X. In Figure 2.1(i), for example, node A is parent of nodes B and E, E is a parent of D and F, and B and D are parents of C. A node with no parents is called a *root* node; node A and node I in Figure 2.1(i) and (ii), respectively, are root nodes.

A sequence of consecutive arrows connecting two nodes X and Y, independent of the direction of the arrows, is known as a *path* between X and Y. For example, in 2.1(i) there are three paths linking nodes A and C: one through the intermediate node B, one through the intermediate nodes E and D, and the third is the path $A - E - F - G -$

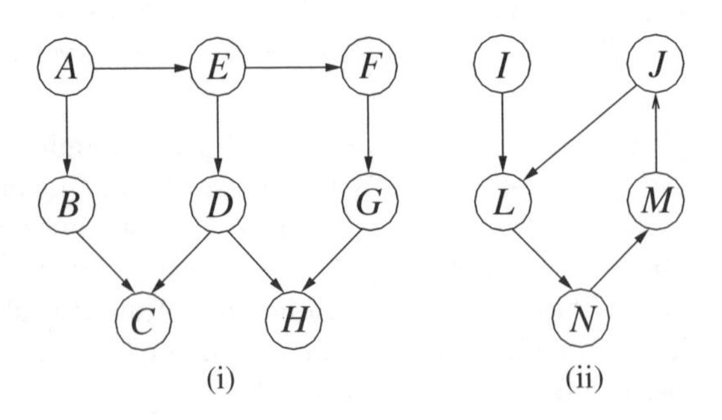

Figure 2.1 A directed acyclic graph (i) and a directed graph with one cycle (ii).

$H - D$; in Figure 2.1(ii) the nodes I and J are linked by the paths $I - L - N - M - J$ and $I - L - J$.

Nodes can have *ancestors* and *descendants*: node X is an ancestor of node Y (and Y is a descendant of X) in the case where there is a unidirectional path from X to Y, linking intermediate nodes. In Figure 2.1(i), A is a parent of B and E and an ancestor of every other node in the graph. In Figure 2.1(ii), node L is an ancestor of itself. Hereafter, only directed graphs which do not contain *cycles* and which are *connected* will be considered. A cycle is said to exist if a node is an ancestor, and also descendant of itself and a graph is connected if there exists at least one path between every two nodes. Both graphs in Figure 2.1 are connected:

- a connected directed graph with no cycles is called a *directed acyclic graph* (acronym DAG).

A formal definition of Bayesian networks will be given in Section 2.2.2. Meanwhile, consider a Bayesian network as a DAG in which:

- nodes represent *random variables*, where the random variable may be either discrete, with a finite set of *mutually* exclusive states which themselves can be categorical, discrete or continuous;

- arcs represent direct *relevance* relationships among variables; for each variable X with parents $Y_1, Y_2, \ldots, Y_n$, there is associated a *conditional probability table* $Pr(X \mid Y_1, Y_2, \ldots, Y_n, I)$, where I denotes, as usual, the background knowledge, which is all the relevant knowledge that does not explicitly appear under the form of nodes in the graph. If X is a root node, then its table reduces to probabilities $Pr(X \mid I)$, unconditional on other nodes in the graph.

Let X be a variable with n states $x_1, \ldots, x_n$. A *state* is a possible value associated with a random variable. If X is a root node then the (un)conditional probability table $Pr(X \mid I)$ will be an n-table (a table with n entries) containing the probability distribution $\{Pr(X = x_i),\ i = 1, \ldots, n\}$, with $\sum_{i=1}^{n} Pr(X = x_i) = 1$. For notational simplicity, explicit mention of background information I has been omitted. Also, when the context is sufficiently clear that there will be no confusion in so doing, the subscript i is omitted from the notation and one writes $Pr(X = x)$ and $\sum Pr(X = x) = 1$.

Let Y be a variable with m states y: if Y is a parent of X then the conditional probability table $Pr(X \mid Y)$ will be an $n \times m$ table containing all the probability assignments $Pr(X = x \mid Y = y)$. For example, suppose that variables B and D in Figure 2.1(i) each have two possible states and variable C has three. Then the conditional probability table for C (Table 2.1) will contain 12 entries $Pr(C = c_i \mid B = b_j, D = d_k) = p_{ijk}$, with $(i = 1, 2, 3; j = 1, 2; k = 1, 2)$.

Notice that the notation for the states of the variables as used here may vary, in the course of the book, with the domain of application. It may happen, for example, that the name of a variable, say E, also characterises a state of that variable (*e.g.*, Section 4.1). On other occasions, the states of a variable may be described quite differently from the name of the variable: for instance, an individual's genotype, denoted gt, has states describing the components of that individual's DNA profile (*i.e.*, genotypes, as discussed in Section 5.2). Sometimes, names and states of variables will be denoted in a way which conforms to the

Table 2.1 Conditional probability table for child C with parents B and D where B has two states (b_1, b_2), C has three states (c_1, c_2, c_3), and D has two states (d_1, d_2).

		B :	b_1		b_2	
		D :	d_1	d_2	d_1	d_2
C :	$Pr(C = c_1 \mid B = b_j, D = d_k)$		p_{111}	p_{112}	p_{121}	p_{122}
	$Pr(C = c_2 \mid B = b_j, D = d_k)$		p_{211}	p_{212}	p_{221}	p_{222}
	$Pr(C = c_3 \mid B = b_j, D = d_k)$		p_{311}	p_{312}	p_{321}	p_{322}

original literature in which the topic of interest was first described, (*e.g.*, Section 4.2 on nodes with indexed lower case letters).

Bayesian networks have a built-in computational architecture for computing the effect of evidence on the states of the variables. This architecture:

- updates probabilities of the states of the variables, on learning new evidence;

- utilises probabilistic independence relationships, both explicitly and implicitly represented in the graphical model, to make computation more efficient.

No algorithms are available to cope with feedback cycles, and this is the reason for the requirement that the network must be a *DAG*. Computational tractability has been one of the points raised against the use of probability theory in the development of artificial reasoning systems. It was claimed that probability would not be good in artificial intelligence (*AI*) because the complexity of probability calculations grows exponentially with the number of variables, and neither the human mind nor the most powerful computer imaginable can execute exponential-time calculations.

It is known today that computation in Bayesian networks is, in general, very hard, in the sense that it is at least as complex as a particular class of decision problems, called NP-problems (acronym for *Non Polynomial*), for which no algorithm has yet been found that can solve them in polynomial time. Most people believe that none exists (Cooper 1990). But it is also known that, although the general problem is *NP-hard*, for certain types of graphs, called *junction trees*, there are efficient algorithms for computing the probability distributions of nodes exactly, apart from rounding errors. (The type of graph for which this is possible is defined in Section 2.2.5.) These algorithms allow the users to handle problems with both discrete and continuous random variables which are already too complex for unaided common sense reasoning. (Discrete random variables have been defined in Section 1.1.9; continuous random variables are discussed in Chapter 10.)

The human mind is good in selecting those features of reality that are important but poor at aggregating the features. Human experts are good in building the model, but they are not so good in reasoning through the model. A computer program is not good in building the model, but it is very good in performing calculations. It is acknowledged that Bayesian networks do not describe how the human mind works. It is claimed only that in simple cases, they provide intuitively reasonable answers, and that they are better than human minds in performing some more complex reasoning tasks. The goal is not to replace human experts, the goal is to help them.

2.1.3 A graphical model for relevance

A type of graphical model, namely, probability trees (*see* Figures 1.1 and 1.2), for representing the possible alternative scenarios which can be built from a given set of propositions has already been discussed in Section 1.1.7. This kind of graphical representation can be used only for very small sets of propositions because the number of branches grows exponentially: for n binary variables the tree has 2^n branches. Moreover, relevance relationships amongst propositions are not imbedded in the structure of the graph. Numerical assignments to the branches are needed to show the relevance relationships amongst the propositions. In Section 1.2.6, for instance, the relevance of H_1 for R can be indicated only by the fact that the probability assigned to the branch labelled R which follows the branch labelled H_1 is different from the probability assigned to the branch labelled R which follows the branch labelled $\bar{H}_1$. Directed graphs provide an economic and powerful model for visualising dependencies between variables.

From Section 1.1.9, it is known that a proposition can be associated with a binary random variable. With an abuse of notation, nodes of the graph which stand for propositions will be labelled with the same letters as used for corresponding binary random variables. It is interesting to know that the person who first foresaw the possibility of transforming propositions into numerical values, Gottfried Wilhelm Leibniz, was a scholar of mathematics and law. In a paper written in 1669, when he was only 23 years old, dealing with the issue of 'conditional rights', that is, rights which hold only if certain conditions happen to be true, he proposed to assign the number 0 to *jus nullum* (a person has no rights to the possession of a certain asset), the number 1 to *jus purum* (a person has unconditional rights to the possession of a certain asset), and a fractional number to *jus conditionale* (a person has rights to the possession of a certain asset which depend on the occurrence of certain conditions) (Leibniz 1930, p. 420).

Consider two nodes H_1 and R, interpreted as propositions. It is judged that proposition H_1 is relevant for R so, a directed arc may be drawn from H_1 and R. The tree in Figure 1.3 can thus be represented in the *DAG* of Figure 2.2 with associated conditional probability values p, q, and r, represented in Tables 2.2 and 2.3.

The reader can appreciate how parsimonious a directed graph is, compared with a probability tree, even for the simplest possible case: instead of three random nodes and six

Figure 2.2 A Bayesian network for the probability tree shown in Figure 1.3 where H_1 and R each have two values, t (*true*) and f (*false*).

Table 2.2 Probability table of the node H_1 (Figure 2.2).

H_1:	t	f
$Pr(H_1)$	p	$1 - p$

Table 2.3 Probability table of the node R (Figure 2.2).

	H_1:	t	f
R :	$Pr(R = t \mid H_1)$	q	r
	$Pr(R = f \mid H_1)$	$1 - q$	$1 - r$

Figure 2.3 A Bayesian network for the coroner's report.

Table 2.4 Probability table of node H (Figure 2.3).

H:	H_1	H_2	H_3
$Pr(H = H_i)$	p_1	p_2	p_3

Table 2.5 Probability table of node R (Figure 2.3).

	H:	H_1	H_2	H_3
R :	$Pr(R = t \mid H)$	q_1	q_2	0
	$Pr(R = f \mid H)$	$1 - q_1$	$1 - q_2$	1

arcs, we have only two nodes and one arc and, above all, the relevance of H_1 for R is clearly visualised.

Another advantage of representing relevance by directed graphs is that a node can also represent a *cluster of propositions*. For example, a variable H, 'Hypothesis', may be defined, which has as its possible values the triplet composed of the truth-values of the propositions H_1 (Sir Charles Baskerville died by accident), H_2 (Sir Charles Baskerville died by a criminal act), and H_3 (Sir Charles Baskerville committed suicide). The possible states of the variable H will be $\{1, 0, 0\}$ (H_1 is true), $\{0, 1, 0\}$ (H_2 is true), or $\{0, 0, 1\}$ (H_3 is true). The Bayesian network shown in Figure 2.3 depicts Watson's beliefs (1.11) and (1.12) of the preceding chapter, where the conditional probability tables (Tables 2.4 and 2.5) are simplified by taking, with an abuse of notation, H_1, H_2, H_3 as the possible states of H.

2.1.4 Conditional independence

The fundamental concept in building Bayesian networks is conditional independence. It usually happens that it must be judged whether a given proposition B is relevant for another proposition A in a context that includes more than background knowledge. For instance, in the Baskerville case, Watson has to judge if the footprints are relevant to the crime hypothesis, given knowledge of the coroner's report. It is convenient to formulate the concept of conditional independence in terms of variables.

Let X, Y and Z be random variables and let $Pr(X \mid Y, Z, I)$ denote 'the degree of belief that a particular value of variable X is observed, given that a particular value of variable Y is observed, *and* given that a particular value of variable Z is observed, plus background information I'. Then

- X is conditionally independent of Y given Z if and only if $Pr(X \mid Y, Z, I) = P(X \mid Z, I)$ for all the states of X, Y and Z.

Statisticians use the concept of conditional independence to build computationally efficient statistical models. Here is a simple example.

It is required to forecast the outcomes of three successive tosses of a coin that will be thrown by another person. The result will be made available. The coin can be either a mint (fair) coin, a coin with two tails or a coin with two heads. Denote the cluster of hypotheses with the variable H and the possible outcomes with binary variables X_1, X_2 and X_3, which can take the values $X_i = h$ (the outcome of toss number i is heads) and $X_i = t$ (the outcome of toss number i is tails). It is obvious that the outcome of each toss depends upon which hypothesis is true, and it is also believed that the outcome of each toss in the sequence is relevant for the degree of belief in the outcome of the next toss. These dependence relationships can be represented by means of a directed graph, such as shown in Figure 2.4(i).

This figure may be simplified by taking advantage of the conditional independence structure of the problem domain. Clearly, if it were known which hypothesis were true, the outcome of each toss would not be relevant for the degree of belief concerning the next one. Absence of a direct link between two nodes means that there is no direct dependency between them: they are not directly relevant to each other. The arcs between the X_i variables may be dropped.

The graph in Figure 2.4(ii) depicts explicitly *all* the *direct* dependency relationships and *only* them. This graph is a faithful representation of the dependence structure of the problem, in the sense that *all* the *indirect* dependency relationships are implicitly represented. This can be demonstrated by an intuitive analysis.

The outcome of one toss is relevant for the expectation of the outcome of the next toss because it changes the relative degree of belief about the hypotheses, and this, in turn, changes the expectations for the other tosses. Suppose, for instance, that the outcome of the first toss is 'heads': this evidence changes the expectation of the next two tosses because

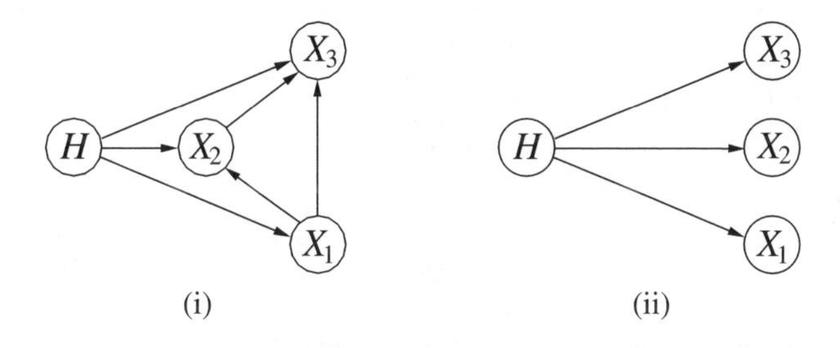

Figure 2.4 A *DAG* (i) and a parsimonious *DAG* (ii) for the coin-tossing problem.

it falsifies the hypothesis 'two tails'. The effect of the evidence acts directly upon node H, and only indirectly upon nodes X_2 and X_3, through the direct influence that node H exerts upon them. Imagine that Figures 2.4(i) and (ii) are physical nets where nodes are computing units and information can go from one unit to the others only by travelling along the arcs. Then evidence from unit X_1 is transmitted to units X_2 and X_3 by the computations executed by unit H, and channels $X_1 \rightarrow X_2$, $X_1 \rightarrow X_3$ and $X_2 \rightarrow X_3$ are superfluous (see Figure 2.4(i)). On the other hand, if the value of the variable H is fixed, then the flow of information between X_1, X_2, and X_3 through H is blocked. Therefore, the arcs $H \rightarrow X_1$, $H \rightarrow X_2$ and $H \rightarrow X_3$ are sufficient to represent faithfully the flow of information that travels through the net (see Figure 2.4(ii)).

These considerations can be generalised to cover all the possible ways by which information can be transmitted through a variable in a directed graph, obtaining a graphical decision rule to check whether any two variables are independent, given the knowledge of another variable or set of variables of the graph.

2.1.5 Graphical models for conditional independence: d-separation

There are only three possible connections by which information can travel through a variable in a directed graph: *diverging*, *serial*, and *converging* connections.

The case of *diverging connections* has already been discussed in the coin-tossing problem. Basic statistical models can be represented graphically by diverging connections. A diverging connection (*see* Figure 2.5 (i)) is an appropriate graphical model whenever it is believed that Z is relevant for both X and Y and that X and Y are conditionally independent given Z. This means that:

- if the state of Z is known, then knowledge also of the state of X does not change the belief about the possible states of Y (and vice versa);
- if the state of Z is not known, then knowledge of the state of X provides information about the possible states of Y (and vice versa).

Information may be transmitted through a diverging connection unless the value of the middle variable is fixed. It is then said that X and Y are d-separated (directionally separated) given Z.

Serial connections typically apply to 'chain' reasoning: a Markov Chain is a classical example. A serial connection (*e.g.*, Figure 2.5(ii)) is an appropriate graphical model whenever it is believed that A is relevant for C, that C is relevant for E, and that A and E are *conditionally independent* given C. This means that:

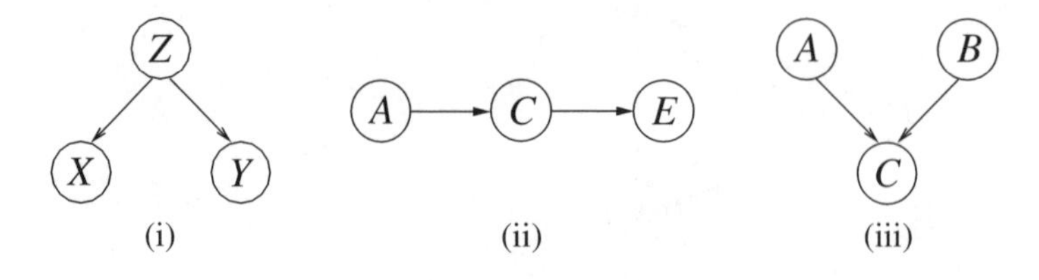

Figure 2.5 A diverging connection (i), a serial connection (ii) and a converging connection (iii).

- if the state of C is known, then knowledge also of the state of A does not change the belief about the possible states of E (and vice versa);

- if the state of C is not known, then knowledge of the state of A provides information about the possible states of E (and vice versa).

Information may be transmitted through a serial connection unless the state of the middle variable is known and it is then said that A and E are d-separated given C.

For example, let A in Figure 2.5(ii) be the proposition 'The suspect is the offender', C the proposition 'The blood stain found on the crime scene comes from the suspect', and E the proposition 'The suspect's blood sample and the bloodstain from the crime scene share the same DNA profile'. Then A is relevant for C and C for E, but there might be an explanation of the occurrence of the event described by C different from the occurrence of the event described by A and, given that C occurred, then the match does not depend upon which explanation of C is true.

Converging connections describe a slightly more sophisticated kind of reasoning. The ability to cope with this kind of reasoning is a real asset of Bayesian networks. A converging connection (*see* Figure 2.5 (iii)) is an appropriate graphical model whenever it is believed that A and B are both relevant for C, A is not relevant for B, but it does become relevant if the state of C is known. In other words, it is believed that A and B are unconditionally independent but conditionally dependent given C. This means that:

- if the state of C is known, then knowledge of the state of A provides information about the possible states of B (and vice versa);

- if the state of C is not known, then knowledge of the state of A provides no information about the possible states of B (and vice versa): the flow of information is blocked if the state of the middle variable is unknown.

Consider the following example. Let A and C in Figure 2.5(iii) be the propositions 'The suspect is the offender' and 'The bloodstain found on the crime scene comes from the suspect', respectively, and B the proposition 'The blood stain found on the crime scene comes from the offender'. Knowledge that either the event described by A or the event described by B occurred does not provide information about the occurrence of the other: the fact that the source of the bloodstain is the offender has no bearing at all, taken alone, on the culpability of the suspect. But, if C is true, then A and B become obviously related.

Converging connections have another peculiarity: to open the channel, it is not necessary that the state of the middle variable is known for certain, but it suffices that there is some information, even though it is not such as to achieve certainty. It is customary in the literature about Bayesian networks to call information that fixes the state of a variable *hard evidence*; otherwise it is called *soft evidence*. In other words, hard evidence bearing on a given variable allows its true state to be known, whereas soft evidence allows only the assignment of probability values lower than 1 (the value associated with certainty) to the states of the variable. Bayes' theorem deals with hard evidence, whereas formula (1.24), the so-called Jeffrey's rule, deals with soft evidence.

- Information may only be transmitted through a converging connection if there is some evidence (hard or soft) bearing on the middle variable or there is some evidence bearing on one of its descendants.

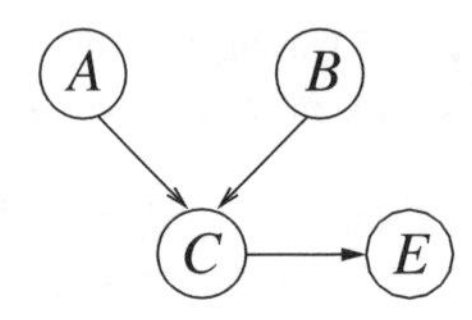

Figure 2.6 A *DAG* for 'one-trace' scenarios. A: suspect is the offender; B: blood stain at crime scene comes from offender; C: bloodstain at crime scene comes from suspect; E: suspect's blood sample and crime stain share same DNA profile.

The last condition can be easily understood by an example. Put together Figures 2.5(ii) and (iii) with the nodes interpreted as above, and a directed graph for the 'one-trace' scenario (Garbolino and Taroni 2002, p. 151), depicted by Figure 2.6, is obtained. A formal development of the likelihood ratios associated with this Bayesian network will be presented in Section 4.1.

Knowing that the DNA profile of the bloodstain matches the profile of the blood sample of the suspect, i.e., knowing the state of child E of C, opens the channel between A and B because this information, in turn, changes the degree of belief in C, even though this degree is still lower than certainty, allowing for the possibility of a random match.

2.1.6 A decision rule for conditional independence

Following the rules given in the preceding paragraph it is possible to build up step by step a Bayesian network for a given set of variables such that, if variables X and Y are conditionally independent given Z, then X and Y are d-separated given Z in the network.

With the same rules it is also possible to decide for any pair of variables X and Y in a given Bayesian network whether they are independent given another variable Z. A *decision rule* for d-separation can be formulated in general, for subsets of variables in a *DAG*, as follows (Pearl 2000, pp. 16–17).

> Let $\mathbf{X}$, $\mathbf{Y}$ and $\mathbf{Z}$ be disjoint subsets of variables in a DAG. Then $\mathbf{X}$ and $\mathbf{Y}$ are d-separated given $\mathbf{Z}$ if and only if every path between a variable in $\mathbf{X}$ and a variable in $\mathbf{Y}$ contains:
>
> - either a serial connection $\rightarrow Z \rightarrow$, or a diverging connection $\leftarrow Z \rightarrow$, such that the middle node Z belongs to $\mathbf{Z}$;
> - or a converging connection $\rightarrow W \leftarrow$ such that the middle node W does not belong to $\mathbf{Z}$ and no descendant of W belongs to $\mathbf{Z}$.

As noted above, nodes A and B are not d-separated given $\mathbf{Z} = \{E\}$ in Figure 2.6 because E is a descendant of C. In Figure 2.1(i), nodes B and D are d-separated given $\mathbf{Z} = \{A, E\}$, because there are two paths going between B and D, namely, $B \leftarrow A \rightarrow E \rightarrow D$ and $B \rightarrow C \leftarrow D$: in the first there is a diverging connection at A (and a serial connection at E) ; in the second, there is a converging connection at C and neither C belongs to the set $\{A, E\}$ nor any descendant of C belongs to $\{A, E\}$ (in fact, C has no descendants). On the other hand, B and D are not d-separated given $\mathbf{Z} = \{A, C\}$ because, in this case, the middle node C of the converging connection belongs to the set $\mathbf{Z} = \{A, C\}$.

It can be proved for Bayesian networks that:

- if sets of nodes **X** and **Y** are d-separated given **Z**, then they are conditionally independent given **Z**.

Therefore, it is not necessarily the case that the graphical structure represents all the independence relationships which hold among the variables, but the graphical structure ensures that, at least, no false relationship of independence can be read from the graphical structure and that all the dependencies are represented.

2.1.7 Networks for evidential reasoning

Both in Graphs 2.4(ii) and 2.6 we can distinguish between *hypothesis nodes* and *evidence nodes*. In the statistical model the hypothesis node is **H**, the 'parameter' variable, and the evidence nodes are the variables X_i describing the possible results of empirical observations. In the 'one-trace' scenario the hypothesis nodes are A, B and C, namely, the propositions which describe singular events whose occurrence is only hypothesised and the evidence node is E, the proposition describing an observed event. The arcs in these graphs are said to go top-down, from hypotheses to evidence.

Relevance is a symmetric relation (*see* Section 1.1.4). One may wonder whether graphs with the direction of arcs reversed, going bottom-up, from evidence to hypotheses, would not provide models for dependency which are as good as those in Figure 2.4(ii) and 2.6. The answer is that it is usually easier to construct top-down graphs. This is because Bayesian inference goes top-down, from hypotheses to evidence (*see* Section 1.4.3):

- Top-down arcs go from *explanans* to *explanandum*.

Scholars of evidence as David Schum, Terence Anderson and William Twining have argued that what is to be sought in evidential reasoning are justified or warranted beliefs. As Schum wrote (Schum 1994, p. 81, note 4):

> (····) people whose opinions we are trying to influence will wish to know why we believe we are entitled to reason from one stage to another (····) For each reasoning stage or step identified we have to make an assertion about what we believe provides the ground for or gives us license to take each one of these steps. Among some logicians and legal scholars such assertions are called *generalisations*. Other writers refer to such assertions as warrants. Whether we call them generalisations or warrants, the intent is the same; they are assertions we make about why we believe we are entitled to reason from one stage to another.

The question 'Why do we believe we are entitled to make an inference from evidence E to hypothesis H?' is not to be confused with the question, 'Why do we believe, to a certain degree, that H is true?'. The answer to the latter question is: 'Because we have observed that E is true and E is relevant for the truth of H'. However, the former question asks 'Why do we believe that E is relevant for H?' The answer to this question is: 'Because H is a *possible explanatory fact* for E'.

This is an important point related to the issue of 'subjective' *versus* 'objective' probabilities which has been already raised in Section 1.3.1. Acknowledgement in the evaluation of evidence that one has to trust one's personal beliefs does not mean that one is left merely with personal feelings that something is relevant for someone else. One is asked to explain *why* it is believed that an observed event E is relevant for another, unobserved, event H. The answer points to a potential explanation of E whose *explanans* contains H *plus* scientific laws and common sense generalisations which provide the required warranties of the inference. Scientific laws and common sense generalisations are 'objective' in the sense that there is a widespread agreement about them.

- E is 'objectively' relevant for H if H is an explanatory fact in a potential explanation of E.

Anderson and Twining associate *probandum* with *explanans*, and *probans* with *explanandum.* The question to be addressed by the 'proponent of evidence' is, according to them (Anderson and Twining 1998, p. 71):

> *Does the evidentiary fact point to the desired conclusion* (not as the only rational inference, but) *as the inference* (or explanation) *most plausible or most natural out of the various ones that are conceivable*? Or (to state the requirement more weakly), is the desired conclusion (not, the most natural, but) a natural or plausible one among the various conceivable ones? [...] How probable is the *Probandum* as the explanation of this *Probans*? [Italics by the authors]

And the task of the 'opponent of evidence' is to provide an alternative explanation (Anderson and Twining 1998, p. 73):

> That fact has been admitted in evidence, but its force may now be diminished or annulled by showing that some explanation of it other than the proponent's is the true one. Thus every sort of evidentiary fact may call for treatment in a second aspect, by the opponent, viz.: *What are the other possible inferences which are available for the opponent as explaining away* the force of the fact already admitted. [Italics by the authors]

The quest for explanations breaks the symmetry of relevance relationships assigning a vantage point to top-down directed graphs: the *probans* is relevant for the *probandum* and vice versa, but it is only the latter that explains the former. In the coin-tossing problem of Figure 2.4(ii), the state of X_i is relevant for the possible states of H, but it is the state of H that explains the state of X_i.

Consider again the 'one-trace scenario' of Figure 2.6. The appropriate conditional probabilities tables are given in Tables 2.6, 2.7 and 2.8. Notice that some of the relationships in the network are deterministic, i.e., they have assigned probability values of one or zero.

Consider the link between the variables C and E: the state of E provides information about the possible states of C, but it is the state of C that explains the state of E. Also intermediate hypotheses are to be explained by means of high-level hypotheses. Node C is

Table 2.6 Unconditional probabilities assigned to the nodes A and B of Figure 2.6.

A:	t	f
$Pr(A)$	a	$1-a$

B:	t	f
$Pr(B)$	r	$1-r$

Table 2.7 Conditional probability table for the node C of Figure 2.6.

	A:	t		f	
	B:	t	f	t	f
C:	$Pr(C = t \mid A, B)$	1	0	0	p
	$Pr(C = f \mid A, B)$	0	1	1	$1 - p$

Table 2.8 Conditional probability table for the node E of Figure 2.6.

	C:	t	f
E:	$Pr(E = t \mid C)$	1	γ
	$Pr(E = f \mid C)$	0	$1 - \gamma$

the intermediate hypothesis whose state can be explained by the states of A and B. Node B represents the factor that has been called *relevance* (Stoney 1991) because its probability $Pr(B \mid I)$ encodes the information available for the belief that the bloodstain is relevant for the case.

In the conditional probability tables taken from Garbolino and Taroni (2002, p. 151), assignments of probabilities 1 and 0 are straightforward, probability γ is the random-match probability and p, the probability that the stain would have been left by the suspect even though he was innocent, is to be assigned on the basis of the alternative explanation proposed by the defence and the information related to that hypothesis. If the defence agrees that C is true, then, of course, this scenario is no longer appropriate.

2.1.8 Relevance and causality

Explanations, especially those involving generalisations based on common sense, are often purportedly *causal explanations*. The fact that warrants for evidence are usually sought, and they make up scenarios or histories that follow a chronological order, leads, using common sense, to an understanding of the order as a causal order; events which are part of the *probandum* occur *before* events which constitute the *probans* and, normally, *causes* occur before their effects.

> Constructing a network representation of an inference problem is a purely subjective judgmental task, one likely to result in a different structural pattern by each person who performs it. In the contemporary literature on inference networks, there is now some controversy concerning what people should and do attend to when such tasks are performed. (· · · ·) this controversy concerns whether such structuring always involves the tracking of causal relations. (Schum 1994, p. 175)

Many authors, among them Salmon (1998) (mentioned in Section 1.4.1), consider that statistical relevance has not a genuine explanatory import, and that only causal relevance has. A consequence of this position is that a probabilistic explanation would be legitimate if it were possible to identify the *causal mechanism* underlying it. Pearl has forcefully argued for the advantages of building *DAG* models around causal relationships, claiming that they are more meaningful, more accessible, and felt by people as more reliable than mere probabilistic relevance relationships (Pearl 2000, p. 21; 25):

> (····) conditional independence judgments are accessible (hence reliable) only when they are anchored onto more fundamental building blocks of our knowledge, such as causal relationships.
>
> (····) Causal relationships are more stable than probabilistic relationships. We expect such difference in stability because causal relationships are *ontological*, describing objective physical constraints on our world, whereas probabilistic relationships are epistemic, reflecting what we know or believe about the world. (····) The element of stability (of mechanisms) is also at the heart of the so-called explanatory accounts of causality, according to which causal models (····) aim primarily to provide an "explanation" or "understanding" of how data are generated.

In the context of evidential reasoning, it can turn out to be very difficult, if not impossible, to substitute some epistemic relationships with nodes and arcs exhibiting a recognisable genuine causal mechanism.

For instance, an intuitive interpretation of the path from node A to node E through node C in the graph in Figure 2.6 as a 'mechanism' suggests itself easily to the mind, but the arc from node B to node C seems to represent the influence of an epistemic condition necessary to 'open' the channel from A to C rather than a causal 'mechanism' acting between B and C. One could argue that A and B fit the definition of *inus* conditions in John Mackie's *regularity theory of causation* (Mackie 1974) and that, therefore, there is a kind of causal explanation after all.

An *inus* condition for some effect is an insufficient but non-redundant part of an unnecessary but sufficient condition. In our example, A and B are *inus* conditions for C, for

- they are both insufficient: $Pr(C \mid \bar{A}, B, I) = Pr(C \mid A, \bar{B}, I) = 0$;
- they are jointly sufficient: $Pr(C \mid A, B, I) = 1$;
- they are not necessary: $Pr(C \mid \bar{A}, \bar{B}, I) > 0$.

The answer is that, even though this particular explanation can be interpreted as a causal explanation according to Mackie's theory, there is no obligation to try to figure out causal explanations in the style of Mackie for any argument in evidential reasoning.

Evidential analysis should take a neutral stance, as long it is possible, on the controversies about what a causal relationship is and what constitutes a causal explanation. Therefore, there is agreement with Schum's viewpoint that the capability of providing a causal explanation is not a necessary condition for justifying the existence of a relevance relationship (Schum 1994, pp. 141; 178):

> In many inference tasks the hypotheses under consideration represent possible causes for various patterns of observable effects. In such cases, we may be

> able to trace what we regard as a causal linkage from some ultimate or major hypothesis to events or effects that are observable. If such linkage can be established, evidence of these effects would be relevant in an inference about these hypothesised causes. But there are many instances in which evidence can be justified as relevant on hypotheses of interest when links in a chain of reasoning may indicate no causal connection.
>
> (⋯) To require that the arcs of an inferential network signify direct causal influences is to hamstring the entire enterprise of applying conventional probability theory in the analyses of inferential networks.

The strategy of building up a Bayesian network following the path of Bayesian inference (*see* Section 1.4.3) is sufficient to break the symmetry of probabilistic relevance and the only requirement to be satisfied is that the d-separation property holds:

- if sets of variables $\mathbf{X}$ and $\mathbf{Y}$ are conditionally independent given the set $\mathbf{Z}$ in the chosen explanatory pattern, then $\mathbf{X}$ and $\mathbf{Y}$ must be d-separated given $\mathbf{Z}$ in the Bayesian network representing that pattern.

The explanatory pattern may be a causal pattern, according to one's favourite theory of causation, and it is true that 'causal' explanations which make use of 'causal' common sense generalisations are usually felt as more intuitive and therefore more 'objective' and acceptable (but this is not always the case if recourse has to be made to sophisticated scientific laws which express some 'causal' relationships that are not intuitive). It is better if causal explanations can be provided, but it is not necessary that they be so.

2.1.9 The *Hound of the Baskervilles* revisited

Doctor Watson wishes to draw a *DAG* which represents the explanatory inferences supporting the evidential reasoning in the Baskerville case. The propositions in question are: 'Sir Charles Baskerville died by accident' (H_1), 'Sir Charles Baskerville died by a criminal act' (H_2), 'Sir Charles Baskerville committed suicide' (H_3), 'The proximate cause of Sir Charles Baskerville's death was a heart attack' (R), and 'A gigantic hound was running after Sir Charles Baskerville' (F). Therefore the nodes will be H, with three possible states $\{H_1, H_2, H_3\}$, R and F, with two possible states each. (*see* Figure 2.7)

There will be an arc pointing from node H to node R, because it is easy to provide a probabilistic explanation of Sir Charles Baskerville's death by heart attack: the propositions H_i constitute a partition of the general class of all the possible death causes and $Pr(R \mid H_i) \neq Pr(R \mid H_j)$ for $i \neq j$. The level of details of the partition depends

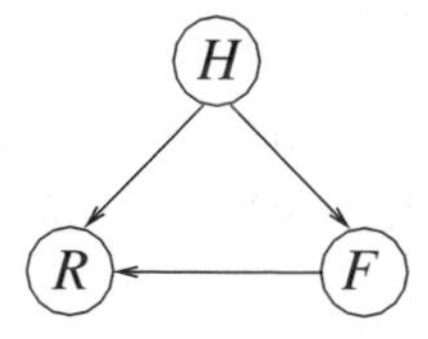

Figure 2.7 A Bayesian network for the Baskerville case.

upon the context in which the explanation is used and the reason why the explanation is requested.

As regards the potential explanations of F, Watson might articulate his opinion (1.21) by arguing that the probability of F given that an intentional action was carried out is greater than the probability of F given that no intentional action was carried out, and that the probability of F is the same in case that hypothesis H_1 is true and in case hypothesis H_3 is true. The occurrence of a heart attack does not explain the presence of a hound any more than Sir Charles Baskerville's will to commit suicide would explain it. In both cases, the only possible explanation is in terms of the frequency of the presence of stray dogs on the moor.

Finally, according to his professional opinion, Watson believes that a stray hound running after a person with a weak heart can explain the occurrence of a heart attack. For the case he committed suicide, there would be no connection at all between the two events. Therefore, there will be an arc pointing from node F to node R. The numerical values assigned in the conditional probability tables, Tables 2.9 and 2.10, are Dr Watson's responsibility and they are coherent with the qualitative assessments (1.11), (1.21) and (1.22).

This simple story contains an instance of the phenomenon called *asymmetric independence*, that occurs when variables are independent for some but not all of their values. The arc from F to R is necessary to encode the dependency of R from F, given H_1 and H_2, but the full *DAG* cannot represent the independence of R from F, given H_3. This asymmetric independence relationship contained in the story, at least as told by Watson, can be read only in the values of the conditional probability table. In order to be able to read it directly from a graphical model, local networks can be drawn, each representing part

Table 2.9 Probabilities for the node H (left) and conditional probabilities for F, given H(right) of Figure 2.7. Node H concerns the method by which Sir Charles Baskerville died with three states, death by accident, H_1, death by a criminal act, H_2, and death by suicide, H_3. Node F represents the event that a gigantic hound was running after Sir Charles Baskerville with two states, true, t, and false, f.

H:	H_1	H_2	H_3
$Pr(H)$	0.98	0.01	0.01

	H:	H_1	H_2	H_3
F:	$Pr(F = t \mid H)$	0.01	0.05	0.01
	$Pr(F = f \mid H)$	0.99	0.95	0.99

Table 2.10 Conditional probability table for the node R, given H and F of Figure 2.7. Node R represents the event that the proximate cause of Sir Charles Baskerville's death was a heart attack with two states, true, t, and false, f.

	H:	H_1		H_2		H_3	
	F:	t	f	t	f	t	f
R:	$Pr(R = t \mid H, F)$	0.4	0.1	0.4	0.001	0	0
	$Pr(R = f \mid H, F)$	0.6	0.9	0.6	0.999	1	1

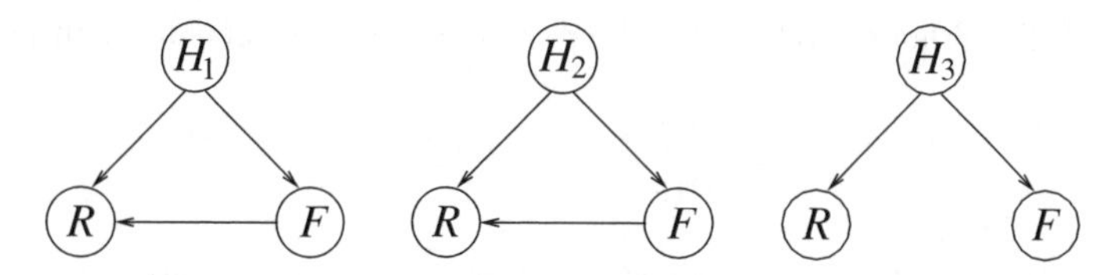

Figure 2.8 The local networks for the Baskerville case. Three states are proposed for the method by which Sir Charles Baskerville died, death by accident, H_1, death by a criminal act, H_2, and death by suicide, H_3.

of the problem. Methods to deal with Bayesian networks with asymmetric independencies have been proposed in Geiger and Heckerman (1991), Friedman and Goldszmidt (1996) and Mahoney and Laskey (1999).

Following for instance, the *Bayesian Multinet* method of Geiger and Heckerman, a DAG is associated with each H_i and the DAG is the Bayesian network given that one of the values in H holds. The Bayesian Multinet of Figure 2.7 is the set of the local networks as given in Figure 2.8, together with a marginal probability distribution on H.

Local networks represent what is also called *context-specific independence*. This example is computationally simple and it can work with the full probability distribution of the net (*see* Figure 2.7). However, *context-specific* independence can be exploited to reduce the burden of model specification and to improve the efficiency of the inference.

To work out the example, first verify that the numerical values in the conditional probability tables, Tables 2.9 and 2.10, are coherent with the qualitative judgment (1.12) (numbers are rounded off to the third decimal):

$$\begin{aligned}
Pr(R \mid H_1, I) &= Pr(R \mid H_1, F, I)Pr(F \mid H_1, I) + \\
&\qquad Pr(R \mid H_1, \bar{F}, I)Pr(\bar{F} \mid H_1, I) = 0.103, \\
Pr(R \mid H_2, I) &= Pr(R \mid H_2, F, I)Pr(F \mid H_2, I) + \\
&\qquad Pr(R \mid H_2, \bar{F}, I)Pr(\bar{F} \mid H_2, I) = 0.021, \\
Pr(R \mid H_3, I) &= 0\,.
\end{aligned}$$

and with (1.17):

$$Pr(F \mid H_1, R, I) = \frac{Pr(R \mid H_1, F, I)Pr(F \mid H_1, I)}{Pr(R \mid H_1, I)} = 0.039\,,$$

$$Pr(F \mid H_2, R, I) = \frac{Pr(R \mid H_2, F, I)Pr(F \mid H_2, I)}{Pr(R \mid H_2, I)} = 0.952\,.$$

The likelihood ratio is slightly in favour of H_2:

$$V = \frac{Pr(F \mid H_1, I)}{Pr(F \mid H_2, I)} \times \frac{Pr(R \mid H_1, F, I)}{Pr(R \mid H_2, F, I)} = \frac{0.01}{0.05} \times \frac{0.4}{0.4} = 0.2\,.$$

The posterior odds, although lower than the prior odds, are still decidedly in favour of H_1:

$$V \times \frac{0.98}{0.01} = 0.2 \times 98 = 19.6\,.$$

Conditional on learning Dr Mortimer's story, Watson's common sense assigns a posterior probability of about 95% to the hypothesis that poor Sir Charles Baskerville has been victim of a tragic fatality.

2.2 Reasoning with Bayesian networks

2.2.1 *'Jack loved Lulu'*

The Bayesian network for the Baskerville case, at least for the initial step of the inquiry, is quite simple and, as a matter of fact, it has been possible to execute immediately the required calculations. Consider a more complex network, obtained by combining a 'one-trace scenario' with an 'eye-witness scenario' (Figure 2.9 and Table 2.11).

A young girl, Lulu, has been found murdered at her home with many knife wounds. The knife has not been found. Some bloodstains have been recovered on the scene of the crime which do not share Lulu's DNA profile. A friend of hers, Jack, has been seen near Lulu's house around the time of the murder by John. There is some evidence that Jack was badly in love with Lulu. This information is considered as background information for the evaluation of the probability of the root node J. John has stated that he also was in love with Lulu. A blood sample has been taken from Jack. The fact that John bore witness as he did is an event that must be explained, and the explanatory facts for it are that the event John related occurred and that John is a reliable witness, under the common sense generalisation that 'reliable witnesses usually say the truth'. It is not known for certain that Jack loved Lulu, otherwise that would be part of the background information also and it would not be a node of the graph. Instead, it is a hypothesis based on information which is not totally reliable and a hypothesis that can furnish an alternative explanation of Jack's presence near Lulu's house. It is also assumed that the fact that John himself fell in love with Lulu can prejudice his reliability as a witness.

The network could be enlarged by taking into account the testimony, and its reliability, of some witnesses that Jack loved Lulu and by also taking into account the analysis of eye-witness reliability through consideration of the variables involved in the evaluation of testimonial evidence (Schum 1994, pp. 100–114; 324–344). Readers can find in Kadane and Schum's work on Sacco and Vanzetti evidence (Kadane and Schum 1996) good examples of how complicated a real case analysis can grow.

In what follows, the 'black box' of the computational architecture of Bayesian networks is examined and it is shown that the architecture works according to the principles of the logic of uncertainty that we have outlined in Chapter 1. Formal proofs are not given of the mathematical results exploited because, for readers interested in practical applications of Bayesian networks and who are going to use commercial software already available, it is more important to understand why the answers provided by the 'black box' may be trusted, rather than to know in any detail the mathematics of the inference engine.

Table 2.11 Definitions of the nodes used in the 'Lulu' Bayesian network in Figure 2.9.

A:	Jack stabbed Lulu.
B:	Bloodstain at crime scene comes from offender.
C:	Bloodstain at crime scene comes from Jack.
E:	Jack's blood sample and crime stain share the same DNA profile.
F:	Jack was in a certain place f near the house where Lulu lived shortly after the time the crime was committed.
W:	John says that Jack was in place f shortly after the time when the crime was committed.
J:	Jack loved Lulu.
D:	John was jealous of Jack.

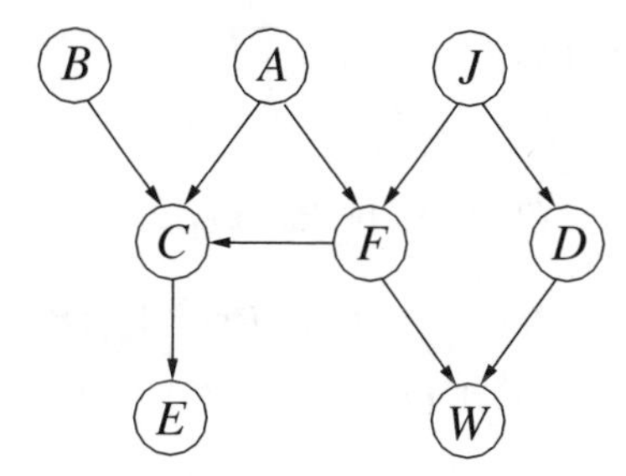

Figure 2.9 The 'Lulu' Bayesian network. The descriptions of the nodes are given in Table 2.11

2.2.2 The Markov property

The multiplication law allows us to decompose a joint probability distribution with n variables as a product of $n - 1$ conditional distributions and a marginal distribution:

$$Pr(X_1, \ldots, X_n) = \left[\prod_{i=2}^{n} Pr(X_i \mid X_1, \ldots, X_{i-1})\right] Pr(X_1).$$

The following fact holds in any Bayesian network and it constitutes the fundamental property of Bayesian networks, also known in the literature as the *Markov property*:

a DAG with a joint probability distribution $Pr()$ over its variables is a Bayesian network if and only if

- for every variable X in the *DAG*, and every set $\mathbf{Y}$ of variables such that it does not include the set $\mathbf{DE}(X)$ of descendants of X, X is conditionally independent from $\mathbf{Y}$ given the set $\mathbf{PA}(X)$ of its parents:

$$Pr(X \mid \mathbf{PA}(X), \mathbf{Y}) = Pr(X \mid \mathbf{PA}(X)).$$

The joint probability distribution associated with the *DAG* in Figure 2.9 is:

$$Pr(A, B, C, D, E, F, J, W) \tag{2.1}$$

and it is easy to check that the Markov property holds. For example, B is d-separated from $\{A, F, J, D, W\}$ given the *null set* $\oslash$ (root nodes have the null set as the only parent), because the path to A contains a converging connection at C and C does not belong to $\oslash$. On the other hand, B is not d-separated from E, given $\oslash$, because the connection at C is serial. Consider two more examples: F is d-separated from $\{B, D\}$ given $\{A, J\}$ because there is a diverging connection at A and a converging connection at C, and neither C nor E belong to $\{A, J\}$; D is d-separated from $\{A, B, C, E, F\}$ given $\{J\}$ because there is a diverging connection at J and a converging connection at W.

From the Markov property the so-called *chain rule* for Bayesian networks follows immediately:

- a Bayesian network can be factorised as the product, for all variables in the network, of their probabilities conditional on their parents only:

$$Pr(X_1, \ldots, X_n) = \prod_{i=1}^{n} Pr(X_i \mid \mathbf{PA}(X_i)).$$

Therefore, the joint distribution (2.1) can be factorised as

$$\begin{aligned} &Pr(A)Pr(B)Pr(C \mid A, B, F)Pr(E \mid C)Pr(F \mid A, J) \\ &\quad Pr(J)Pr(D \mid J)Pr(W \mid D, F). \end{aligned} \tag{2.2}$$

The task of computing a given joint probability distribution becomes very rapidly intractable because its complexity increases exponentially with the number of variables. In a simple problem with only eight binary variables like the 'Lulu' network the joint distribution contains $2^8 = 256$ values. The Markov property provides a method by which the computational problem of probability calculus may be handled. Although it does not guarantee a tractable task, the method is very efficient. The key lies in the observation that, although the full joint probability distribution of the 'Lulu' problem needs memory space for 2^8 values, the computation of the biggest factor in (2.2) needs only memory space for 2^4 values.

In order to be able to apply Bayes' rule in the 'Lulu' case, it is necessary to calculate the marginal distributions of the evidence nodes E and W from the conditional probabilities in (2.2). Consider, first, the task of computing the marginal probability distribution of E:

$$Pr(E) = \sum_{C} Pr(E \mid C)Pr(C). \tag{2.3}$$

The marginal probability distribution of C is given by

$$Pr(C) = \sum_{ABF} Pr(C \mid A, B, F)Pr(A, B, F). \tag{2.4}$$

The variable B is d-separated from $\{A, F\}$ given $\oslash$, therefore we can rewrite (2.4) as:

$$Pr(C) = \sum_{ABF} Pr(C \mid A, B, F)Pr(A, F)Pr(B). \tag{2.5}$$

The final goal is the marginal probability distribution $Pr(A, F)$,

$$Pr(A, F) = \sum_{J} Pr(A, F, J) = \sum_{J} Pr(F \mid A, J)Pr(A)Pr(J). \tag{2.6}$$

A subset of nodes of a graph is called a *cluster*. The determination of a value for (2.6) (also known as a *sub-task*) requires working with the cluster $\{A, F, J\}$ and $2^3 = 8$ probability values have to be stored. Execution of sub-task (2.5) requires working with the cluster $\{A, B, C, F\}$ and the storage of a further $2^4 = 16$ probability values. Finally, the cluster of sub-task (2.3) is $\{C, E\}$, with the requirement to store $2^2 = 4$ probability values.

The marginal probability distribution of W may be computed in an analogous manner.

$$Pr(W) = \sum_{DF} Pr(D, F, W) = \sum_{DF} Pr(W \mid D, F)Pr(D, F). \tag{2.7}$$

The next sub-task is the computation of the marginal distribution $Pr(D, F)$. The variable D is d-separated from F given J, *i.e.*, D is conditionally independent of F given J. Therefore:

$$\begin{aligned} Pr(D, F) &= \sum_{J} Pr(D, F, J) = \sum_{J} Pr(D \mid F, J)Pr(F, J) \\ &= \sum_{J} Pr(D \mid J)Pr(F, J). \end{aligned} \tag{2.8}$$

The marginal distribution $P(F, J)$ can be obtained from the probability values which have already been stored in memory when (2.6) was executed:

$$Pr(F, J) = \sum_{A} Pr(A, F, J) = \sum_{A} Pr(F \mid A, J)Pr(J)Pr(A). \tag{2.9}$$

This means that sub-task (2.9) handles the same cluster as (2.6) and that extra memory space is not needed for this computation. Sub-task (2.8) works with cluster $\{D, F, J\}$ and adds $2^3 = 8$ probability values to the store, while the cluster of sub-task (2.7) is $\{W, F, D\}$, and $2^3 = 8$ more values are added to the store.

Thus, $(8 + 16 + 4 + 8 + 8) = 44$ probability values have been computed instead of $2^8 = 256$, quite a remarkable saving of computational resources, and the goals have been achieved working with five clusters:

$$\{A, F, J\},\ \{A, B, C, F\},\ \{C, E\},\ \{D, F, J\}, \{D, F, W\}. \tag{2.10}$$

Let us call the set of clusters (2.10) the *Markov domain* of the *DAG* in Figure 2.9.

2.2.3 *Divide and conquer*

The flow of computations (2.3)–(2.9) is illustrated by Figure 2.10, where rectangular boxes are inputs and outputs, and oval boxes indicate the computations that are performed. Care has to be taken with these symbols to note the context; in other contexts, oval boxes denote discrete variables and rectangular boxes denote continuous variables (*see* Chapter 10).

The basic idea for the implementation of the flow of computations by an expert system can be summarised as follows:

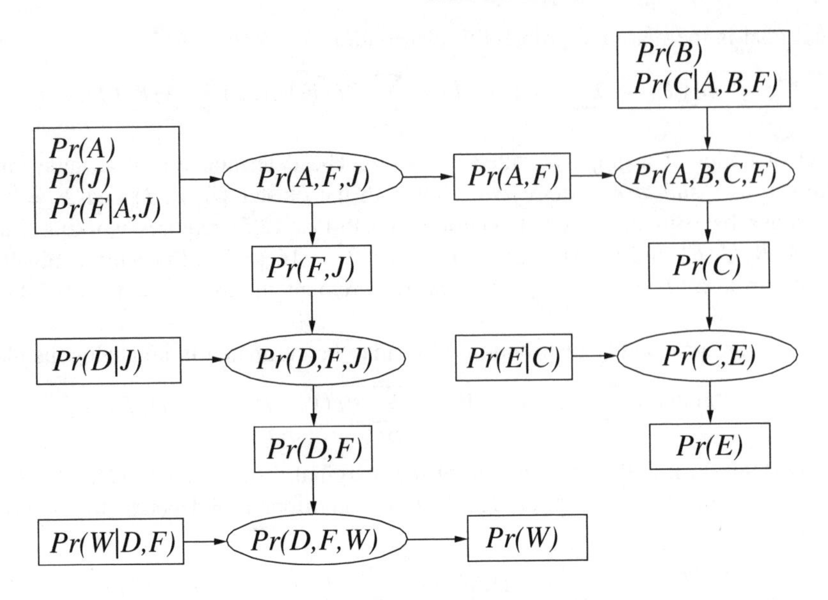

Figure 2.10 Flow diagram of the computation of $Pr(E)$ and $Pr(W)$.

A new graph is constructed which has subsets of nodes of the original *DAG* as its nodes: such a so-called *cluster graph* satisfies the two properties that

(i) for every pair of clusters there exists a unique path connecting them;

(ii) every node that belongs to two clusters also belongs to every cluster in the path between them.

These two properties together ensure a correct propagation of information because the new graph has the same topology as that depicted in Figure 2.10; there is only one path between any two oval boxes in Figure 2.10 and the oval box on the path between $Pr(A, F, J)$ and $P(D, F, W)$ does contain the variable F.

The cluster graph will become a permanent part of the expert system knowledge representation, a part that will change only if the original *DAG* changes. Each cluster stores its local joint distribution which can be used to answer queries for probabilities of any node in the cluster. In the following sections, the 'Lulu' network is used to give an overview of the procedure for the construction of the appropriate cluster graph and of the local propagation algorithm which passes information between pairs of adjacent clusters.

There is another desirable property the cluster graph should satisfy, beyond those already mentioned. The number of nodes contained in each cluster can be taken as a measure of computational complexity, and it is desirable to use clusters with as few nodes as possible. Unfortunately, the general problem of finding clusters with minimal size is NP-hard, as is the general problem of probabilistic computation in a Bayesian network. The following procedure, although it does not guarantee to find the minimal solution, is among the most efficient.

Clustering (Whittaker 1990):

- transform the *DAG* into an undirected graph, one with the arrowheads removed, and with links between parents of common children, the *moral graph* (the definition of which will be given in Section 2.2.4);
- turn the moral graph into a *triangulated graph* (the definition of which will be given in Section 2.2.4);
- find the clusters of the triangulated graph which are the *cliques* of the graph (the definition of which will be given in Section 2.2.5);
- connect the cliques to form a *junction*, or *join tree*.

A junction tree satisfies the properties of a cluster graph. The procedure is such to guarantee that the set of nodes of the junction tree, *i.e.*, the set of cliques, is identical with the Markov domain of the original *DAG*.

2.2.4 From directed to triangulated graphs

The first step of clustering is the *moralisation* of the directed graph. The name comes from the fact that parents of the same child are 'married' together.

- A graph obtained by adding a link between parents of common children in a directed graph, and then removing all the arrowheads is called a *moral graph*.

It is obvious that moralising is a mechanical procedure so that it can be executed by a computer.

The second step of clustering is the crux of the matter: *triangulation*.

- An undirected graph is said to be *triangulated* if any cycle of length four or more has at least one *chord*. A chord is a link between two nodes in a cycle that is not contained in the cycle.

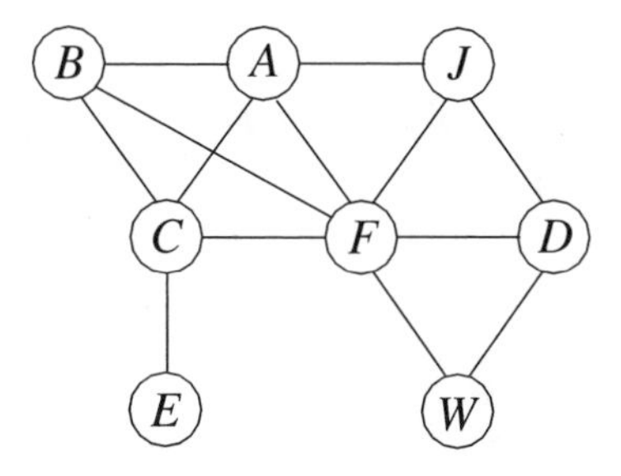

Figure 2.11 The 'Lulu' moral graph.

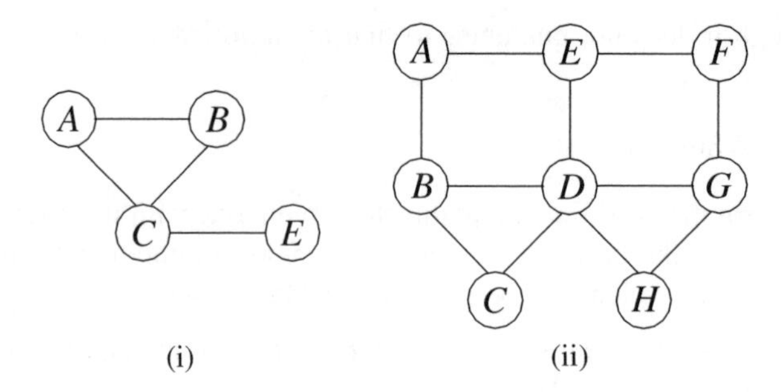

Figure 2.12 (i) Moral graph of the 'one-stain' scenario *DAG* (Figure 2.6) and (ii) moral graph for the *DAG* shown in Figure 2.1(i).

The 'Lulu' moral graph (*see* Figure 2.11) is triangulated, as is the moral graph (*see* Figure 2.12(i)) associated with the 'one-trace' scenario (*see* Figure 2.6).

Not every moral graph is triangulated. An example of a non-triangulated graph is given in Figure 2.12(ii).

Why should the moral graph be triangulated? There is a theorem of Graph Theory saying that an undirected graph has a junction tree if and only if it is triangulated.

- A necessary, and sufficient, condition for completing the procedure of construction of the junction tree is that the moral graph is triangulated.

A non-triangulated graph can be triangulated by adding chords to break the cycles. This process is called *fill-in* or *triangulation*. A cycle can be broken in several ways, so that there are different ways to triangulate a graph. Figure 2.13 shows two ways of triangulating the moral graph in Figure 2.12(ii). Readers are left to check other ways.

The problem that arises in filling-in a non-triangulated graph, with respect to the complexity issue, is that adding chords raises the size of the cliques, therefore it is desirable to add as few chords as possible, in order to obtain clusters with the smallest size. A minimal fill-in has the minimum number of chords needed to triangulate a given graph. The fill-in in Figure 2.13(ii) is minimal but the one in Figure 2.13(i) is not, because removing either the chords $A - D$ or $B - E$ will still leave a triangulated graph. The problem of finding a minimal fill-in is NP-hard, and hence the general problem of finding minimal

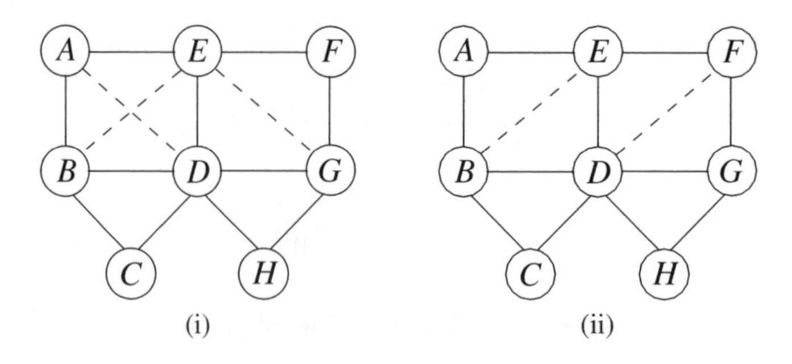

Figure 2.13 Two different triangulations for the moral graph depicted in Figure 2.12(ii).

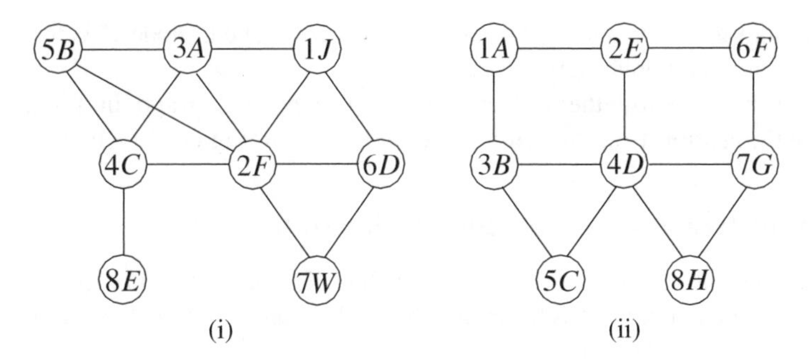

Figure 2.14 A possible ordering by MCS of (i) the 'Lulu' moral graph and (ii) the moral graph presented in Figure 2.12(ii).

clusters is NP-hard. We have efficient algorithms to check whether an undirected graph is triangulated and for triangulating it, if it is not. But, these algorithms do not guarantee a minimum fill-in.

A simple triangulation method makes use of two algorithms, the *maximum cardinality search* (*MCS*), followed by the *elimination algorithm*.

Maximum cardinality search orders nodes of the moral graph in the following manner:

(i) assign number 1 to an arbitrary node;

(ii) assign the next number to an unnumbered node with the largest number of numbered neighbours (the *neighbours* of a node X are all the nodes linked to X in an undirected graph); break ties arbitrarily.

A possible ordering of nodes obtained by applying MCS algorithm to 'Lulu' moral graph (*see* Figure 2.11) is given in Figure 2.14(i).

The *elimination algorithm* visits nodes of the moral graph in reverse numerical order and for each node X it executes the following instructions:

(i) add links so that all neighbours of X are linked to each other;

(ii) remove X together with all its links;

(iii) continue until all nodes have been eliminated.

The original graph together with all the links which have been added by executing the procedure is triangulated. A graph is already triangulated if and only if all the nodes can be eliminated without adding new links.

It is clear by visual inspection that the graph in Figure 2.14(i) is triangulated without adding new links, but a computer needs a mechanical procedure to make the inspection and the elimination algorithm is such a procedure.

Starting from the highest numbered node E, its neighbour is C, node E and its link to C may be eliminated. Then move to node W, its neighbours F and D are already linked, so eliminate node W and its links to nodes F and D. Repeat the procedure with node D, and so on. On arrival at node, C, nodes E and B have already been eliminated; therefore the neighbours of C at this step of the procedure are A and F, and they are linked together.

Eliminate C and go to A, and so on, until node J is reached. Node J can be eliminated because it has no neighbours left.

Readers can check by themselves that the triangulated graph in Figure 2.13(ii) is obtained by elimination from the *MCS* ordering shown in Figure 2.14(ii).

2.2.5 From triangulated graphs to junction trees

There exist efficient exact probabilistic algorithms for particular kind of graphs. These graphs are a particular kind of cluster graphs, called *junction trees*, whose nodes are the *cliques of a triangulated graph*.

- A subset of nodes of an undirected graph is said to be *complete* if all its nodes are pairwise linked. A complete set of nodes is called a *clique* if it is not a subset of another complete set, *i.e.*, if it is maximal.

A mechanical procedure to find the cliques of a triangulated graph is very simple, given that, for every node X of a moral graph, after step (i) of the elimination algorithm has been executed, it holds that X and its neighbours form a complete set.

> *Construction of the set of cliques of a triangulated graph*. For every node X, after step (i) of elimination:
>
> (i) check if the complete set composed by X and its neighbours is maximal;
>
> (ii) if the answer is 'yes', add to the list of cliques.

Apply this algorithm to the 'Lulu' graph (*see* Figure 2.14(i)). On arrival during elimination at node C, the check for the set $\{A, C, F\}$ is negative: it is not a clique because it is a subset of the set $\{A, B, C, F\}$ already added to the list when B was eliminated. On arrival at F, the set $\{F, J\}$ is not a clique because the set $\{A, F, J\}$ was added to the list when A was eliminated. Therefore the set of cliques of the triangulated graph in Figure 2.14(i) is (with node numbers according to *MCS* ordering):

$$\{C_4, E_8\},\ \{D_6, F_2, W_7\},\ \{D_6, F_2, J_1\},$$
$$\{A_3, B_5, C_4, F_2\},\ \{A_3, F_2, J_1\}. \tag{2.11}$$

Therefore, the set of cliques (2.11) of the triangulated graph in Figure 2.14(i) obtained *via* the clustering procedure is exactly the Markov domain (2.10) of the 'Lulu' *DAG*.

The last step of the procedure is the construction of the cluster graph having the cliques as its nodes.

- A connected undirected graph is called a *tree* if for every pair of nodes there exists a unique path linking them. A cluster tree is a *junction tree*, if every node that belongs to two clusters also belongs to every cluster in the path between them. This condition is called the *junction tree property*.

The following algorithm guarantees that the tree of the cliques of a triangulated graph obtained *via MCS* and elimination is a junction tree.

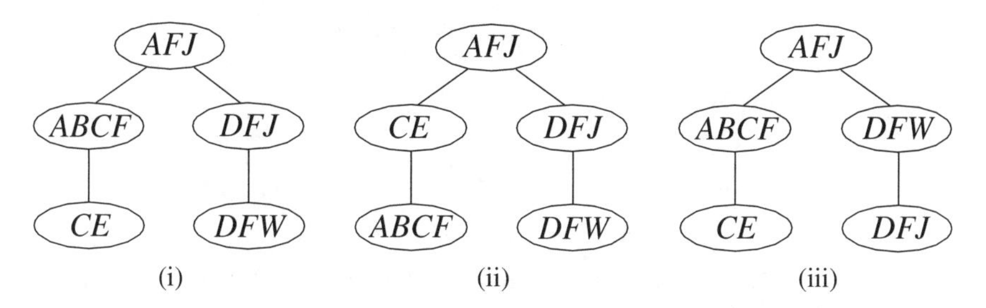

Figure 2.15 (i) is a junction tree; (ii) and (iii) are not junction trees.

Construction of a junction tree.

(i) order the cliques in ascending order according to their highest numbered node;

(ii) go through the cliques in order, choosing for each clique **X** a clique **Y** among the previously numbered cliques with maximum number of common nodes and add an undirected link between **X** and **Y**; break ties arbitrarily.

In Figure 2.15 three different arrangements of the set of cliques (2.11) are given, all of them are trees but only one is a junction tree. Readers can check that the algorithm produces the tree in Figure 2.15(i) and only in that.

Let **X** and **Y** be adjacent cliques of a junction tree. The subset of nodes they have in common is called the *separator* for **X** and **Y** and its graphical rendering consists in associating with the link joining the cliques, a box containing the nodes in the separator.

Call the complement of a clique **X** with respect to the separator **S** for an adjacent clique **Y**, the *residual* of **X** with respect to **S**. For example, the residual of $\{A, B, C, F\}$ with respect to $\{A, F\}$ is $\{B, C\}$, and the residual of $\{A, F, J\}$ with respect to the same separator $\{A, F\}$ is $\{J\}$. The following fact, called the *Markov property for junction trees*, holds:

- the residuals of two adjacent nodes in a junction tree are *conditionally independent given their separator.*

It can be checked in the original *DAG* (*see* Figure 2.9) that $\{B, C\}$ is d-separated from $\{J\}$ given $\{A, F\}$: the path $C \leftarrow F \leftarrow J$ is serial at F and the path $C \leftarrow A \rightarrow F \leftarrow J$ is divergent at A. The residual of $\{F, D, J\}$ with respect to $\{F, J\}$ is $\{D\}$ and it is d-separated from $\{A\}$ given $\{F, J\}$: the path $A \rightarrow F \leftarrow J \rightarrow D$ is diverging at J and the path $A \rightarrow F \rightarrow W \leftarrow D$ is serial at F. Finally, $\{W\}$ is d-separated from $\{J\}$ given $\{D, F\}$ and $\{E\}$ and it is d-separated from $\{A, B, F\}$ given $\{C\}$.

Therefore the joint probability distribution over the junction tree can be written as the product of the conditional probability distributions of the residuals of the cliques, given their separators:

$$
\begin{aligned}
&Pr(A, F, J)Pr(B, C \mid A, F)Pr(D \mid F, J)Pr(W \mid D, F)Pr(E \mid C)\\
&\quad = Pr(A)Pr(J)Pr(F \mid A, J)Pr(B)Pr(C \mid A, B, F)\\
&\qquad Pr(D \mid J)Pr(W \mid D, F)Pr(E \mid C)\,. \qquad (2.12)
\end{aligned}
$$

The factorisation (2.12) of the joint distribution over the junction tree in Figure 2.16 is thus the same as the factorisation (2.2) of the joint distribution over the *DAG* in Figure 2.9.

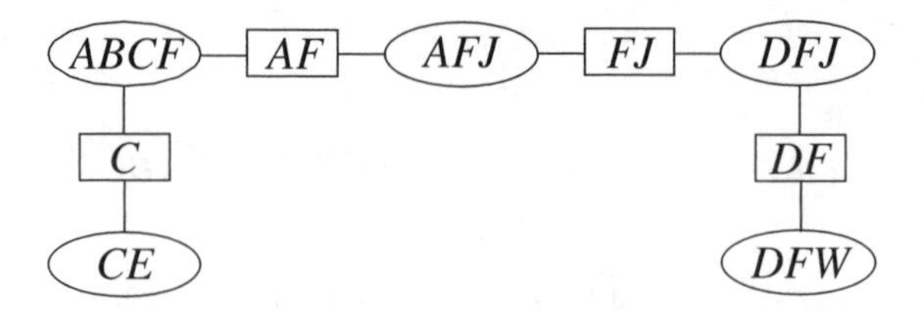

Figure 2.16 The junction tree for graph in Figure 2.14(i) with separators.

2.2.6 *Calculemus*

In this paragraph, it is shown how it is possible to execute probability updating with paper and pencil on the junction tree in Figure 2.16. The aim is to let the reader realise that the junction tree architecture allows coherent probability updating according to the rules of the logic of uncertainty outlined in Chapter 1.

Given the role that Jeffrey's rule (1.24) will play, it is useful to rewrite the general formula here. Let A and B be two propositions, and let $Pr_0(A, B)$ be, by abuse of language, their joint probability distribution given the state of information at time t_0. If the state of information at time t_1 is such that the new probability for B is different from the old probability for B, *i.e.*, $Pr_1(B) \neq Pr_0(B)$, and nothing else is different according to the knowledge available, then the new joint probability distribution is given by

$$Pr_1(A, B) = Pr_0(A, B) \times \frac{Pr_1(B)}{Pr_0(B)}. \tag{2.13}$$

Now, suppose that the conditional probability tables associated with the graph in Figure 2.9 are as shown in Table 2.12 (the numerical values are chosen to make calculations easier). The first task is to compute the initial marginal distributions $Pr_0()$ for all cliques and separators in the tree of Figure 2.16.

It is convenient to execute the computation following the order given in Figure 2.10. An example of the probability table for the clique $\{A, F, J\}$, obtained from values in Table 2.12

Table 2.12 Probability distributions for Figure 2.9.

A:	$Pr_0(A) = 0.01$	J:	$Pr_0(J) = 0.9$
B:	$Pr_0(B) = 0.95$	F:	$Pr_0(F \mid A, J) = 1$
C:	$Pr_0(C \mid A, B, F) = 1$		$Pr_0(F \mid A, \bar{J}) = 1$
	$Pr_0(C \mid A, B, \bar{F}) = 1$		$Pr_0(F \mid \bar{A}, J) = 0.1$
	$Pr_0(C \mid A, \bar{B}, F) = 0$		$Pr_0(F \mid \bar{A}, \bar{J}) = 0.01$
	$Pr_0(C \mid A, \bar{B}, \bar{F}) = 0$	D:	$Pr_0(D \mid J) = 1$
	$Pr_0(C \mid \bar{A}, B, F) = 0$		$Pr_0(D \mid \bar{J}) = 0.6$
	$Pr_0(C \mid \bar{A}, B, \bar{F}) = 0$	W:	$Pr_0(W \mid F, D) = 0.99$
	$Pr_0(C \mid \bar{A}, \bar{B}, F) = 0.1$		$Pr_0(W \mid F, \bar{D}) = 0.99$
	$Pr_0(C \mid \bar{A}, \bar{B}, \bar{F}) = 0.01$		$Pr_0(W \mid \bar{F}, D) = 0.8$
E:	$Pr_0(E \mid C) = 1$		$Pr_0(W \mid \bar{F}, \bar{D}) = 0.01$
	$Pr_0(E \mid \bar{C}) = 0.001$		

Table 2.13 Initial marginal distributions for the clique $\{A, F, J\}$.

$Pr_0(A, F, J) = 0.009$	$Pr_0(\bar{A}, F, J) = 0.089$
$Pr_0(A, \bar{F}, J) = 0$	$Pr_0(\bar{A}, \bar{F}, J) = 0.802$
$Pr_0(A, F, \bar{J}) = 0.001$	$Pr_0(\bar{A}, F, \bar{J}) = 0.001$
$Pr_0(A, \bar{F}, \bar{J}) = 0$	$Pr_0(\bar{A}, \bar{F}, \bar{J}) = 0.098$

Table 2.14 Initial tables of marginal distributions for the separators $\{A, F\}$ (left column) and $\{F, J\}$ (right column).

$Pr_0(A, F) = 0.01$	$Pr_0(F, J) = 0.098$
$Pr_0(A, \bar{F}) = 0$	$Pr_0(F, \bar{J}) = 0.002$
$Pr_0(\bar{A}, F) = 0.09$	$Pr_0(\bar{F}, J) = 0.802$
$Pr_0(\bar{A}, \bar{F}) = 0.9$	$Pr_0(\bar{F}, \bar{J}) = 0.098$

Table 2.15 Initial marginal distributions for the cliques $\{C, E\}$ and $\{D, F, W\}$.

$Pr_0(C, E) = 0.01$	$Pr_0(D, F, W) = 0.09801$	$Pr_0(\bar{D}, F, W) = 0.00099$
$Pr_0(C, \bar{E}) = 0$	$Pr_0(D, \bar{F}, W) = 0.6888$	$Pr_0(\bar{D}, \bar{F}, W) = 0.00039$
$Pr_0(\bar{C}, E) = 0.001$	$Pr_0(D, F, \bar{W}) = 0.00099$	$Pr_0(\bar{D}, F, \bar{W}) = 0.00001$
$Pr_0(\bar{C}, \bar{E}) = 0.989$	$Pr_0(D, \bar{F}, \bar{W}) = 0.1722$	$Pr_0(\bar{D}, \bar{F}, \bar{W}) = 0.03861$

by applying the multiplication rule (with values rounded off at the third decimal place) is given in Table 2.13.

The table of the separators $\{A, F\}$ and $\{F, J\}$ (*see* Table 2.14) is obtained by extending the conversation (formulae (2.6) and (2.9)).

The initial probabilities of the evidence nodes E and W are obtained by extending the conversation from Table 2.15 of the cliques $\{C, E\}$ and $\{D, F, W\}$ (formulae (2.3) and (2.7)):

$$Pr_0(E) = 0.011 \; ;$$

$$Pr_0(W) = 0.788 \,. \tag{2.14}$$

After all the cliques and separator tables have been computed, the Bayesian network is said to be *initialised* and ready to receive new evidence.

Evidence can come under two forms given the model, *i.e.*, given that the structure of the *DAG* is not altered:

- either by a change of some of the conditional probabilities of Table 2.12,
- or by a change in the probabilities of some evidence nodes.

In the first case, the network has to be re-initialised using the new conditional probabilities table. In the last case, it is to be updated following these steps:

- evidence is incorporated by updating the table of the clique containing the evidence node according to rule (2.13);
- evidence is propagated by sequentially updating the tables of all cliques and separators in the tree, according to rule (2.13);
- propagation stops when the tables of all the cliques and separators have been updated.

It is of interest to answer the query: 'What is the probability of proposition A given the knowledge that propositions E and W are true?'. In order to answer this query, it is divided into two queries, which are discussed sequentially.

Consider first the query: 'What is the probability of proposition A given the knowledge that proposition W is true?'. This knowledge about W changes the probability of node W from the value given in (2.14) to the new value

$$Pr_1(W) = 1\,.$$

The marginals of the clique $\{D, F, W\}$ in Table 2.15 are updated using Bayes' theorem,

$$Pr_1(D, F, W) = Pr_0(D, F, W) \times \frac{Pr_1(W)}{Pr_0(W)} = \frac{Pr_0(D, F, W)}{Pr_0(W)}\,.$$

Given that $Pr_1(\bar{W}) = 0$, all the terms in the table containing $\bar{W}$ take the value zero, so new values for the marginals for the clique $\{D, F, W\}$ are given in Table 2.16 (rounding-off to the fourth decimal place).

From Table 2.16, the new table for the separator $\{D, F\}$ may be computed by extending the conversation:

$$Pr_1(D, F) = \sum_W Pr_1(D, F, W)\,.$$

The second term of the sum is null because $Pr_1(\bar{W}) = 0$. Then, the table for the clique $\{D, F, J\}$ may be updated,

$$Pr_1(D, F, J) = Pr_0(D, F, J) \times \frac{Pr_1(D, F)}{Pr_0(D, F)}\,,$$

and the new table for the separator $\{F, J\}$ is computed from

$$Pr_1(F, J) = \sum_D Pr_1(D, F, J)\,,$$

and so on, until evidence has been propagated up to the clique $\{C, E\}$.

Table 2.16 New marginal distribution for the clique $\{D, F, W\}$.

$Pr_1(D, F, W) = 0.1243$	$Pr_1(\bar{D}, F, W) = 0.0012$
$Pr_1(D, \bar{F}, W) = 0.8740$	$Pr_1(\bar{D}, \bar{F}, W) = 0.0005$
$Pr_1(D, F, \bar{W}) = 0$	$Pr_1(\bar{D}, F, \bar{W}) = 0$
$Pr_1(D, \bar{F}, \bar{W}) = 0$	$Pr_1(\bar{D}, \bar{F}, \bar{W}) = 0$

During propagation, the table for the clique $\{A, F, J\}$ has been also updated, and the query: 'What is the probability of proposition A given knowledge that proposition W is true?' is answered by extending the conversation (the value has been rounded off to the third decimal place)

$$Pr_1(A) = \sum_{FJ} Pr_1(A, F, J) = 0.013\,. \tag{2.15}$$

Now consider the query: 'What is the probability of A given knowledge that proposition E is true?'. The initial distribution is $Pr_1()$, the probability distribution at time t_1 that incorporates the knowledge that W is true. The value $Pr_1(E)$ is given by

$$Pr_1(E) = \sum_{C} Pr_1(C, E) = 0.014\ , \tag{2.16}$$

from the table of the clique $\{C, E\}$, which is as follows (values rounded off to the third decimal place)

$$Pr_1(C, E) = 0.013,\ Pr_1(C, \bar{E}) = 0,$$
$$Pr_1(\bar{C}, E) = 0.001,\ Pr_1(\bar{C}, \bar{E}) = 0.986.$$

The new information, again, fixes the probability of node E at the value

$$Pr_2(E) = 1\ ,$$

and another run of computations may be started, updating the table for $\{C, E\}$:

$$Pr_2(C, E) = Pr_1(C, E) \times \frac{Pr_2(E)}{Pr_1(E)}\,.$$

In doing these paper and pencil calculations, those divisions where a zero appears in the denominator are not executed. Simply write the value zero in the new table: so, for example, $P_2(C, \bar{E}) = 0$.

The run stops after the calculation of the new table for $\{D, F, W\}$, and the final probability of A is (value rounded off to the third decimal):

$$Pr_2(A) = \sum_{FJ} Pr_2(A, F, J) = 0.857\,. \tag{2.17}$$

In answering the query: 'What is the probability of proposition A given the knowledge that propositions E and W are true?' updating was done in two stages, first upon W and then upon E. The ordering is not important because, if the calculations are done in the reverse order, *i.e.*, taking first $Pr_1(E) = 1$ and then $Pr_2(W) = 1$ as inputs, the result is the same.

2.2.7 A probabilistic machine

Doing computations by paper and pencil is time; wasting even for a very simple network like 'Lulu'. Fortunately, junction trees support efficient computational algorithms that can be implemented on personal computers. One of these algorithms runs on a commercially available program called HUGIN, that is used in the rest of this book. The algorithm contains the instructions which allow a machine to execute the clustering procedures described in Sections 2.2.4 and 2.2.5 and the calculations informally described in Section 2.2.6.

A node of the junction tree (a clique) is chosen to act as a temporary root node; this choice is said to trigger the direction of message passing. Then messages are passed from all the other nodes (cliques) to the root node. Propagation in this direction is called *collect evidence* (to the root node); propagation from the root to all the other nodes is called *distribute evidence* (from the root node).

- Each non-root node waits to send its message to a given neighbour until it has received messages from all its other neighbours.
- The root waits to send messages to its neighbours until it has received messages from all of them.
- The algorithm stops when both *collect evidence* and *distribute evidence* (in this order) have been performed.

The HUGIN Graphical Interface enables the construction of the 'Lulu' network as Figure 2.17, and the entry of conditional probability values as in Table 2.12.

The program compiles the initial probability distributions that can be displayed using monitor windows as in Figure 2.18.

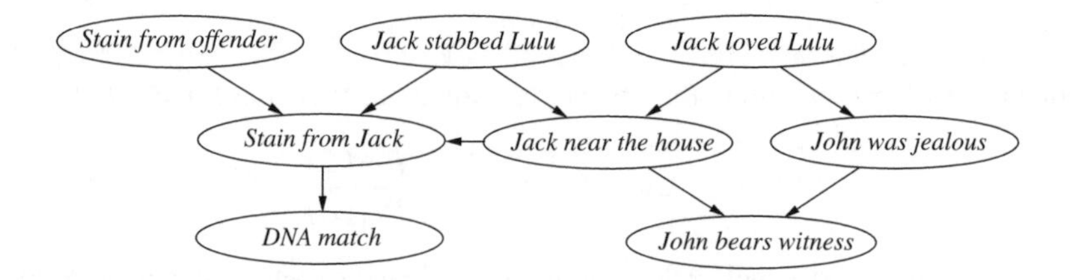

Figure 2.17 HUGIN 'Lulu' network.

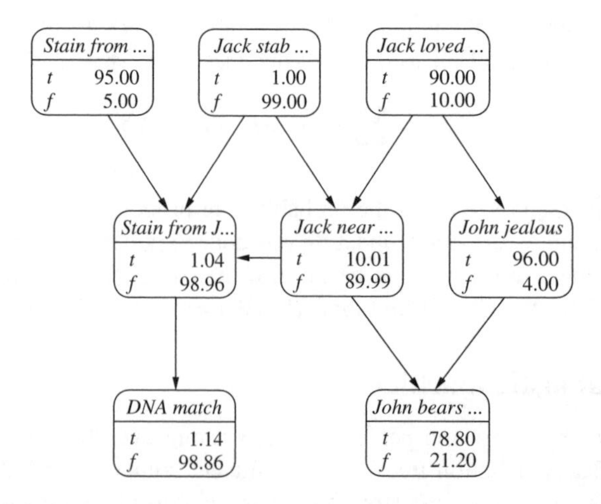

Figure 2.18 HUGIN 'Lulu' initial probability distribution.

After the evidence that W (*i.e.*, that John bears witness) is true has been entered and propagated, the monitor windows report the new probability distribution (*see* Figure 2.19).

HUGIN can deal with several items of evidence at the same time. In Figure 2.20, evidence that W (*i.e.*, that John bears witness) and E (*i.e.*, DNA match) are both true has been entered and propagated. The order of entry does not matter.

Discrete nodes in HUGIN can have more than two possible states, therefore clusters of propositions may be represented as in the case of the Baskerville network of Figure 2.21 where the conditional probability values are those of Tables 2.9 and 2.10.

HUGIN can also perform updating on uncertain evidence (*see* Section 1.2.6). In HUGIN language, it is called *entering likelihood evidence* and an example for the Baskerville network is shown in Figure 2.22.

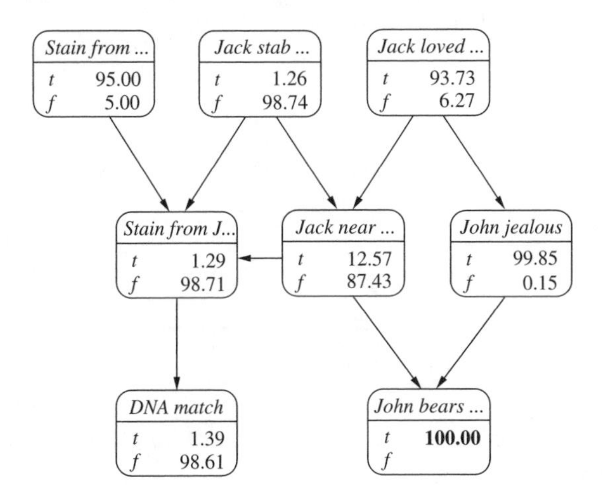

Figure 2.19 HUGIN 'Lulu' distribution with W true.

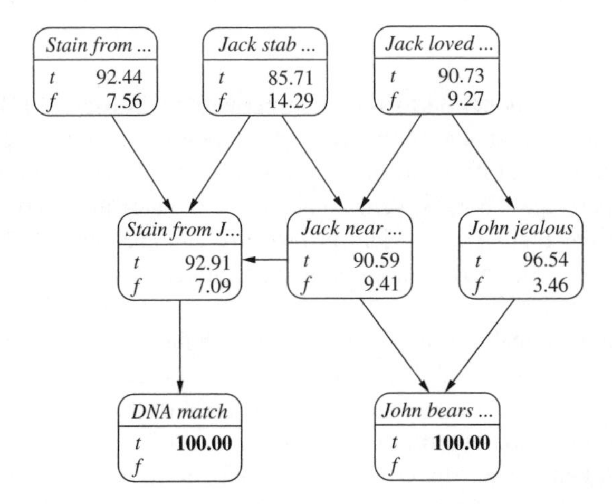

Figure 2.20 HUGIN 'Lulu' distribution with W and E both true.

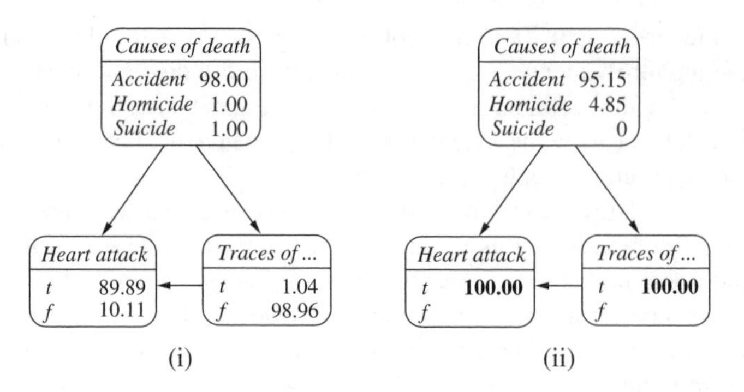

Figure 2.21 (i) HUGIN 'Watson' initial distribution; (ii) HUGIN 'Watson' distribution with 'heart attack' and 'traces of hound' both true.

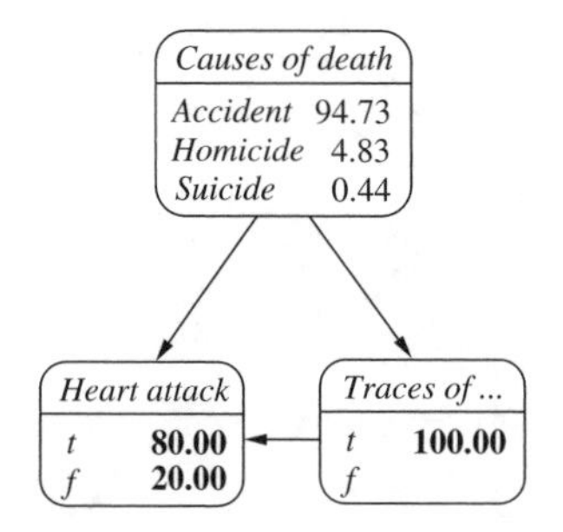

Figure 2.22 HUGIN 'Watson' distribution with 'traces of hound' true and 'heart attack' true with probability 0.8.

2.3 Further readings

2.3.1 General

An accessible general introduction to Bayesian networks is Jensen (2001). More technical treatment of the concepts and algorithms from graph theory used for Bayesian networks can be found in Castillo et al. (1997) and in Cowell et al. (1999). The concept of d-separation is due to Judea Pearl and Thomas Verma (Pearl 1988; Pearl and Verma 1987). A good, informal, proof of the fundamental Markov property (*see* Section 2.2.2) can be found in Charniak (1991). Interested readers can find more technical discussions of the d-separation theorem in Castillo et al. (1997), Cowell et al. (1999) and Pearl (2000). A formal proof of the fact that the joint probability distribution over the junction tree can be written as the product of the conditional probability distributions of the residuals of the cliques, given their separators (*see* Section 2.2.5) can be found in Castillo et al. (1997, p. 228).

Junction tree propagation algorithms were first introduced by Lauritzen and Spiegelhalter (1988) and by Shenoy and Shafer (1990). Readers are referred to Cowell et al. (1999) for a detailed presentation and more sophisticated examples than 'Lulu', both with discrete and continuous variables. The HUGIN architecture for probability updating in junction trees has been first introduced in Jensen et al. (1990). The so-called *'lazy' propagation algorithm* implemented in HUGIN is described in Jensen (2001) and has been introduced by Madsen and Jensen (1999). A free version of HUGIN can be found at `http://www.hugin.com`.

Schum's understanding of a 'generalisation' differs from the understanding given in Section 2.1.7, because the graphs for evidence analysis used in Schum (1994) and Kadane and Schum (1996) are not Bayesian networks but are modelled on Wigmore's reference charts (*see* Section 2.3.2), Wigmore (1937). Those graphs are bottom-up. The examples of 'generalisations' the authors put forward are also bottom-up, so to speak, as this example from Kadane and Schum (1996, p. 14) shows

> Persons who are at the scene of a crime when it is committed are frequently (····) the ones who took part in it.

The appropriate common sense generalisation is: 'If a person takes part in a crime, then he is at the scene of the crime when it is committed'. This may then constitute the premiss of the following *D-N* explanation (*see* Section 1.4.1) that justifies the relevance of the fact that a reliable eyewitness says that they saw the individual in question at the scene of the crime (Garbolino 2001, pp. 1508–1511):

1. If a person participates in a crime then he is at the scene of the crime when it is committed.

2. John Smith participated in the robbery.

3. John Smith was at the scene of the robbery when it was committed.

With regards to causality, the problem with Mackie's regularity approach (*see* Section 2.1.8) is that (Hitchcock 2002, p. 3):

> (····) the regularity approach makes causation incompatible with indeterminism: if an event is not determined to occur, then no event can be a part of a sufficient condition for that event. (····) Many philosophers find the idea of indeterministic causation counterintuitive. Indeed, the word 'causality' is sometimes used as a synonym for determinism. A strong case for indeterministic causation can be made by considering the epistemic warrant for causal claims. There is now very strong empirical evidence that smoking causes lung cancer. Yet the question of whether there is a deterministic relationship between smoking and lung cancer is wide open. The formation of cancer cells depends upon mutation, which is a strong candidate for being an indeterministic process. (····) Thus the price of preserving the intuition that causation presupposes determinism is agnosticism about even our best supported causal claims.

Hitchcock (2002) is a useful introductory survey to the important topic of probabilistic causality. Accessible introductions to the quite technical works of Pearl (2000) and Spirtes et al. (2001) can be found in Cooper (1999), in Scheines (1997) and in Pearl (1999). Other different viewpoints on probabilistic causation and causal modelling are offered in Dawid (2000, 2002) and Williamson (2004).

2.3.2 Bayesian networks in judicial contexts

Since the early 1990s, both legal scholars and forensic scientists have shown an increased interest in the applicability of Bayesian networks in judicial contexts. While lawyers tend merely to be concerned with structuring cases as a whole, forensic scientists focus on the evaluation of selected items of scientific evidence, such as fibres or blood. Below, relevant literature pertaining to the former topic is mentioned. A more detailed discussion of the latter topic is given in the remaining chapters of this book.

Bayesian networks have been proposed for structuring and reasoning about issues of complex and historically important cases. For example, Edwards (1991) provided an alternative analysis of the descriptive elements presented in the Collins case (*People v. Collins*, 68 Cal. 2d 319, 438 P. 2d 33, 66 Cal. Rptr. 497 (1968)). In this case, the prosecution assumed independence between different descriptive elements such as the colour of the getaway car or the colour of the offender's hair. The prosecution multiplied the estimated frequencies of each descriptive element to arrive at a frequency of 1/12.000.000, inviting the jury to conclude that there was a chance of 1 in 12 million that the two accused were innocent. This argument has been criticised extensively in literature, *e.g.*, by Finkelstein and Fairley (1970). Edwards (1991) proposes various dependence relationships among the different descriptive elements. Bayesian networks are used as a graphical representation scheme as well as a means to perform numerical evaluation.

Schum (1994) and Kadane and Schum (1996) worked on a probabilistic analysis of the Sacco and Vanzetti case with an emphasis on the credibility and relevance of evidence given by human sources, *i.e.*, testimony.

Probabilistic case analysis was also proposed for the Omar Raddad case (Levitt and Blackmond Laskey 2001) and more recently, for the O.J. Simpson case (Thagart 2003).

A thorough discussion of Bayesian networks for structuring and analysing legal arguments is given by Robertson and Vignaux (1993). The authors present Bayesian networks as a contribution to the field of fact analysis and point out the advantages of Bayesian networks over previously proposed charting methods, such as Wigmore charts (Wigmore 1913) or route diagrams (Friedman 1986a,b). 'Wigmore charts' are a representational scheme, intended to provide formal support for reaching and defending conclusions based on a '(· · · ·) mixed mass of evidence (· · · ·)' (Wigmore 1913); however, there is no incorporation of an explicit measure of uncertainty, *e.g.*, through probabilities. 'Route diagrams', as proposed by Friedman, are a modification of probability trees (*see* Section 1.1.7), combining nodes that lead to the same final state.

Fenton and Neil (2000) use Bayesian networks to point out a previously unreported pitfall of intuition, called the *jury fallacy*. This kind of fallacy applies to situations in which a prior similar conviction of the defendant is revealed after the jury returns a not guilty verdict. Using Bayesian networks with rather conservative assumptions, the authors show that it is fallacious to argue that a prior similar conviction should decrease the belief that the jury was correct with their not guilty verdict. Arguments of Fenton and Neil (2000) are also discussed by Jowett (2001a,b).

3

Evaluation of scientific evidence

3.1 Introduction

Consider a legal investigation. At the time of the preliminary investigation and that of the trial, the proof of a point at issue often takes the form of a 'technical proof' and relies on the forensic sciences. This technical proof may be used to support the existence or the nature of a criminal act (for example in forgery cases), or to help demonstrate links between elements involved in a criminal act (for example, an aggression between an offender and a victim in a rape case). In the latter case, scientists investigate these links with the help of a comparative analysis of multiple traces (such as DNA evidence or textile fibres) which could have been transferred during the act.

Developments in the last twenty years have greatly increased the range of possible types of traces and of analytical techniques which have been presented to the courts. These developments have allowed the analysis of more numerous traces and of traces of reduced size. However, it is not only the results of the analyses which should be presented to a court, but also the results presented have to be evaluated and interpreted in the context of the case under trial.

Scientific evidence requires considerable care in its interpretation. Emphasis needs to be put on the importance of asking the question 'What do the results mean in this particular case?' (Jackson 2000). Scientists and jurists have to abandon the idea of absolute certainty in order to approach the individualisation process. If it can be accepted that nothing is absolutely certain, then it becomes logical to determine the degree of confidence that may be assigned to a particular belief (Kirk and Kingston 1964).

Therefore, theory, methods and applications of probability and statistics underlie the evaluation of scientific evidence. The trier of fact should use probabilistic reasoning and assessment to understand scientific evidence which is often presented in a numerical way. Forensic scientists should give the court an evaluation that illustrates the convincing force of their results. Such an evaluation inevitably uses probabilities as measures of uncertainty.

Forensic science suggests itself as a case study for evidence interpretation as there is a degree of consensus that forensic scientific evidence should be thought about in probabilistic terms.

Bayesian Networks and Probabilistic Inference in Forensic Science F. Taroni, C. Aitken, P. Garbolino and A. Biedermann

Scientific evidence has, by its nature, a close link to statistical assessment (see Chapter 1), but there is the potential for misinterpretation of the value of statistical evidence when, routinely, such evidence supports a scientific argument at trial. Appropriate interpretative procedures for the assistance of both jurists and scientists have been proposed in scientific and judicial literature and applied practically to ease communication between the forensic and judicial world, and to aid in the correct interpretation of statistical evidence.

This chapter proposes a summary of the probabilistic approach based on Bayes' theorem and the use of the likelihood ratio to assess scientific evidence. Formulae for likelihood ratios are developed and presented in the context of the case under examination through the use of the idea of the 'hierarchy of propositions' (Cook et al. 1998a).

Forthcoming chapters will translate and use these equations in a graphical environment.

3.2 The value of evidence

The interpretation of scientific evidence may be thought of as the assessment of a comparison. This comparison is between evidential material found at the scene of a crime and evidential material found on a suspect, a suspect's clothing or around his environment. Let M_c denote the event of finding crime-related material on a scene, and M_s the event of recovering material, for example, on a suspect. For the remainder of this book, the events M_c and M_s will be taken to refer, by abuse of language, to the material pertaining to these events. Denote the combination by $M = (M_c, M_s)$. As an example, consider a bloodstain. The crime stain is M_c, the receptor or transferred stain form of the evidential material, M_s is the genotype of the suspect and is the source form of the material. Alternatively, suppose glass is broken during the commission of a crime. M_s would be the fragments of glass (the source form of the material) found at the crime scene, M_c would be the fragments of glass (the receptor or transferred particle form of the material) found on the clothing of a suspect and M would be the two sets of fragments.

Observed qualities, such as the genotypes, or results of measurements, such as the refractive indices of the glass fragments, are taken from M. Comparisons are made of the source form and the receptor form. Again, for the ease of notation, denote these by E_c and E_s, respectively, and let $E = (E_c, E_s)$ denote the combined set. Comparison of E_c and E_s is to be made, and the assessment of this comparison has to be quantified. The totality of the evidence is denoted as Ev and is such that $Ev = (M, E)$.

Consider now the odds form of Bayes' Theorem (Section 1.2.3) in the forensic context of assessing the value of some evidence. Replace H_1 in (1.16) by a *proposition* H_p, that the suspect (or defendant if the case has come to trial) is truly guilty. H_2 is replaced by proposition H_d, that the suspect is truly innocent. The evidence under consideration, Ev, may be written as $(E, M) = (E_c, E_s, M_c, M_s)$, the type of evidence and observations of it. In conformity to a predominant part of forensic literature on evidence evaluation, note that, from now on, the term 'proposition' is used instead of 'hypothesis' to designate the ingredients deemed to be relevant for the scenario under study. Also, and for the sake of simplicity, apostrophes (' ') will be omitted when describing verbally the content of a proposition, except when a proposition is part of a sentence.

The odds form of Bayes' Theorem then enables the prior odds (*i.e.*, prior to the presentation of Ev) in favour of guilt to be updated to posterior odds given Ev, the evidence under consideration. This is done by multiplying the prior odds by the likelihood ratio which, in this context, is the ratio of the probabilities of the evidence assuming guilt and assuming innocence of the suspect. With this notation, the odds form of Bayes' Theorem may be written as

$$\frac{Pr(H_p \mid Ev)}{Pr(H_d \mid Ev)} = \frac{Pr(Ev \mid H_p)}{Pr(Ev \mid H_d)} \times \frac{Pr(H_p)}{Pr(H_d)}.$$

Explicit mention of the background information I is omitted in general from probability statements for ease of notation. With the inclusion of I, the odds form of Bayes' Theorem is

$$\frac{Pr(H_p \mid Ev, I)}{Pr(H_d \mid Ev, I)} = \frac{Pr(Ev \mid H_p, I)}{Pr(Ev \mid H_d, I)} \times \frac{Pr(H_p \mid I)}{Pr(H_d \mid I)}.$$

Notice that in the evaluation of the evidence Ev, it is two probabilities that are necessary: the probability of the evidence if the suspect is guilty and the probability of the evidence if the suspect is innocent. For example, it is not sufficient to consider only the probability of the evidence if the suspect is innocent and to declare that a small value of this is indicative of guilt. The probability of the evidence if the suspect is guilty also has to be considered.

Similarly, it is not sufficient to consider only the probability of the evidence if the suspect is guilty and to declare that a high value of this is indicative of guilt. The probability of the evidence if the suspect is innocent is also to be considered.

Consider the likelihood ratio $Pr(Ev \mid H_p)/Pr(Ev \mid H_d)$ further, where explicit mention of I has again been omitted. This equals

$$\frac{Pr(E \mid H_p, M)}{Pr(E \mid H_d, M)} \times \frac{Pr(M \mid H_p)}{Pr(M \mid H_d)}.$$

The second ratio in this expression, $Pr(M \mid H_p)/Pr(M \mid H_d)$, concerns the type and quantity of evidential material found at the crime scene and on the suspect. It may be written as

$$\frac{Pr(M_s \mid M_c, H_p)}{Pr(M_s \mid M_c, H_d)} \times \frac{Pr(M_c \mid H_p)}{Pr(M_c \mid H_d)}.$$

The value of the second ratio in this expression may be taken to be 1. The type and quantity of material at the crime scene is independent of whether the suspect is the criminal or whether someone else is. The value of the first ratio, which concerns the evidential material found on the suspect, given the evidential material found at the crime scene and the guilt or otherwise of the suspect, is a matter for subjective judgement, and it is not proposed to consider its determination further here. Instead, consideration will be concentrated on

$$\frac{Pr(E \mid H_p, M)}{Pr(E \mid H_d, M)}.$$

In particular, for notational convenience, M will be subsumed into I and omitted, for clarity of notation. Then,

$$\frac{Pr(M \mid H_p)}{Pr(M \mid H_d)} \times \frac{Pr(H_p)}{Pr(H_d)}$$

which equals

$$\frac{Pr(H_p \mid M)}{Pr(H_d \mid M)}$$

will be written as

$$\frac{Pr(H_p)}{Pr(H_d)}.$$

Thus

$$\frac{Pr(H_p \mid Ev)}{Pr(H_d \mid Ev)} = \frac{Pr(H_p \mid E, M)}{Pr(H_d \mid E, M)}$$

will be written as

$$\frac{Pr(H_p \mid E)}{Pr(H_d \mid E)}$$

and

$$\frac{Pr(Ev \mid H_p)}{Pr(Ev \mid H_d)} \times \frac{Pr(H_p)}{Pr(H_d)}$$

will be written as

$$\frac{Pr(E \mid H_p)}{Pr(E \mid H_d)} \times \frac{Pr(H_p)}{Pr(H_d)}.$$

The full result is then

$$\frac{Pr(H_p \mid E)}{Pr(H_d \mid E)} = \frac{Pr(E \mid H_p)}{Pr(E \mid H_d)} \times \frac{Pr(H_p)}{Pr(H_d)}, \tag{3.1}$$

or if I is included

$$\frac{Pr(H_p \mid E, I)}{Pr(H_d \mid E, I)} = \frac{Pr(E \mid H_p, I)}{Pr(E \mid H_d, I)} \times \frac{Pr(H_p \mid I)}{Pr(H_d \mid I)}. \tag{3.2}$$

The likelihood ratio is the ratio

$$\frac{Pr(H_p \mid E, I)/Pr(H_d \mid E, I)}{Pr(H_p \mid I)/Pr(H_d \mid I)} \tag{3.3}$$

of posterior odds to prior odds. The likelihood ratio converts the prior odds in favour of H_p into the posterior odds in favour of H_p. This representation also emphasises the dependence of the prior odds on background information.

The likelihood ratio may be thought of as the *value* of the evidence. Evaluation of evidence is taken to mean the determination of a value for the likelihood ratio. This value will be denoted as V.

Consider two competing propositions, H_p and H_d, and background information I. The *value* V of the evidence E is given by

$$V = \frac{Pr(E \mid H_p, I)}{Pr(E \mid H_d, I)}, \tag{3.4}$$

the likelihood ratio that converts prior odds $Pr(H_p \mid I)/Pr(H_d \mid I)$ in favour of H_p relative to H_d into posterior odds $Pr(H_p \mid E, I)/Pr(H_d \mid E, I)$ in favour of H_p relative to H_d.

3.3 Relevant propositions

It is widely accepted for the assessment of scientific evidence, that the forensic scientist should consider different propositions which commonly represent alternatives proposed by the prosecution and the defence, to illustrate their description of the facts under examination. These alternatives are formalised representations of the framework of circumstances. The forensic scientist evaluates the evidence under these propositions. The formulation of the propositions is a crucial basis for a logical approach to the evaluation of evidence (Cook et al. 1998a). This procedure can be summarised by three key principles (Evett and Weir 1998):

- Evaluation is only meaningful when at least one alternative proposition (two or more competing propositions) is addressed, the two propositions are conventionally denoted as H_p and H_d.
- Evaluation of scientific evidence (E) considers the probability of the evidence given the propositions that are addressed, $Pr(E \mid H_p)$ and $Pr(E \mid H_d)$.
- Evaluation of scientific evidence is carried out within a framework of circumstances, denoted as I. The evaluation is conditioned not only by the competing propositions but also by the structure and content of the framework.

Therefore, propositions play a key role in this process.

Generally, propositions are considered in pairs. There will be situations where there will be three or more (*see* Section 5.5). This happens quite often with DNA mixtures, for example, where the number of contributors to the mixture is in dispute (Buckleton et al. 1998; Lauritzen and Mortera 2002). It is generally possible to reduce the number of propositions to two, which will be identified with the respective prosecution and defence positions. Clearly the two propositions must be mutually exclusive. It is tempting to specify that they are exhaustive, but this is not necessary. The simplest way to achieve the exhaustive situation is with the addition of the word 'not' into the first proposition, saying for example: 'Mr C is the man who kicked Mr Z', and 'Mr C is not the man who kicked Mr Z'. However, this gives the court no idea of the way in which the scientist has assessed the evidence with regard to the second proposition. Mr C may not have kicked the victim, but he may have been present at the incident. Analogously, consider the proposition 'Mr B had sexual intercourse with Miss Y', and 'Mr B did not have sexual intercourse with Miss Y'. In fact, if semen has been found on the vaginal swab, then it may be inferred that someone has had sexual intercourse with Miss Y and, indeed, the typing results from the semen would be evaluated by considering the probability given that it came from some other man. It will help the court if this is made plain in the alternative proposition that is specified. So the alternative could be 'Some unknown man, unrelated to Mr B, had sexual intercourse with Miss Y' (with no consideration of relatives).

In summary, the simple use of 'not' to frame the alternative proposition is unlikely to be particularly helpful to the court. In the same sense, it is useful to avoid the use of misleading words like 'contact' to describe the type of action in the propositions. In fact, there is a danger in using such a vague word. As stated by Evett et al. (2002b), the statement that a suspect has been in recent contact with broken glass could mean many things. There is a clear need to specify correctly the propositions in a framework of circumstances for a transparent approach to the consideration of the evidence.

Identification of the propositions to be considered is not an easy task for the scientist. A fruitful approach to assist the scientist has been proposed by Cook et al. (1998a). Practically speaking, the propositions that are addressed in a judicial case, depend on (a) the circumstances of the case, (b) the observations that have been made and (c) the available background data. A classification (called *hierarchy*) of these propositions into three main categories or levels has been proposed, notably the source level (level I), the activity level (level II), and the crime level (level III).

3.3.1 Source level

The assessment of the level I category (the source) depends on analyses and measurements done on the recovered and the control samples. The value of a trace (or a stain) under source level propositions (such as 'Mr X's pullover is the source of the recovered fibres', and 'Mr X's pullover is not the source of the recovered fibres', so that another clothing is the source of the trace) does not need to take account of anything other than the analytical information obtained during examination. The probability of the evidence under the first proposition (numerator) is considered from a careful comparison between two samples (the recovered and the control) assuming that they have come from the same population. The probability of the evidence under the second proposition (denominator) is considered by a comparison of the characteristics of the control and recovered samples in the context of a relevant population of alternative sources.

3.3.1.1 Notation

As usual, let E be the evidence, the value of which has to be assessed. Let the two propositions to be compared be denoted as H_p and H_d. The likelihood ratio, V, is then

$$V = \frac{Pr(E \mid H_p)}{Pr(E \mid H_d)}.$$

The propositions will be stated explicitly for any particular context. In examples concerning DNA profiles and a proposition that a stain of body fluids did not come from a suspect, it will be understood that the origin of the stain was from some person unrelated to, and not sharing the same sub-population of, the suspect (for evaluations in these contexts *see* Balding 2005 and Section 5.12).

The influence of background information I also has to be assessed. The likelihood ratio is then

$$V = \frac{Pr(E \mid H_p, I)}{Pr(E \mid H_d, I)}. \tag{3.5}$$

3.3.1.2 Single sample

Consider a scenario in which a blood stain has been left at the scene of the crime by the person who has committed the crime. A suspect has been identified, and it is desired to establish the strength of the link between the suspect and the crime stain. A comparison between the profile of the stain and the profile of a sample given by the suspect is made by a forensic scientist. The two propositions to be compared are as follows:

- H_p : the crime stain comes from the suspect;
- H_d : the crime stain comes from some (unrelated) person other than the suspect.

The scientist's results, denoted by E, may be divided into two parts (E_c, E_s) as follows:

- E_s: the DNA profile, Γ, of the suspect;
- E_c: the DNA profile, Γ, of the crime stain.

A general formulation of the problem is given here.

The scientist knows, in addition, from data previously collected (population studies) that profile Γ occurs in $100\gamma\%$ of some relevant population, Ψ say.

The value to be attached to E is given by

$$V = \frac{Pr(E \mid H_p, I)}{Pr(E \mid H_d, I)}.$$

This can be simplified.

$$\begin{aligned} V &= \frac{Pr(E \mid H_p, I)}{Pr(E \mid H_d, I)} \\ &= \frac{Pr(E_c, E_s \mid H_p, I)}{Pr(E_c, E_s \mid H_d, I)} \\ &= \frac{Pr(E_c \mid E_s, H_p, I)Pr(E_s \mid H_p, I)}{Pr(E_c \mid E_s, H_d, I)Pr(E_s \mid H_d, I)}. \end{aligned}$$

Now, E_s is the evidence that the suspect's profile is Γ. Here, it is assumed that a person's profile is independent of whether he was at the scene of the crime (H_p) or not (H_d). Thus,

$$Pr(E_s \mid H_p, I) = Pr(E_s \mid H_d, I)$$

and so

$$V = \frac{Pr(E_c \mid E_s, H_p, I)}{Pr(E_c \mid E_s, H_d, I)}.$$

If the suspect was not at the scene of the crime (H_d is true), then the evidence (E_c) about the profile of the crime stain is independent of the evidence (E_s) about the profile of the suspect. Thus,

$$Pr(E_c \mid E_s, H_d, I) = Pr(E_c \mid H_d, I)$$

and

$$V = \frac{Pr(E_c \mid E_s, H_p, I)}{Pr(E_c \mid H_d, I)}. \tag{3.6}$$

Note that the assumption that the suspect's characteristics are independent of whether he is the origin of the crime stain should not be made lightly. In a DNA context, the question of how many other individuals among the population of possible culprits might also be expected to share this DNA type is of great relevance. The selection of a genotype (the suspect's genotype) will increase the probability of selection of an identical one coming from the offender (proposition H_d). This question is complicated by the phenomenon of genetic correlations due to shared ancestry (Balding and Nichols 1997). Simple calculation of profile frequencies (assuming Hardy–Weinberg equilibrium) is not sufficient for the evaluation of evidence when there are dependencies amongst different individuals involved in the case under examination, such as the suspect and the perpetrator (the real source of the recovered sample) featured in the proposition, H_d.

The more common source of dependency is a result of membership in the same population and the possession of similar evolutionary histories. The mere fact that populations are finite in size means that two people taken at random from a population have a non-zero chance of having relatively recent common ancestors. Disregarding this correlation of alleles in the calculation of the value of the evidence results in an exaggeration of the strength of the evidence against the person whose genotype is being compared. For the sake of simplicity, in the proposed development above and in what follows, independence is assumed. This assumption is relaxed in Section 5.12.

Moreover, some crime scenes for example, are likely to transfer materials to an offender's clothing. If the characteristics of interest relate to such materials and the offender is later identified as a suspect, the presence of such materials is not independent of his presence at the crime scene. If the legitimacy of the simplifications are in doubt then the original expression is the one which should be used.

The background information, I, may be used to assist in the determination of the relevant population from which the criminal is supposed to have come. For example, consider an example where I may include an eyewitness description of the criminal as Chinese. This is valuable because the frequency of DNA profiles can vary between ethnic groups and affect the value of the evidence.

First, consider the numerator $Pr(E_c \mid H_p, E_s, I)$ of the likelihood ratio. This is the probability that the crime stain is of profile Γ given the suspect was the source of the crime stain *and* the suspect has DNA profile Γ *and* all other information, including, for example, an eyewitness account that the criminal is Chinese. This probability is just 1 since if the suspect was the source of the stain and has profile Γ, then the crime stain is of profile Γ, assuming as before that all innocent sources of the crime stain have been eliminated. Thus, $Pr(E_c \mid H_p, E_s, I) = 1$.

Now consider the denominator $Pr(E_c \mid H_d, I)$. Here the proposition H_d is assumed to be true; for example, the suspect was not the source of the crime stain. I is also assumed known. Together I and H_d define the relevant population.

Suppose I provides no information about the criminal, this will affect the probability of his blood profile being of a particular type. For example, I may include eyewitness evidence that the criminal was a tall, young male. However, a DNA profile is independent of all three of these qualities, so I gives no information affecting the probability that the DNA profile is of a particular type.

It is assumed that the suspect was not the source of the stain. Thus, the suspect is not the criminal. The relevant population is deemed to be Ψ. The criminal is an unknown member of Ψ. Evidence E_c is to the effect that the crime stain is of profile Γ. This is to say that an unknown member of Ψ is Γ. The probability of this is the probability that a person drawn at random from Ψ has a profile Γ, which is γ. Thus,

$$Pr(E_c \mid H_d, I) = \gamma.$$

The likelihood ratio V is then

$$V = \frac{Pr(E_c \mid H_p, E_s, I)}{Pr(E_c \mid H_d, I)} = \frac{1}{\gamma}. \tag{3.7}$$

This value, $1/\gamma$, is the value of the evidence of the profile of the blood stain when the criminal is a member of Ψ.

There can be uncertainty around the relevance of the evidence. Because of the sensitivity of DNA profiling technology, it is now possible to envisage situations in which it is not necessarily the case that a particular profile actually came from what was observed as a discernible region of staining. In such cases, it may be necessary to address what are termed *sub-level I* propositions. In a DNA context, level I propositions such as 'The semen came from the suspect', and 'The semen came from some other man', have to be replaced by 'DNA came from the suspect', and 'DNA came from some other person' (Evett et al. 2002b). The information available and the context of the case influence the choice of proposition. However, the value V still remains $1/\gamma$.

Extensions of the likelihood ratio for evidence assessed at the 'source level' were applied to deal with more sophisticated situations involving, for example, mixtures (which is a stain (trace) that contains a mixture of genetic material from more than one person, *see* Section 5.8) and N-traces (cases involving several stains and several offenders), notably the two-trace problem (*see* Section 4.3).

3.3.2 Activity level

The next level (level II) in the hierarchy of propositions is related to an activity. This implies that the definitions of the propositions of interest have to include an action. Such propositions could be for example, 'Mr X assaulted the victim', and 'Mr X did not assault the victim' (some other man assaulted her, and Mr X is not involved in the offence), or 'Mr X sat on the car driver's seat', and 'Mr X never sat on the car driver's seat'. The consequence of this activity (the assault or the sitting on a driver's seat) is the contact (between the two people involved in the assault, or the contact between the driver and the seat of the car) and consequently a transfer of material (*i.e.*, fibres in this example). The scientist needs to consider more detailed information about the case under examination relative to the transfer and persistence of the fibres on the receptor (the victim's pullover, for example). Circumstances of the case (*e.g.*, the distance between the victim and the criminal, the strength of the contact and the modus operandi) are needed to be able to answer relevant questions like 'Is this the sort of trace that would be seen if Mr X were the man who assaulted the victim?' or 'Is this the sort of trace that would be seen if Mr X were not the man who assaulted the victim?'. The assessment of evidence under level I propositions requires little in the way of circumstantial information. Only I, the background information, is needed. This could be useful to define the relevant population for use in the assessment of the rarity of the characteristic of interest. Activity (level II) propositions cannot be addressed without a framework of circumstances. The importance of this will be discussed in the pre-assessment approach (Section 3.4) when the expert is required to examine the scenarios of the case and to verify that all relevant information for the proper assessment of the evidence is available. The main advantage of level II over level I propositions is that the evaluation of evidence under activity (level II) propositions does not strictly depend on the recovered material; for example, it is possible to assess the fact that no fibres have been recovered. It is clearly important to assess the importance of the absence of material (such absence of material is evidence of interest).

3.3.2.1 Notation

This situation involves a stain at the crime scene, which could not be identified as coming from the criminal. The complication introduced when it cannot be assumed that the transferred particle form of the evidence is associated with the crime is made more explicit when transfer is in the direction from the scene to the criminal.

A crime has been committed during which the blood of a victim has been shed. A suspect has been identified. A single blood stain of genotype Γ has been found on an item of the suspect's clothing. The suspect's genotype is *not* Γ. The victim's genotype is Γ. There are two possibilities:

- T_0: the blood stain came from some background source;
- T_n: the blood stain was transferred during the commission of the crime.

As before, there are two propositions to consider:

- H_p: the suspect assaulted the victim;
- H_d: the suspect did not assault the victim; *i.e.*, he is not involved in any way whatsoever with the victim.

The evidence E to be considered is that a single blood stain has been found on the suspect's clothing and that it is of genotype Γ. The information that the victim's genotype is Γ is to be considered as part of the relevant background information I. This is a scene-anchored perspective (Stoney 1991). The value of the evidence is then

$$V = \frac{Pr(E \mid H_p, I)}{Pr(E \mid H_d, I)}.$$

Consider the numerator first and event T_0 initially. Then, the suspect and the victim have been in contact (H_p) and no blood has been transferred to the suspect. This is an event with probability $Pr(T_0 \mid H_p)$. Also, a stain of genotype Γ has been transferred by some other means; an event with probability $Pr(B, \Gamma)$ where B refers to the event of a transfer of a stain from a source (the background source) other than the crime scene.

Consider T_n. Blood has been transferred to the suspect, an event with probability $Pr(T_n \mid H_p)$; given T_n, H_p and the genotype Γ of the victim, it is certain that the group of the transferred stain is Γ. This also implies that no blood has been transferred from a background source.

The probability that no blood has been transferred by other means also has to be included. Let $t_0 = Pr(T_0 \mid H_p)$ and $t_1 = Pr(T_n \mid H_p)$ denote the probabilities of no stain or of one stain being transferred during the course of a crime. Let b_0 and $b_{1,1}$, respectively, denote the probabilities that a person from the relevant population will have zero blood stains or one group of one blood stain on his clothing. The probabilities are taken with respect to an object, here a person but not necessarily always so, which has received the evidence. Such a body is a *receptor*. Let γ denote the probability that a stain acquired innocently on the clothing of a person from the relevant population will be of genotype Γ. This probability may be different from the relative frequency of Γ amongst the general population. Then $Pr(B, \Gamma) = \gamma b_{1,1}$. The numerator can be written as (Evett 1984)

$$t_0 \gamma b_{1,1} + t_1 b_0.$$

The first term accounts for T_0, the second for T_n.

Now consider the denominator. The suspect and the victim were not in contact (so the presence of the stain is explained by chance alone). The denominator then takes the value $Pr(B, \Gamma)$, which equals

$$\gamma b_{1,1}.$$

The value of the evidence is thus

$$\begin{aligned} V &= \frac{\{t_0 \gamma b_{1,1} + t_1 b_0\}}{\{\gamma b_{1,1}\}} \\ &= t_0 + \frac{t_1 b_0}{\gamma b_{1,1}}. \end{aligned} \tag{3.8}$$

Consider the probabilities that have to be estimated:

- t_0: no stain transferred during the commission of a crime;
- t_1: one stain transferred during the commission of a crime;
- b_0: no stain transferred innocently;
- $b_{1,1}$: (one group of) one stain transferred innocently;
- γ: the frequency of genotype Γ amongst stains of body fluids on clothing.

The first four of these probabilities relate to what has been called *extrinsic evidence*, the fifth to *intrinsic evidence* (Kind 1994). The estimation of probabilities for extrinsic evidence is subjective and values for these are matters for a forensic scientist's personal judgement. The estimation of probabilities for intrinsic evidence may be determined by observation and measurement. Extrinsic evidence for this example is the evidence of transfer of a stain. Intrinsic evidence is the result of the frequency of the profile. More generally, extrinsic evidence can be physical attributes (the number, position and location of stains) and intrinsic evidence can be descriptors of the stains (profiles).

Note that in general, t_0 is small in relation to $t_1 b_0 / \{b_{1,1} \gamma\}$ and may be considered negligible.

Let Γ be a DNA profile that has a match probability of approximately 0.01 amongst Caucasians in England. Assume that the distribution of DNA profiles among stains on clothing is approximately the distribution among the relevant population. This assumption is not necessarily correct. Then $\gamma = 0.01$. Consider the results of a survey of men's clothing from which it appears reasonable that $b_0 > 0.95$, $b_{1,1} < 0.05$ and suppose $t_1 > 0.5$, then, irrespective of the value of t_0 (except that it has to be less than $(1 - t_1)$; *i.e.*, less than 0.5 in this instance),

$$V > \frac{0.5 \times 0.95}{0.05 \times 0.01} = 950.$$

This value indicates very strong evidence to support the proposition that the suspect assaulted the victim. The evidence is at least 950 times more likely if the suspect assaulted the victim than if he did not.

Notice that such a result is considerably different from $1/\gamma$ (= 100 in the numerical example). This latter result would hold if $(t_1 b_0 / b_{1,1})$ were approximately 1, which may mean that unrealistic assumptions about the relative values of the transfer probabilities would have to be made.

In conclusion, the evaluation of the evidence has to take into account the possibility of transfer of evidence from a source other than the suspect. Transfer and background probabilities are relevant.

Extensions to cases involving

1. evidence left by the offender (not the suspect),
2. two recovered stains of the same type (typically groups of glass fragments) and a control group,
3. one recovered and two control groups,
4. two recovered and two control groups,
5. a generalisation involving n stains, k groups and k (or $m \neq k$) offenders and
6. cross-transfer (two-way transfer)

have been proposed (*e.g.*, Aitken and Taroni 2004).

3.3.3 Crime level

Level III, the so-called 'crime level', is close to the activity level. At level III, the propositions are really those of interest to the jury. Non-scientific information, such as whether a crime occurred, or whether an eyewitness is reliable, plays an important role in the decision. In routine work, forensic scientists generally use the source level to assess scientific evidence, notably for DNA evidence. Evidence under the activity level propositions requires that an important body of circumstantial information is available to the scientist. Unfortunately, this is often not the case because of a lack of interaction between the scientists and investigators. There are limitations in the use of a source level evaluation in a criminal investigation compared with an activity level evaluation. The lower the level at which the evidence is assessed, the lower is the relevance of the results in the context of the case discussed in the courts. For ease of simplicity, note that even if the value, V, of the evidence is such that it adds considerable support to the proposition that the stain comes from the suspect, this does not help determine whether the stains had been transferred during the criminal action or for an innocent reason. Consequently, there is often dissatisfaction if the scientist is restricted to level I propositions.

A formal development of the likelihood ratio under 'crime level' propositions shows that two additional parameters are of interest. The first concerns material that may be *relevant* (Stoney 1991, 1994). Crime material that came from the offender is said to be relevant in that it is relevant to the consideration of suspects as possible offenders. The second parameter concerns the recognition that if the material is not relevant to the case, then it may have arrived at the scene from the suspect for innocent reasons.

3.3.3.1 Notation

A likelihood ratio has been developed for a scenario involving k offenders. Consider a crime that has been committed by k offenders. A single blood stain is found at the crime scene in a position where it may have been left by one of the offenders. A suspect is found and he gives a blood sample. The suspect's sample and the crime stain are of the same profile Γ with relative frequency γ amongst the relevant population from which the criminals have come. As before, consider two propositions:

- H_p: the suspect is one of the k offenders,
- H_d: the suspect is not one of the k offenders.

Notice the difference between these propositions and those of the previous sections on source or activity propositions. At the source level, the propositions referred to the suspect being, or not being, the donor of the blood stain found at the crime scene. Now, the propositions are stronger, namely that the suspect is, or is not, one of the offenders. The value V of the evidence is

$$V = \frac{Pr(E_c \mid E_s, H_p)}{Pr(E_c \mid H_d)} \tag{3.9}$$

where E_c is the profile Γ of the crime stain and E_s is the profile Γ of the suspect, where Γ has frequency γ in the relevant population, the population from which the offenders have come.

A link is needed between what is observed, the stain at the crime scene, and the propositions, the suspect is or is not one of the offenders. The link is made in two steps.

The first step is the consideration of a proposition that the crime stain came from one of the k offenders (rather than H_p that the suspect is one of the k offenders) and the alternative proposition that the crime stain did not come from any of the k offenders. These propositions are known as *association propositions*.

Assume that the crime stain came from one of the k offenders. The second step is then the consideration of a proposition that the crime stain came from the suspect and the alternative proposition that the crime stain did not come from the suspect. These propositions are known as *intermediate association propositions* (source level propositions).

Development of these two pairs of propositions indicates that other factors have to be considered. These are the relevance, first, and the innocent acquisition (for the receptor), second. The evaluation of these factors may be done by partitioning the expressions in the numerator and the denominator of the likelihood ratio. There are two types of subjective probabilities of interest, those of

- innocent acquisition, usually denoted as p: It is a measure of belief that evidence has been acquired in a manner unrelated to the crime and

- relevance, usually denoted as r: this is the probability of relevance. In this context, it is the probability that the stain recovered from the crime scene is connected with the crime; it has been left by one of the offenders.

3.3.3.2 Association propositions

Consider the following

- G : the crime stain came from one of the k offenders,
- $\bar{G}$: the crime stain did not come from any of the k offenders.

The value, V, of the evidence may now be written, by extending the conversation, as

$$V = \frac{Pr(E_c \mid H_p, G, E_s)Pr(G \mid H_p, E_s) + Pr(E_c \mid H_p, \bar{G}, E_s)Pr(\bar{G} \mid H_p, E_s)}{Pr(E_c \mid H_d, G)Pr(G \mid H_d) + Pr(E_c \mid H_d, \bar{G})Pr(\bar{G} \mid H_d)}.$$

In the absence of E_c, the evidence of the profile of the crime stain, knowledge of H_p and of E_s does not affect our belief in the truth or otherwise of G. This is what is meant by *relevance* in this context. Thus

$$Pr(G \mid H_p, E_s) = Pr(G \mid H_p) = Pr(G)$$

and

$$Pr(\bar{G} \mid H_p, E_s) = Pr(\bar{G} \mid H_p) = Pr(\bar{G}).$$

Let $Pr(G) = r,\ Pr(\bar{G}) = (1 - r)$ and call r the *relevance term, i.e.*, relevance is equated to the probability that the stain had been left by one of the offenders. The higher the value of r, the more relevant the stain becomes. Thus

$$V = \frac{Pr(E_c \mid H_p, G, E_s)r + Pr(E_c \mid H_p, \bar{G}, E_s)(1 - r)}{Pr(E_c \mid H_d, G, E_s)r + Pr(E_c \mid H_d, \bar{G}, E_s)(1 - r)}. \quad (3.10)$$

3.3.3.3 Intermediate association propositions

In order to determine the component probabilities of the likelihood ratio developed introducing association propositions, intermediate association propositions are used.

- F: the crime stain came from the suspect,
- $\bar{F}$: the crime stain did not come from the suspect.

Now consider the following four conditional probabilities:

(a) $Pr(E_c \mid H_p, G, E_s)$

This is the probability that the crime stain would be of profile Γ if it had been left by one of the offenders (G), the suspect had committed the crime (H_p) and the suspect is of profile Γ.

$$Pr(E_c \mid H_p, G, E_s) = Pr(E_c \mid H_p, G, F, E_s)Pr(F \mid H_p, G, E_s) + Pr(E_c \mid H_p, G, \bar{F}, E_s)Pr(\bar{F} \mid H_p, G, E_s).$$

Here $E_c = E_s = \Gamma$, the profile of the crime stain and of the suspect, and $Pr(E_c \mid H_p, G, F, E_s) = 1$. In the absence of E_c, F is independent of E_s and so

$$Pr(F \mid H_p, G, E_s) = Pr(F \mid H_p, G) = 1/k,$$

where it is assumed that there is nothing in the background information I to distinguish the suspect, given H_p, from the other offenders as far as blood shedding is considered.

In a similar manner, $Pr(\bar{F} \mid H_p, G, E_s) = (k-1)/k$. Also,

$$Pr(E_c \mid H_p, G, \bar{F}, E_s) = Pr(E_c \mid H_p, G, \bar{F}) = \gamma,$$

since if $\bar{F}$ is true, E_c and E_s are independent and one of the other offenders left the stain (since G holds). Thus

$$Pr(E_c \mid H_p, G, E_s) = \{1 + (k-1)\gamma\}/k.$$

(b) $Pr(E_c \mid H_p, \bar{G}, E_s)$

This is the probability that the crime stain would be of profile Γ if it had been left by an unknown person who was unconnected with the crime. (This is the implication of assuming $\bar{G}$ to be true.) The population of people who may have left the stain is not necessarily the same as the population from which the criminals are assumed to have come. Thus, let

$$Pr(E_c \mid H_p, \bar{G}, E_s) = \gamma'.$$

where γ' is the probability of profile Γ amongst the population of people who may have left the stain (the prime $'$ indicating it may not be the same value as γ, which relates to the population from which the criminals have come).

Consider now that the suspect is innocent and H_d is true.

(c) $Pr(E_c \mid H_d, G, E_s) = Pr(E_c \mid H_d, G) = \gamma$,

the frequency of Γ amongst the population from which the criminals have come. There is no need to partition these probabilities to consider F and $\bar{F}$ as the suspect is assumed not to be one of the offenders and G is that the stain was left by one of the offenders.

(d)

$$Pr(E_c \mid H_d, \bar{G}, E_s) = Pr(E_c \mid H_d, \bar{G}, F, E_s)Pr(F \mid H_d, \bar{G}, E_s) + Pr(E_c \mid H_d, \bar{G}, \bar{F}, E_s)Pr(\bar{F} \mid H_d, \bar{G}, E_s).$$

If F is true, $Pr(E_c \mid H_d, \bar{G}, F, E_s) = 1$. Also $Pr(F \mid H_d, \bar{G}, E_s) = Pr(F \mid H_d, \bar{G})$. This is the probability p of innocent acquisition, that the stain would have been left by the suspect even though the suspect was innocent of the offence. Here, it is assumed that the propensity to leave a stain is independent of the profile of the person who left the stain. Hence $Pr(F \mid H_d, \bar{G}) = p$ and $Pr(\bar{F} \mid H_d, \bar{G}, E_s) = Pr(\bar{F} \mid H_d, \bar{G}) = 1 - p$. Also $Pr(E_c \mid H_d, \bar{G}, \bar{F}) = \gamma'$. Thus

$$Pr(E_c \mid H_d, \bar{G}, E_s) = p + (1-p)\gamma'.$$

Substitution of the above expressions into (3.10) gives

$$V = \frac{[r\{1 + (k-1)\gamma\}/k] + \{\gamma'(1-r)\}}{\gamma r + \{p + (1-p)\gamma'\}(1-r)}$$
$$= \frac{r\{1 + (k-1)\gamma\} + k\gamma'(1-r)}{k[\gamma r + \{p + (1-p)\gamma'\}(1-r)]}. \quad (3.11)$$

Values of V for varying values of r and p are presented in Aitken and Taroni (2004).

A practical example of the use of the above is given by Evett et al. (1998b) in the context of footmarks evaluation with $k = 1$, its formal analysis using Bayesian networks is presented in Section 4.1.1.

Extensions of this probabilistic result have been presented for scenarios involving two stains and one offender as developed in Section 4.3.

3.4 Pre-assessment of the case

The evaluation process should start when the scientist first meets the case. It is at this stage that the scientist thinks about the questions that are to be addressed and the outcomes that may be expected. The scientist should attempt to frame propositions and think about the value of evidence that is expected (Evett et al. 2002b). However, there is a tendency to consider evaluation of evidence as a final step of a casework examination, notably at the time of preparing the formal report. This is so even if an earlier interest in the process enables the scientist to make better decisions about the allocation of resources. For example, consider a case of assault involving the possible cross-transfer of textile fibres between a victim and an assailant. The scientist has to decide whether to look first for potential transferred fibres on the victim's pullover or for fibres on the suspect's pullover. If traces compatible with the suspect's pullover are found on the victim's pullover, then the expectation of the detection of traces from the victim's pullover on the suspect's pullover has to be assessed. This includes the possibility of reciprocal transfer. Should the scientist have expectations? How can they be quantified? If so, what is the interpretative consequence when those expectations are or are not met (presence or absence of evidence)? Matters to be considered include the following:

- the appropriate nature of the expectations;
- the quantification of the expectations;
- the interpretation of the presence or absence of evidence, through the success or failure of the expectations.

The scientist requires an adequate appreciation of the circumstances of the case so that a framework may be set up for consideration of the kind of examinations that may be carried out and what may be expected from them (Cook et al. 1998b), in order that a logical decision can be made.

Such a procedure of pre-assessment can be justified on a variety of grounds. Essentially it is justified because the choice of level for the propositions for the evaluation of scientific evidence is carried out within a framework of circumstances, and these circumstances have to be known before any examination is made in order that relevant propositions may be proposed (*e.g.*, at the activity instead of source level). This procedure may inform a discussion with the customer before any substantial decision is reached (*e.g.*, about expense). Moreover, this process provides a basis for consistency of approach by all scientists who are thereby encouraged to consider carefully, factors such as circumstantial information and data that are to be used to evaluate the evidence and to declare in the final report.

The scientist should proceed by considering an estimation of the probability of whatever evidence will be found given each proposition. Consider, for example, a case where a window is smashed and assume that the prosecution and defence propose the following alternatives: 'The suspect is the man who smashed the window', and 'The suspect was not present when the window was smashed'. The examination of the suspect's pullover will reveal a quantity Q of glass fragments, where Q can be classified, for example, as one of the following states {*no, few* or *many*}. So:

- the first question asked for assessment of the numerator of the likelihood ratio is 'What is the probability of finding a quantity Q of matching glass fragments if the suspect is the man who smashed the window?'
- the second question asked for assessment of the denominator of the likelihood ratio is 'What is the probability of finding a quantity Q of matching glass fragments if the suspect was not present when the window was smashed?'

The scientist is asked initially to assess six different probabilities:

1. the probability of finding *no* matching glass fragments if the suspect is the man who smashed the window;
2. the probability of finding *few* matching glass fragments if the suspect is the man who smashed the window;
3. the probability of finding *many* matching glass fragments if the suspect is the man who smashed the window;
4. the probability of finding *no* matching glass fragments if the suspect was not present when the window was smashed;
5. the probability of finding *few* matching glass fragments if the suspect was not present when the window was smashed;
6. the probability of finding *many* matching glass fragments if the suspect was not present when the window was smashed.

These probabilities may not be easy to derive because of a lack of information available to the scientist (Cook et al. 1998a). For example, it will be very difficult to assess transfer probabilities if the scientist has no answer to questions like the following.

- Was the window smashed by a person or by a vehicle? The fact that a window was smashed by a person or by a vehicle changes the amount of glass fragments the scientist expects to be transferred.
- How (*modus operandi*) was the window smashed? If it was smashed by a person, then was that person standing close to it? Was a brick thrown through it? Information

about the way a window is smashed is important because it provides information on the amount of glass potentially projected. Information of the distance between the person who smashed the window and the window offers relevant information on the amount of glass fragments the scientist will expect to recover.

Where there is little information about the time of the alleged offence and the time of the investigators seizing the clothes, the lapse of time between the offence and the collection of evidence cannot be precisely estimated. It is also difficult to assess the probability of persistence of any transferred glass fragment. Therefore, if the scientist has a small amount of information about the case under examination, then assessment has to be restricted to level I (or sub-level I) propositions (Section 3.3.1.2).

Consider another brief example presented in Evett et al. (2002a). A cigarette end is recovered from the scene of a burgled house. None of the family in the house smokes. A suspect is apprehended. The cigarette end is submitted for DNA profiling (because the suspect denied ever being anywhere near to the house that was burglared) and the results are compared with the suspect's DNA. Various outcomes are possible.

- a *Match*, where a single profile from the cigarette end is attributable to the same genotype as that of the suspect;
- a *Mixture/Match*, in which a mixed profile from the cigarette end includes alleles that are present in the suspect's profile;
- a *Difference*, which includes either a single profile that is a different genotype from that of the suspect, or a two-persons mixture that differs from the suspect's DNA profile;
- a *Non-profile*, where no DNA profile is detected.

If propositions at the activity level are proposed (*e.g.*, H_p, the suspect is the person who smoked the cigarette, and H_d, some other unknown person smoked the cigarette), then the assessment of the outcomes depends on probabilities the scientist specifies for the possibility that the DNA entered (or not) the process by innocent means, the fact that the person who smoked the cigarette left (or not) sufficient DNA to give a profile, or that the DNA of a third person entered (or not) the process.

Circumstantial evidence and information of laboratory quality are necessary for such assessments. It will then be possible to propose a pre-assessment table expressing a likelihood ratio for each of the four outcomes. It is clear that it will be difficult to assign precise values to different parameters of interest for the evaluation of the outcomes. Sensitivity analyses (*see* Chapter 9) can be considered.

The case pre-assessment process can be summarised by the following steps:

- collection of information that the scientist may need about the case
- consideration of the questions that the scientist can reasonably address, and, consequently, the level of propositions for which the scientist can reasonably choose to assess evidence
- identification of the relevant parameters that will appear in the likelihood ratio

- assessment of the strength of the likelihood ratio expected given the background information
- determination of the examination strategy
- conduct of tests and observation of outcomes
- evaluation of the likelihood ratio and report of the value.

A practical procedure using Bayesian networks is presented in Chapter 8.

Examples considering the pre-assessment of various types of scientific evidence are presented in the literature. Cook et al. (1998b) present pre-assessment through the example of a hypothetical burglary involving potential glass fragments (an unknown quantity, Q, of recovered fragments). Stockton and Day (2001) consider an example involving signatures in questioned documents. Champod and Jackson (2000) consider a burglary case involving fibres evidence. Booth et al. (2002) discuss a drug case. A cross-transfer (also called 'two-way transfer') case involving textiles is presented by Cook et al. (1999) where it is shown how the pre-assessment can be updated when a staged approach is taken. The results of the examination of one of the garments are used to inform the decision about whether the second garment should be examined.

3.5 Evaluation using graphical models

3.5.1 Introduction

Bayesian networks have been found to be useful in assisting human reasoning in a variety of disciplines in which uncertainty plays an important role. In forensic science too, Bayesian networks have been proposed as a method of formal reasoning that could assist forensic scientists to understand the dependencies that may exist between different aspects of evidence. For example, Aitken and Gammerman (1989) were among the first to suggest the use of graphical probabilistic models for the assessment of scenarios involving scientific evidence. Ideas expressed by these authors have further been developed by authors such as Dawid and Evett (1997) and Garbolino and Taroni (2002). Although these studies provide clear examples of the relevance of Bayesian networks for assisting the evaluation of evidence, explanations, if any, as to how practitioners should use the method to build their own models, are mostly brief and very general.

The problem is well posed in Dawid et al. (2002), where the authors note that finding an appropriate representation of a case under examination is crucial for several reasons (viability, computational routines, etc.), and that the graphical construction is to some extent an art form, but one that can be guided by scientific and logical considerations.

While no explicit guidelines exist as to how one should proceed in setting up a Bayesian network, there are some concepts considered useful for eliciting sensible network structures. These are discussed below.

3.5.2 Aspects of constructing Bayesian networks

Consider a two-stain scenario described by Evett (1987). A crime has been committed by two men, each of whom left a bloodstain at the crime scene. A hypothetical characteristic Γ

of the stains is analysed. This characteristic has some variability so that the various possible values may be denoted as $\Gamma_1, \Gamma_2, \Gamma_3, \ldots, \Gamma_n$. Suppose that one of the two crime stains is of type Γ_1 and the other is of type Γ_2. Some time later, a suspect is apprehended as a result of information completely unrelated to the blood evidence. A blood sample provided by the suspect is found to be of type Γ_1. Notice that this scenario assumes that there is no evidence in the form of injuries and the evidence is confined solely to the results of the blood analysis. It is further assumed that the evidential material is relevant to the offence and that there are exactly two offenders.

A forensic scientist may be asked to evaluate the evidence with respect to the following two propositions:

- H_p: The suspect was one of the two men who committed the crime.
- H_d: The suspect was not one of the two men who committed the crime.

It has already been noted that the value of the evidence is appropriately expressed in the form of a likelihood ratio V. In the case of the two-stain scenario, the evidence E consists of two distinct pieces of information, which are as follows:

- E_1: The bloodstains at the crime scene are of type Γ_1 and Γ_2.
- E_2: The suspect's blood is of type Γ_1.

A general expression of the likelihood ratio is as follows (Evett 1987):

$$V = \frac{Pr(E \mid H_p)}{Pr(E \mid H_d)} = \frac{Pr(E_1 \mid E_2, H_p) \times Pr(E_2 \mid H_p)}{Pr(E_1 \mid E_2, H_d) \times Pr(E_2 \mid H_d)} . \tag{3.12}$$

Consider the conditioning of E_2 on H_p and on H_d on the right-hand side of (3.12). It appears to be a reasonable assumption that the probability of the suspect's blood being type Γ_1 does not depend on whether he was one of the two men who committed the crime. So, one may set $Pr(E_2 \mid H_p) = Pr(E_2 \mid H_d)$ and the likelihood ratio becomes:

$$V = \frac{Pr(E_1 \mid E_2, H_p)}{Pr(E_1 \mid E_2, H_d)} . \tag{3.13}$$

3.5.3 Eliciting structural relationships

An advantage of constructing Bayesian networks on the basis of an existing formula is that the number and definition of the nodes are already given. In the case of the two-trace scenario there are the following nodes: H, E_1 and E_2. For the moment, assume that these variables have binary states.

As an aid to visualise the potential dependencies about which one will need to decide, consider a complete undirected graph over the variables of interest (Figure 3.1(i)). Now, a probabilistic formula may not only help in defining relevant variables, but also can contain expressions that may guide one in defining the number of edges as well as their direction. For example, the term $Pr(E_1 \mid E_2, H_p)$ in (3.12), to be read as 'the probability of the two blood stains at the crime scene are of type Γ_1 and Γ_2 given the suspect's blood is of type Γ_1 and the suspect is one of the two men who committed the crime', indicates that E_1

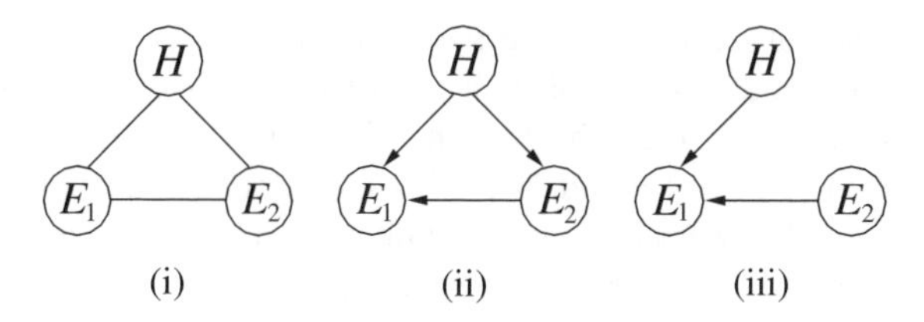

Figure 3.1 (i) Complete undirected graph over the variables E_1, E_2 and H; (ii) and (iii) possible network structures for the two-trace transfer problem.

is conditioned on both H and E_2. An appropriate graphical expression of this conditioning would be a converging connection at the node E_1 with parental variables H and E_2.

The question remains as to whether there should be a link of some kind between H and E_2. Consider again (3.12). This suggests that there should be an arc pointing from H and E_2. This would lead to a graphical structure as shown in Figure 3.1(ii). However, if it is understood that the probability of E_2, the suspect's blood being of type Γ_1, is the same whether he is or is not one of the two men who committed the crime, then the link $H \rightarrow E_2$ carries no inferential force. In other words, if $Pr(E_2 \mid H_p) = Pr(E_2 \mid H_d)$, knowledge about E_2 does not affect H along $H \rightarrow E_2$. Thus, the stated assumption serves as a formal argument for justifying the omission of the link $H \rightarrow E_2$ (*see* Figure 3.1(iii)).

Notice that this does not mean that there is no influence between H and E_2. The converging connection at the node E_1 implies that H and E_2 become dependent as soon as knowledge about E_1 becomes available (Section 2.1.5).

3.5.4 Level of detail of variables and quantification of influences

An inference problem may be approached on different levels of detail and this may result in structural differences in Bayesian networks. It is possible to illustrate this in case of the two-trace problem, which has, until now, been studied on the binary level.

Consider the variable E_1 first. The discovery of two crime-related traces of type Γ_1 and Γ_2 at the scene of a crime is only one of several possible outcomes. The stains may have been of some other type. For simplicity, aggregate the other possible types $\Gamma_3, \ldots, \Gamma_n$ by Γ_x. The frequency of Γ_i is $\gamma_i, i = 1, \ldots, n$ with $\gamma_x = \gamma_3 + \cdots + \gamma_n$. The possible states (*i.e., genotypes*) for the node E_1 can then be defined as follows: $\Gamma_1 - \Gamma_1, \Gamma_1 - \Gamma_2, \Gamma_1 - \Gamma_x, \Gamma_2 - \Gamma_2, \Gamma_2 - \Gamma_x, \Gamma_x - \Gamma_x$. The suspect is a potential contributor to the pair of recovered stains, so it appears reasonable to assume the following possible values for the node E_2: Γ_1, Γ_2, or Γ_x. The third variable of interest, H, is concerned with the major proposition. The suspect may or may not be one of the two criminals, so H may be assumed to be a binary node.

For a numerical specification of the Bayesian network, unconditional probabilities are required for the nodes H and E_2, whereas conditional probabilities are required for the node E_1. The probabilities assigned to the node H are the prior beliefs held by some evaluator, for example, an investigator or a court of law, about the two alternative propositions. The prior probabilities of the node E_2 could be the relative frequencies of the characteristics Γ_1, Γ_2 and Γ_x in a relevant database. Note that the frequency γ_x of Γ_x is $\sum_{i=3}^{n} \gamma_i$ and that $\sum_{i=1}^{n} \gamma_i = 1$. A more detailed study is required for the probabilities assigned to the node E_1:

- If H_p is true, then at least one of the two recovered stains must be of the same type as the suspect's blood. For example, if the suspect is one of the two criminals and his blood is of type Γ_1, then the only possible outcome for the pair of recovered stains are $\Gamma_1 - \Gamma_1$, $\Gamma_1 - \Gamma_2$ and $\Gamma_1 - \Gamma_x$. The second man may have left a stain of type Γ_1, Γ_2 or Γ_x, respectively. Assuming the characteristics of a man's blood to be independent of his tendency to criminal activity, then the probability of the second man's blood being of, for example, type Γ_2 can be estimated by the relative frequency γ_2 in a relevant population. Thus a probability such as $Pr(E_1 = \Gamma_1 - \Gamma_2 \mid E_2 = \Gamma_1, H_p)$ is obtained by $1 \times \gamma_2$. Analogous considerations apply to all other conditional probabilities that assume H_p to be true.

- If H_d is true, that is, the suspect is not one of the two criminals, then the characteristics of his blood are surely irrelevant for determining the characteristics of the pair of recovered stains. In other words, the stains have been left by two unknown men. These two individuals may be regarded as drawn randomly from the population. If the two stains have different characteristics, that is, $\Gamma_i - \Gamma_j$, then the first man may have left the stain of type Γ_i, and the second man may have left the stain of type Γ_j, and vice versa. Therefore, the probability of observing two stains of different type, given H_d, is $2\gamma_i\gamma_j$. On the other hand, the probability of observing two stains of the same type, for example, $\Gamma_i - \Gamma_i$, is γ_i^2.

A summary of the probabilities assigned to the node E_1 can be found in Table 3.1. As may be seen, there are far more probabilities than that used in the initial likelihood ratio formula (3.13). The reason for this is that (3.13) accounts only for a specific situation, notably the two stains being of type Γ_1 and Γ_2 and the suspect's blood being of type Γ_1, whereas the Bayesian network accounts for all possible combinations of outcomes for the nodes E_1 and E_2.

As a numerical example, imagine a scenario in which $\gamma_1 = \gamma_2 = 0.05$ and $Pr(H_p) = Pr(H_d) = 0.5$. Figure 3.2(i) shows the numerically specified Bayesian network in its initial state. Notice that the values displayed in the nodes H and E_2 are given by the unconditional probabilities assigned to these nodes. In common with commercially and academically

Table 3.1 Probabilities assigned to the node E_1, conditional on E_2 and H for possible values of E_1, E_2 and H in Figure 3.1. Profile frequencies for Γ_1, Γ_2 and Γ_x are γ_1, γ_2 and γ_x. Propositions are H_p, the suspect was one of the two men who committed the crime, and H_d, the suspect was not one of the two men who committed the crime.

	H:	H_p			H_d		
	E_2:	Γ_1	Γ_2	Γ_x	Γ_1	Γ_2	Γ_x
E_1:	$\Gamma_1 - \Gamma_1$	γ_1	0	0	γ_1^2	γ_1^2	γ_1^2
	$\Gamma_1 - \Gamma_2$	γ_2	γ_1	0	$2\gamma_1\gamma_2$	$2\gamma_1\gamma_2$	$2\gamma_1\gamma_2$
	$\Gamma_1 - \Gamma_x$	γ_x	0	γ_1	$2\gamma_1\gamma_x$	$2\gamma_1\gamma_x$	$2\gamma_1\gamma_x$
	$\Gamma_2 - \Gamma_2$	0	γ_2	0	γ_2^2	γ_2^2	γ_2^2
	$\Gamma_2 - \Gamma_x$	0	γ_x	γ_2	$2\gamma_2\gamma_x$	$2\gamma_2\gamma_x$	$2\gamma_2\gamma_x$
	$\Gamma_x - \Gamma_x$	0	0	γ_x	γ_x^2	γ_x^2	γ_x^2

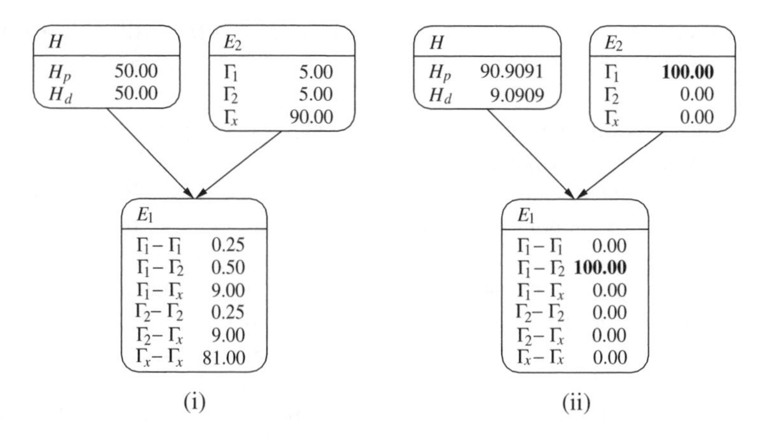

Figure 3.2 Bayesian network for a two-trace transfer scenario: (i) initial state of the model; probability values for E_1 are calculated using (3.14). The probabilities displayed in the nodes H and E_2 are the unconditional probabilities specified for these nodes. (ii) state of the model after instantiation of E_1 and E_2. Node H has two states, H_p, the suspect was one of the two men who committed the crime, H_d, the suspect was not one of the two men who committed the crime. Node E_2 has three states, the suspect's profile is Γ_1, Γ_2 or Γ_x (something other than Γ_1 or Γ_2). Node E_1 has six states, the profiles of the bloodstains at the crime scene are one of the six possible pairs obtainable from Γ_1, Γ_2 and Γ_x.

available Bayesian network software systems, probabilities displayed in the pictorial representations in this book will take the form of numbers between 0 and 100.

For determining the initial state of the variable E_1, all possible parental configurations must be considered. The initial probability of, for example, E_1 being in state $\Gamma_1 - \Gamma_2$ is thus obtained by:

$$\sum_{i=1}^{x} Pr(E_1 = \Gamma_1 - \Gamma_2 \mid E_2 = \Gamma_i, H_p) \times Pr(E_2 = \Gamma_i) \times Pr(H_p) +$$
$$\sum_{i=1}^{x} Pr(E_1 = \Gamma_1 - \Gamma_2 \mid E_2 = \Gamma_i, H_d) \times Pr(E_2 = \Gamma_i) \times Pr(H_d) \ . \tag{3.14}$$

Use of the probability estimates γ_1 and γ_2 from Table 3.1, both replaced by 0.05, and $Pr(H_p) = Pr(H_d) = 0.5$, in (3.14) verifies the value of 0.50 in Figure 3.2 for $\Gamma_1 - \Gamma_2$.

$$\begin{aligned} Pr(E_1 = \Gamma_1 - \Gamma_2) &= (1 \times \gamma_2) \times \gamma_1 \times 0.5 + (\gamma_1 \times 1) \times \gamma_2 \times 0.5 \\ &\quad + (0) \times \gamma_x \times 0.5 + (2 \times \gamma_1 \times \gamma_2) \times \gamma_1 \times 0.5 \\ &\quad + (2 \times \gamma_1 \times \gamma_2) \times \gamma_2 \times 0.5 + (2 \times \gamma_1 \times \gamma_2) \times \gamma_x \times 0.5 \\ &= (0.05 \times 0.05 \times 0.5) + (0.05 \times 0.05 \times 0.5) \\ &\quad + (0.005 \times 0.05 \times 0.5) + (0.005 \times 0.05 \times 0.5) \\ &\quad + (0.005 \times 0.9 \times 0.5) \\ &= 0.005. \end{aligned}$$

Note that probability values in figures are expressed on a 0 to 100 % scale; the value of 0.50 in Figure 3.2 corresponds to a probability for state $\Gamma_1 - \Gamma_2$ of 0.005 on a scale of 0 to 1.

The Bayesian network may now be used to calculate the posterior probability of H given an observed set of evidence. Recall the scenario described at the beginning of Section 3.5.2. There are two stains of type Γ_1 and Γ_2 and the suspect's blood is of type Γ_1. These observations are communicated to the model by instantiating the nodes E_1 and E_2. This situation is shown in Figure 3.2(ii). The instantiation of the nodes E_1 and E_2 affects the node H, and the probability of H_p has increased. Notice that the probabilities displayed for H_p and H_d are posterior probabilities given the observed set of evidence. In their extended form, H_p and H_d may be written as $Pr(H_p \mid E_1 = \Gamma_1 - \Gamma_2, E_2 = \Gamma_1)$ and $Pr(H_d \mid E_1 = \Gamma_1 - \Gamma_2, E_2 = \Gamma_1)$, respectively.

The value of the observed set of evidence may be measured by the likelihood ratio. On the basis of the probabilities defined in Table 3.1, one can find that (3.13) reduces to $1/2\gamma_1$. As γ_1 was set to 0.05, a likelihood ratio of 10 is indicated. The same value can be deduced from the Bayesian network shown in Figure 3.2. Here, equal prior probabilities are assumed for the node H. In such a situation, the likelihood ratio equals the ratio of the posterior probabilities of H, namely, $0.91/0.091 = 10$.

3.5.5 Derivation of an alternative network structure

Different Bayesian networks may be proposed for considering questions surrounding the same situation. This is illustrated in the case of the Bayesian network discussed in Section 3.5.2, which may be used for the evaluation of a two-stain scenario.

Some of the conditional probabilities of the node E_1 in the Bayesian network shown in Figure 3.2 need to be calculated using frequency statistics, in particular, when conditioning on the alternative hypothesis H_d (*see* Table 3.1). Such calculations may be time consuming and, to some degree, prone to error. An interesting question, therefore, is whether there exists a more convenient Bayesian network structure, allowing one to avoid calculations when completing probability tables.

The use of so-called 'compound' probabilities, for example, in the probability table for the node E_1 (*see* Section 3.5.4 for the definition of this node), suggests the presence of assumed intermediate states that have not been represented explicitly. An essential aspect of an approach to find an alternative Bayesian network structure consists of examining the definitions of nodes more closely. Such analysis may reveal hidden propositions that, if represented as distinct nodes, may avoid the use of composed probabilities.

Consider this in the case of the node E_1. In Section 3.5.4, this node was defined as the characteristic of the two recovered stains, that is, $\Gamma_1 - \Gamma_1$, $\Gamma_1 - \Gamma_2$, $\Gamma_1 - \Gamma_x$, $\Gamma_2 - \Gamma_2$, $\Gamma_2 - \Gamma_x$, $\Gamma_x - \Gamma_x$. One possibility to provide a more detailed description of the node E_1 would be to adopt nodes describing the characteristics that each of the recovered stains may assume. For example, let T_1 and T_2 each with states Γ_1, Γ_2 and Γ_x denote the possible characteristics of stain 1 and stain 2 respectively. It is reasonable to postulate that both T_1 and T_2 determine the state of E_1. Consequently, a converging connection with $T_1 \rightarrow E_1 \leftarrow T_2$ is appropriate. A practical consequence of this is that the probability table associated with the node E_1 can now be completed logically using values 0 and 1 (*see* Table 3.2).

Table 3.2 Probabilities assigned to the node E_1, where T_1 and T_2 are the characteristics of stains 1 and 2, respectively, and E_1 is the combination of T_1 and T_2.

	T_1:	Γ_1			Γ_2			Γ_x		
	T_2:	Γ_1	Γ_2	Γ_x	Γ_1	Γ_2	Γ_x	Γ_1	Γ_2	Γ_x
E_1:	$\Gamma_1 - \Gamma_1$	1	0	0	0	0	0	0	0	0
	$\Gamma_1 - \Gamma_2$	0	1	0	1	0	0	0	0	0
	$\Gamma_1 - \Gamma_x$	0	0	1	0	0	0	1	0	0
	$\Gamma_2 - \Gamma_2$	0	0	0	0	1	0	0	0	0
	$\Gamma_2 - \Gamma_x$	0	0	0	0	0	1	0	1	0
	$\Gamma_x - \Gamma_x$	0	0	0	0	0	0	0	0	1

From T_1 and T_2, that is, the characteristics of stain 1 and 2, it is then necessary to construct an argument to E_2 and H. These two variables designate the characteristics of the suspect's blood (E_2) and the propositions according to which the suspect is or is not one of the two men who committed the crime (H). A feasible way to achieve this is as follows:

- Knowledge about the characteristics of the suspect's blood (E_2) may tell one something about the characteristics of the crime stains (T_1 and T_2). However, knowledge about which of the two stains, if any, has been left by the suspect, is relevant here. Therefore, an additional variable F is adopted to represent the uncertainty about which stain has been left by the suspect. The states 'none' (f_0), 'stain 1' (f_1) and 'stain 2' (f_2) are assumed for this variable. Both E_2 and F are chosen as parental variables for T_1 and T_2. Notice that f_0, f_1 and f_2 are mutually exclusive events. This is an implicit assumption that the suspect has left at most one trace.

- The proposition concerning which stain, if any, was left by the offender (F) depends on whether the suspect is one of the two offenders (H). Accordingly, the node H is chosen as a parental variable for the node F.

The Bayesian network shown in Figure 3.3 represents these considerations.

The assignment of values to the probability tables of the nodes T_1, T_2 and F is quite straightforward. Consider the following:

- Nodes T_1 and T_2: If the suspect has left none of the two stains (f_0), then his blood profile characteristics (E_2) are not relevant for assessing the probability of the stains being of type Γ_1, Γ_2 or Γ_x. The stains are then regarded as coming from two unknown men drawn randomly from the population. The probability of the stains being of type Γ_1, Γ_2 or Γ_x is just given by the relative frequency of these characteristics, *i.e.*, γ_1, γ_2 and γ_x. There are further situations in which a stain is considered as originating from a man drawn randomly from the population, notably: if (*a*) the stain considered is T_1 and the suspect has left stain 2 (f_2), and (*b*) the stain considered is T_2 and the suspect has left stain 1 (f_1). In these cases, the probability tables of the nodes T_1 and T_2 contain again the relative frequency of the characteristic considered irrespective of the suspect's blood type. However, the characteristic of the suspect's blood becomes important whenever one considers a stain assumed to be left by the

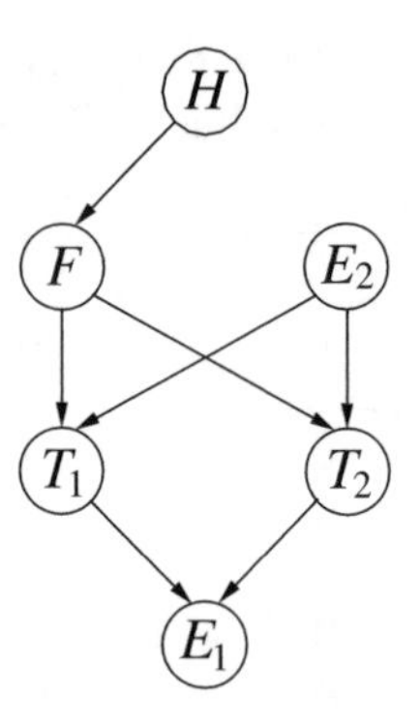

Figure 3.3 An alternative Bayesian network structure for a two-trace transfer scenario, where T_1 and T_2 are characteristics of stains 1 and 2, respectively, and F represents the uncertainty about which stain has been left by the suspect.

suspect. For example, if the suspect has left stain 1 (f_1) and he is of type Γ_1, then certainly the stain 1 is of type Γ_1, formally written $Pr(T_1 = \Gamma_1 \mid f_1, E_2 = \Gamma_1) = 1$. Stated otherwise, a stain left by the suspect cannot have a characteristic that is different from the suspect: $\sum_{i=1,2,x} \sum_{j=1,2} Pr(T_j = \Gamma_i \mid f_j, E_2 \neq \Gamma_i) = 0$.

- Node F: According to the scenario outlined at the beginning of Section 3.5.2, it is assumed that there are two offenders each of whom left exactly one stain. Notice also that both stains are regarded as being left by the offenders. If the suspect is one of the two offenders, then he may have left either stain 1 or stain 2 and these two possibilities are taken to be equally likely. Consequently, $Pr(f_1 \mid H_p) = Pr(f_2 \mid H_p) = 0.5$ and $Pr(f_0 \mid H_p) = 0$. Given H_d, that is, the suspect is not one of the two offenders, neither of the two stains has been left by the suspect: $Pr(f_0 \mid H_d) = 1$ and $\sum_{i=1,2} Pr(f_i \mid H_d) = 0$.

Tables 3.3 and 3.4 provide a summary of all the conditional probabilities assigned.

Table 3.3 Probabilities assigned to the nodes T_1 and T_2, given uncertainty F about the stain left by the suspect: neither, stain 1 or stain 2, with probabilities f_0, f_1, f_2, the profiles Γ_1, Γ_2 and Γ_x of suspect E_2 and T_1 and T_2 are the characteristics of stains 1 and 2, respectively.

	F:	f_0			f_1			f_2		
	E_2:	Γ_1	Γ_2	Γ_x	Γ_1	Γ_2	Γ_x	Γ_1	Γ_2	Γ_x
T_1:	Γ_1	γ_1	γ_1	γ_1	1	0	0	γ_1	γ_1	γ_1
	Γ_2	γ_2	γ_2	γ_2	0	1	0	γ_2	γ_2	γ_2
	Γ_x	γ_x	γ_x	γ_x	0	0	1	γ_x	γ_x	γ_x
T_2:	Γ_1	γ_1	γ_1	γ_1	γ_1	γ_1	γ_1	1	0	0
	Γ_2	γ_2	γ_2	γ_2	γ_2	γ_2	γ_2	0	1	0
	Γ_x	γ_x	γ_x	γ_x	γ_x	γ_x	γ_x	0	0	1

Table 3.4 Probabilities f_0, f_1 and f_2 assigned to the node F, the uncertainty about which stain has been left by the suspect for, neither, stain 1 or stain 2, for propositions H_p, the suspect was one of two who committed the crime and H_d, the suspect was not one of two who committed the crime.

	H:	H_p	H_d
F:	f_0	0	1
	f_1	0.5	0
	f_2	0.5	0

In order to see whether the proposed Bayesian network yields the same results as those obtained with the basic network discussed in the previous section, consider again the numerical example with $\gamma_1 = \gamma_2 = 0.05$ and $Pr(H_p) = Pr(H_d) = 0.5$. A Bayesian network initialised with these values is shown in Figure 3.4(i).

Consider the scenario from Section 3.5.2 where the two stains recovered are of type Γ_1 and Γ_2 and the suspect's blood is of type Γ_1. These observations are entered in the Bayesian network by instantiating the nodes E_1 and E_2 respectively. Figure 3.4(ii) shows the state

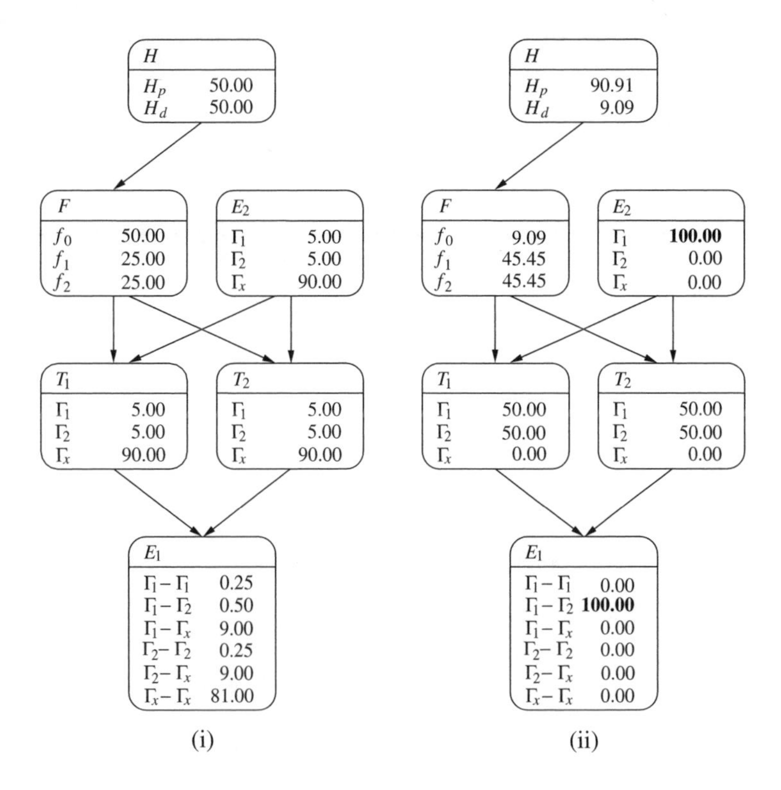

Figure 3.4 (i) Initial state of an alternative Bayesian network for a two-trace transfer scenario; (ii) state of the Bayesian network after instantiation of the nodes E_1 and E_2. The definitions of the nodes are to be found in the text.

of the Bayesian network after entering this evidence and propagating its effect. As may be seen from the values displayed in the node H, the value of the observations corresponds to a likelihood ratio of 10. This result is in agreement with the result presented in Section 3.5.4.

Using the proposed Bayesian network, it is easy to verify that a likelihood ratio of $1/2\gamma_i$ is obtained in all situations in which the suspect's blood is of type Γ_i and the two recovered stains are of types $\Gamma_i - \Gamma_j$ (with $i \neq j$).

4

Bayesian networks for evaluating scientific evidence

At the end of the previous chapter, aspects concerning the construction and evaluation of Bayesian networks, given a particular inferential problem, have been introduced. In this and later chapters, these ideas will be extended in order to provide a more systematic study of the evaluation of evidence in various forensic disciplines. Transfer evidence (*i.e.*, fibres, glass fragments), marks (*e.g.*, shoeprints, firearms/toolmarks) as well as DNA evidence will be considered.

4.1 Issues in one-trace transfer cases

One of the most general scenarios involving scientific evidence is one where a single stain is found at the scene. Suppose the stain is blood and its DNA profile is different from the victim's profile. The stain is found at a position where it may have been left by the offender. As part of the background knowledge it is assumed, for simplicity that there is only one offender. The crime stain and a blood sample provided by a suspect share the same DNA profile.

Assume that the evidence is evaluated at the crime level. The propositions representing the prosecution and defence positions may then be formulated as follows:

- H_p: the suspect is the offender,
- H_d: the suspect is not the offender.

As mentioned in Section 3.3.3, a link is needed between these propositions and what is observed, that is, the stain at the crime scene. This connection is operated in two steps.

The first step consists of considering a proposition according to which the crime stain came from the offender and the alternative proposition that the crime stain did not come from the offender.

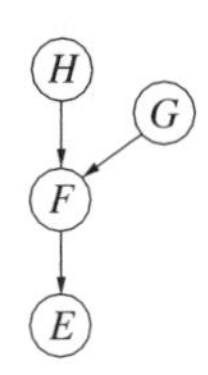

Figure 4.1 A Bayesian network describing the 'one-trace' scenario. Each node has two states. H: the suspect is (H_p) or is not (H_d) the offender. G: the crime stain did or did not come from the offender. F: the crime stain did or did not come from the suspect. E: the blood sample from the suspect and the blood stain found at the crime scene share the same profile. [Reproduced with permission from Elsevier.]

Assume that the crime stain came from the offender. The second step is then the consideration of a proposition that the crime stain came from the suspect and the alternative proposition that the crime stain did not come from the suspect. Let H be the node representing the ultimate probandum (*i.e.*, 'the suspect is/is not the offender'). Given the considerations of the previous paragraph, we also define the following nodes:

- node G: the crime stain came from the offender,
- node F: the crime stain came from the suspect.

The evidence, node E, is defined as 'the suspect's blood sample and the blood stain found at the scene of the crime share the same DNA profile'. Notice that all nodes are binary.

An appropriate graphical representation of the logical relationships among these propositions is shown in Figure 4.1, discussed earlier in Section 2.1.7.

In addition to the probabilistic dependencies represented in Figure 4.1, the following assumptions concerning particular values of the conditional probability matrix of the network are reasonable.

1. If the crime stain came from the suspect, then the suspect's blood certainly matches the crime stain: $Pr(E \mid F) = 1$. Note that no distinction is made here between a *true* match and a *reported* match as proposed by Thompson et al. (2003) or Aitken et al. (2003). This point is discussed in Section 5.11.2.

2. If the crime stain did not come from the suspect, then the probability of the crime stain matching the suspect's blood is given by the random match probability γ of the DNA profile in the relevant population: $Pr(E \mid \bar{F}) = \gamma$ (Balding and Nichols 1994).

3. If the suspect is the offender and the crime stain came from the offender, then certainly the stain came from the suspect: $Pr(F \mid G, H_p) = 1$. Notice, however, that for other evidence types, such as shoe marks, values different from 1 may be possible (*see* Section 4.2.5).

4. Conversely, if the suspect is the offender and the stain did not come from the offender, then the crime stain did not come from the suspect: $Pr(F \mid \bar{G}, H_p) = 0$.

5. If the suspect is not the offender and the stain came from the offender, then certainly the crime stain did not come from the suspect: $Pr(F \mid G, H_d) = 0$.

As one may see, the directed graph of Figure 4.1 is, in a sense, a 'degenerate Bayesian network': this is primarily because of the network containing conditional probabilities equal to zero and one, which itself is a direct consequence of the particular choice of the variables. The choice of such variables fits exactly the purpose of minimising the number of probabilities that are needed and whose values are different from certainty (1) and impossibility (0). One such probability is the random match probability γ, which may be evaluated on the basis of statistical data. Further probabilities different from zero and one are $Pr(H_p)$, $Pr(G)$, and $Pr(F \mid \bar{G}, H_d)$:

- $Pr(H_p)$ can be thought of as the probability that the suspect is the offender given the non-scientific evidence pertaining to the case. Analogous to the node H in the example presented in Section 3.5.2, the values assigned to the node H can represent the (prior) beliefs held by some evaluator, such as an investigator or court of law, before consideration of the scientific evidence.

- $Pr(G)$ is the *relevance* term, the probability, on the basis of the information available, that the blood stain is relevant for the case.

- $Pr(F \mid \bar{G}, H_d)$ is the probability that the stain originates from the suspect, given that he is not the offender and the stain is not relevant to the case. In other words, this is the probability that the stain would have been left by the suspect even though he was innocent of the offence.

4.1.1 Evaluation of the network

A Bayesian network consisting of a qualitative network specification, together with an assignment of values to probability tables, enables the evaluation of target probabilities. The likelihood ratio is of utmost importance to forensic scientists. The numerator and denominator are given by $Pr(E \mid H_p)$ and $Pr(E \mid H_d)$, respectively (*see* Section 3.2).

First, consider the numerator of the likelihood ratio, $Pr(E \mid H_p)$. For the evaluation of $Pr(E \mid H_p)$, the uncertainty in relation to F is also relevant, so it is necessary to write

$$Pr(E \mid H_p) = Pr(E \mid F, H_p)Pr(F \mid H_p) + Pr(E \mid \bar{F}, H_p)Pr(\bar{F} \mid H_p). \tag{4.1}$$

Assumptions (1) and (2) from Section 4.1 and the result that F 'screens off' E from H, which means $Pr(E \mid F, H_p) = Pr(E \mid F)$, reduce (4.1) to

$$Pr(E \mid H_p) = Pr(F \mid H_p) + \gamma Pr(\bar{F} \mid H_p). \tag{4.2}$$

The likelihood of the hypothesis, given that there is a match, is equal to the sum of the likelihood of the hypothesis, given that the stain came from the suspect, and of the likelihood of the hypothesis, given that the stain did not come from the suspect, multiplied by the random match probability.

Then one can calculate the likelihood of H_p given F by 'extending the conversation' (Lindley 1991) to G, that is, the proposition that the stain is relevant and it did not come from an extraneous source:

$$Pr(F \mid H_p) = Pr(F \mid G, H_p)Pr(G \mid H_p) + Pr(F \mid \bar{G}, H_p)Pr(\bar{G} \mid H_p). \tag{4.3}$$

As $Pr(F \mid \bar{G}, H_p) = 0$ (*see* assumption (4) in Section 4.1), (4.3) reduces to

$$Pr(F \mid H_p) = Pr(F \mid G, H_p)Pr(G \mid H_p). \tag{4.4}$$

According to assumption (3), F is certain if both H and G are true. In addition, H and G are probabilistically independent[1]. Thus, (4.4) can further reduced to

$$Pr(F \mid H_p) = Pr(G). \tag{4.5}$$

As may now be seen, the first term in (4.2) is simply the 'relevance' probability. A second major term in (4.2) is the likelihood of H_p given $\bar{F}$. Considering again assumptions (3) and (4) from Section 4.1, it can be shown that $Pr(\bar{F} \mid H_p)$ is $Pr(\bar{G})$[2],

$$\begin{aligned} Pr(\bar{F} \mid H_p) = Pr(\bar{F} \mid G, H_p)Pr(G \mid H_p) + \\ Pr(\bar{F} \mid \bar{G}, H_p)Pr(\bar{G} \mid H_p) = Pr(\bar{G}). \end{aligned} \tag{4.6}$$

Using these considerations, notably the results of (4.5) and (4.6), the numerator of the likelihood ratio (4.2) can be written as follows:

$$Pr(E \mid H_p) = Pr(G) + Pr(\bar{G})\gamma. \tag{4.7}$$

Consider now the denominator of the likelihood ratio, *i.e.*, $Pr(E \mid H_d)$. In analogy to (4.1), uncertainty in relation to F enters the consideration, so one can write

$$Pr(E \mid H_d) = Pr(E \mid F, H_d)Pr(F \mid H_d) + Pr(E \mid \bar{F}, H_d)Pr(\bar{F} \mid H_d). \tag{4.8}$$

Given that F 'screens off' E from H, that is, $Pr(E \mid F, H_d) = Pr(E \mid F)$, and assumptions (1) and (2), the formula for the denominator can be written as

$$Pr(E \mid H_d) = Pr(F \mid H_d) + \gamma Pr(\bar{F} \mid H_d), \tag{4.9}$$

that is, the terms of the sum are the same as in (4.2) except for the substitution of the alternative proposition H_d.

The likelihood of H_d given F, and given $\bar{F}$, can be calculated as before by extending the conversation, obtaining $Pr(F \mid H_d) = Pr(F \mid \bar{G}, H_d)Pr(\bar{G} \mid H_d)$ and $Pr(\bar{F} \mid H_d) = Pr(G \mid H_d) + Pr(\bar{F} \mid \bar{G}, H_d)Pr(\bar{G} \mid H_p)$.

Therefore considering, as given here, that G and H are dependent only conditionally on F, V is given by

$$\begin{aligned} V &= \frac{Pr(E \mid H_p)}{Pr(E \mid H_d)} \\ &= \frac{Pr(G) + Pr(\bar{G})\gamma}{Pr(F \mid \bar{G}, H_d)Pr(\bar{G}) + [Pr(G) + Pr(\bar{F} \mid \bar{G}, H_d)Pr(\bar{G})]\gamma}. \end{aligned} \tag{4.10}$$

Assume that $Pr(G) = r$, and $Pr(F \mid \bar{G}, H_d) = p$; the former is the 'relevance' probability, whereas the latter is the probability that the crime stain was left innocently by

[1]For a graphical illustration of this, consider Figure 4.1 where no direct link exists between H and G.

[2]Remember that $Pr(\bar{F} \mid G, H_p) = 0$ if $Pr(F \mid G, H_p) = 1$, and $Pr(\bar{F} \mid \bar{G}, H_p) = 1$ if $Pr(F \mid \bar{G}, H_p) = 0$.

someone who is now a suspect. Here, it is assumed that the propensity to leave a stain is independent of the blood profile of the person leaving the stain (Aitken 1995). A (simplified) version of Evett's formula (Evett 1993):

$$\begin{aligned} V &= \frac{r + (1-r)\gamma}{p(1-r) + r\gamma + (1-p)(1-r)\gamma} \\ &= \frac{r + (1-r)\gamma}{r\gamma + (1-r)[p + (1-p)\gamma]} \end{aligned} \tag{4.11}$$

has been obtained. Note that to assess the relevance, the scientist has to answer the following question: 'if we find a trace, how sure can one be that it is relevant to the case of interest'? If the personal assessment of the relevance reaches its maximum, $r = 1$, then V is reduced to its simplest form, $1/\gamma$.

Note that the conditional probabilities associated with the links in Figure 4.1 are fixed and independent of the critical epistemic probabilities r and p, the Bayesian network can have a fairly wide inter-subjective acceptance, leaving room for disagreement in the evaluation of r and p.

The Bayesian network shown in Figure 4.1 can be helpful in justifying (4.11) by providing an intuitive model for it, it shows pictorially the dependence and independence assumptions made by the scientist, and the probability assessments needed to evaluate Evett's formula. Moreover, it is a standard model, although the simplest, that can be used in any 'one-trace' case, for instance, the trace could be fibres or fingermarks.

4.2 When evidence has more than one component: footwear marks evidence

In Section 4.1, a scenario has been considered where the evidence was thought of in terms of a 'match'. Remind that no distinction was made between a *true* and a *reported match*. Although being a viable strategy in many situations, the so-called 'match-approach' involves a considerable simplification of reality. For example, when comparing a crime mark and a test mark of some kind, only rarely will a 'perfect' correspondence be observed. In practice, there will usually be similarities as well as differences so that the concept of a 'match' may be a more or less appropriate expression to use (Friedman 1996).

Another way to look at evidence is to consider observations in terms of *sets*. One may distinguish between a set of observations relating to the crime mark and a set of observations relating to the test mark, each of which may itself consist of multiple sets of observations. When comparing footwear marks, for example, a distinction may be made between observable traits relating to a shoe's manufacturing characteristics, such as size or sole pattern, and acquired features, such as wear pattern. Such distinct partitions of observations have also been called *components* (Evett et al. 1998b).

If one decides to consider findings in terms of distinct components, it becomes essential to address the questions of how *(a)* the evidential value of each component is evaluated, and *(b)*, the evidential values of distinct components are logically combined. Evett et al. (1998b) proposed a probabilistic approach for this in the context of footwear marks evidence. The aim here is to discuss the use of Bayesian networks as a means not only to deepen the understanding of this approach but also to clarify the underlying assumptions.

4.2.1 General considerations

First, consider the problem at the source level. An appropriate pair of propositions to represent the prosecution and defence alternatives could be as follows:

F: the mark was left by shoe X,

$\bar{F}$: the mark was left by some unknown shoe.

Shoe X is the shoe of a suspect. Shoe X is used to make test prints and the observations made on them are denoted as x. Likewise, y denotes the observations made on the crime mark. For now, focus solely on the sets of observations x and y without distinguishing between features of manufacture and features that are acquired. The likelihood ratio is then

$$V = \frac{Pr(x, y \mid F)}{Pr(x, y \mid \bar{F})}. \tag{4.12}$$

Applying the third law of probability, V becomes

$$V = \frac{Pr(y \mid x, F)}{Pr(y \mid x, \bar{F})} \frac{Pr(x \mid F)}{Pr(x \mid \bar{F})}. \tag{4.13}$$

It appears reasonable to assume that, in the absence of knowledge of y, uncertainty about x is not affected by the truth or otherwise of F: $Pr(x \mid F) = Pr(x \mid \bar{F})$. In addition, if the crime mark was not made by shoe X, then knowledge about x can be considered as irrelevant[3] for assessing the probability of y: $Pr(y \mid x, \bar{F}) = Pr(y \mid \bar{F})$. (4.13) now reduces to

$$V = \frac{Pr(y \mid x, F)}{Pr(y \mid \bar{F})}. \tag{4.14}$$

The conditioning of y on x and F is explicit in (4.14). A Bayesian network over the three variables x, y and F would thus involve a converging connection as shown in Figure 4.2.

4.2.2 Addition of further propositions

Until now, the problem of footwear marks has been considered at the source level. With an extension of the consideration to the crime level, the following pair of propositions may be adopted.

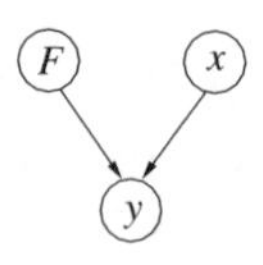

Figure 4.2 A Bayesian network for evaluating footwear mark evidence at the source level. Node F has two states, the mark was left by shoe X and the mark was left by some unknown shoe. Node x is concerned with observations made on reference prints obtained from shoe X. Node y is associated with the observations made on the crime mark. [Adapted from Biedermann and Taroni (2005), reproduced with permission from Elsevier.]

[3]Notice that such an assumption is not valid in the context of DNA evidence (Balding and Nichols 1994), as developed in Section 5.12. The assumption has also been questioned for other kinds of evidence (Aitken and Taroni 1997).

H_p: the suspect is the offender,

H_d: some unknown person is the offender.

As noted in Section 4.1, if one seeks to address propositions at the crime level, then relevance is a factor of which account has to be taken. Therefore, two propositions are introduced:

G: the footwear mark was left by the offender,

$\bar{G}$: the footwear mark was left by someone other than the offender.

Figure 4.3 shows an extended model of the network shown in Figure 4.2. As may be seen, the variables G and H have been introduced as two distinct parental nodes for the variable F.

Note that the presence of an arc is just as informative as the absence of an arc. By adding the nodes G and H, further dependence and independence properties have been added to the network. These properties may be justified by the following assumptions.

1. In the absence of the knowledge of F, the probability that the footwear mark was left by the offender is not dependent on whether or not the suspect is the offender: $Pr(G \mid H_p) = Pr(G \mid H_d) = Pr(G)$. In analogy to the network discussed in Section 4.1, $Pr(G)$ denotes the *relevance* term.

2. Besides x, only F is directly relevant for assessing the probability of y. Notably, given F or $\bar{F}$, the probability of y is independent of G and H. In other words, F 'screens off' y from G and H.

4.2.3 Derivation of the likelihood ratio

The numerator of the likelihood ratio is concerned with the probability of the observations on the crime mark, given the observations on the test mark and given that the suspect is the offender: $Pr(y \mid x, H_p)$.

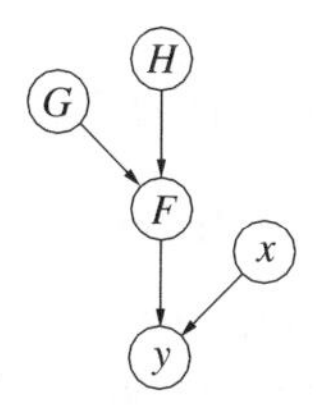

Figure 4.3 A Bayesian network for evaluating footwear mark evidence at the crime level. Node F has two states, the mark was left by shoe X and the mark was left by some unknown shoe. Node x is associated with the observations made on prints made by shoe X. Node y is associated with the observations made on the crime mark. Node G has two states, the footwear mark was left by the offender and the footwear mark was left by someone other than the offender. Node H has two states, the suspect is the offender and some unknown person is the offender. [Adapted from Biedermann and Taroni (2005), reproduced with permission from Elsevier.]

Considering uncertainty in relation to F, the numerator of V, $Pr(y \mid x, H_p)$ in the context of this discussion, becomes

$$Pr(y \mid x, H_p) = Pr(y \mid x, F, H_p)Pr(F \mid x, H_p) + Pr(y \mid x, \bar{F}, H_p)Pr(\bar{F} \mid x, H_p). \quad (4.15)$$

According to Assumption (2) in Section 4.2.2, knowledge about H would not change the belief in y if F were already known, that is, that the mark was left by shoe X. So, $Pr(y \mid x, F, H_p)$ can be reduced to $Pr(y \mid x, F)$. In the absence of knowledge about y, uncertainty about F is not affected by knowledge about x, so, $Pr(F \mid x, H_p)$ can be written as $Pr(F \mid H_p)$. (4.15) is now

$$Pr(y \mid x, H_p) = Pr(y \mid x, F)Pr(F \mid H_p) + Pr(y \mid x, \bar{F})Pr(\bar{F} \mid H_p). \quad (4.16)$$

Next the likelihood of H_p given F may be derived. As in Section 4.1.1, 'extend the conversation' to G and obtain

$$Pr(F \mid H_p) = Pr(F \mid G, H_p)Pr(G \mid H_p) + Pr(F \mid \bar{G}, H_p)Pr(\bar{G} \mid H_p). \quad (4.17)$$

If the suspect is the offender and the crime mark did not come from the offender, then the crime stain was not made by shoe X (assuming no other person than the suspect to be wearing shoe X) so, $Pr(F \mid \bar{G}, H_p) = 0$. From this assumption and the probabilistic independence between H and G, (4.17) reduces to

$$Pr(F \mid H_p) = Pr(F \mid G, H_p)Pr(G). \quad (4.18)$$

Similar arguments apply to the likelihood of H_p given $\bar{F}$, which in its extended form is

$$Pr(\bar{F} \mid H_p) = Pr(\bar{F} \mid G, H_p)Pr(G \mid H_p) + Pr(\bar{F} \mid \bar{G}, H_p)Pr(\bar{G} \mid H_p). \quad (4.19)$$

From the assumption $Pr(F \mid \bar{G}, H_p) = 0$ made above, it follows that $Pr(\bar{F} \mid \bar{G}, H_p) = 1$. Again, G and H are probabilistically independent, so (4.19) becomes

$$Pr(\bar{F} \mid H_p) = Pr(\bar{F} \mid G, H_p)Pr(G) + Pr(\bar{G}). \quad (4.20)$$

Application of (4.18) and (4.20) to (4.16), gives

$$Pr(y \mid x, H_p) = Pr(y \mid x, F)Pr(F \mid G, H_p)Pr(G) + Pr(y \mid x, \bar{F})[Pr(\bar{F} \mid G, H_p)Pr(G) + Pr(\bar{G})]. \quad (4.21)$$

The denominator is concerned with the probability of y given H_d and x. Considering again that F 'screens off' y from H, the denominator can be written as

$$Pr(y \mid x, H_d) = Pr(y \mid x, F)Pr(F \mid H_d) + Pr(y \mid x, \bar{F})Pr(\bar{F} \mid H_d). \quad (4.22)$$

Evaluation of $Pr(F \mid H_d)$ and $Pr(\bar{F} \mid H_d)$ separately gives

$$Pr(F \mid H_d) = Pr(F \mid G, H_d)Pr(G \mid H_p) + Pr(F \mid \bar{G}, H_d)Pr(\bar{G} \mid H_d) \quad (4.23)$$

and

$$Pr(\bar{F} \mid H_d) = Pr(\bar{F} \mid G, H_d)Pr(G \mid H_p) + Pr(\bar{F} \mid \bar{G}, H_d)Pr(\bar{G} \mid H_d). \quad (4.24)$$

Assume the following:

1. If the suspect is not the offender and the crime mark was left by the offender, then the probability of the crime mark being made by shoe X is zero: $Pr(F \mid G, H_d) = 0$. It is therefore assumed that the true offender, if he is not the same person as the suspect, could not have worn shoe X.

2. If the suspect is not the offender and the crime mark was not made by the offender, then the probability of the crime mark being made by shoe X is also zero , $Pr(F \mid \bar{G}, H_d) = 0$. This is an expression of the belief that the suspect could not have left the crime mark for innocent reasons. This probability was denoted as p in Section 4.1.

These assumptions imply that both $Pr(\bar{F} \mid G, H_d)$ and $Pr(\bar{F} \mid \bar{G}, H_d)$ are 1. Application of these assumptions to (4.23) and (4.24) gives

$$Pr(F \mid H_d) = 0, \text{ and} \tag{4.25}$$

$$Pr(\bar{F} \mid H_d) = Pr(G) + Pr(\bar{G}) = 1\,. \tag{4.26}$$

This simplifies the denominator considerably and it can now be written as

$$Pr(y \mid x, H_d) = Pr(y \mid x, \bar{F})\,. \tag{4.27}$$

The combination of (4.21) and (4.27) gives the overall likelihood ratio

$$V = \frac{\begin{array}{c} Pr(y \mid x, F)Pr(F \mid G, H_p)Pr(G) \\ +Pr(y \mid x, \bar{F})\{Pr(\bar{F} \mid G, H_p)Pr(G) + Pr(\bar{G})\} \end{array}}{Pr(y \mid x, \bar{F})}\,. \tag{4.28}$$

This result can be simplified if the following assumptions are made.

1. $Pr(G)$ is the *relevance* term, abbreviated by r. Arguably, $Pr(\bar{G}) = 1 - r$.

2. $Pr(F \mid G, H_p)$ is the probability that the suspect was wearing shoe X, given that he was the offender and that he left the footwear mark. This parameter is abbreviated by w. Its complement, $Pr(\bar{F} \mid G, H_p)$ is therefore $1 - w$.

3. $Pr(y \mid x, \bar{F}) = Pr(y \mid \bar{F}) = \gamma$.

Application of these assumptions to (4.28) yields:

$$V = \frac{Pr(y \mid x, F)rw + \gamma[(1 - w)r + (1 - r)]}{\gamma}\,. \tag{4.29}$$

This can further be simplified to

$$V = rw\frac{Pr(y \mid x, F)}{\gamma} + (1 - rw)\,. \tag{4.30}$$

V, deduced from the Bayesian network shown in Figure 4.3, is thus in agreement with the formula proposed by Evett et al. (1998b).

For the purpose of illustration, consider the formulae for V for the one-trace scenario, that is, (4.11), and for the footwear mark scenario, that is, (4.30). These two formulae might appear to have very little in common if judged by their general appearance. However, comparison of the corresponding Bayesian networks (Figures 4.1 and 4.3), illustrates a strong resemblance. This may be taken as an indication of the coherence of the proposed probabilistic solutions for these two scenarios.

4.2.4 Consideration of distinct components

Until now, y and x have been considered as global assignments for the observations made on the crime and test marks respectively. Figure 4.3 provides a graphical summary of this.

As noted at the beginning of Section 4.2, observations made on a footwear mark, y, may be partitioned into distinct components. A component y_m may be defined with the aim of describing those traits that originate from the features of manufacture of the shoe that left the mark y. In the same way, y_a refers to those aspects of a footwear mark that are thought to originate from the acquired features of the shoe that produced the mark y. With $x = (x_m, x_a)$, an analogous notation can be used to describe observations made on prints obtained from the suspect's shoe under controlled conditions. The terms (y_m, y_a) and (x_m, x_a) will express a mark's or print's characteristics that originate from, respectively, the manufacturing and acquired features of the source of the mark or print.

Graphically, a logical extension from y and x to (y_m, y_a) and (x_m, x_a) could be achieved by 'duplicating' the nodes y and x. Such a Bayesian network is shown in Figure 4.4(i).

Notice, however, that the network shown in Figure 4.4(i) is, to some degree, a minimal representation, providing room for possible extensions. For example, one may consider the following:

- The nodes with subscripts m and a denote the *true* presence of manufactured and acquired features. Thus, no distinction is made between the observation of a characteristic and a characteristic itself. Distinct observational nodes for each component may be adopted, allowing the scientist to consider the potential of observational error. Such a network is shown in Figure 4.4(ii) where observational nodes labelled with a ' ' ' are present.

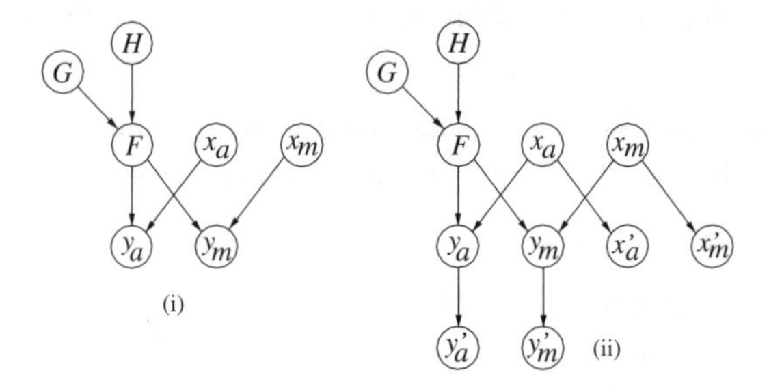

Figure 4.4 Bayesian networks for evaluating footwear mark evidence, partitioning the observations into components. Node F has two states, the mark was left by shoe X and the mark was left by some unknown shoe. Nodes x_m and x_a are associated with the features of the control prints that originate from the manufacturing (x_m) and acquired (x_a) characteristics of the shoe X. Node y_m and y_a denote features on the crime mark deemed to be associated with the manufacturing (x_m) and acquired (x_a) features of the shoe at the origin of the crime mark. The nodes labelled with a ' ' ' represent observations relating to the features covered by the respective parental node. Node G has two states, the footwear mark was left by the offender and the footwear mark was left by someone other than the offender. Node H has two states, the suspect is the offender and some unknown person is the offender. [Adapted from Biedermann and Taroni (2005), reproduced with permission from Elsevier.]

- The absence of an arc between y_m and y_a assumes independence between manufacture and wear. As noted by Evett et al. (1998b), the validity of such an assumption depends on the type of wear and the way in which it has been described.

The formula for V that corresponds to the Bayesian network shown in Figure 4.4 is an extension of (4.30):

$$V = rwV_mV_a + (1 - rw). \tag{4.31}$$

The parameters V_m and V_a represent the following ratios:

$$V_m = \frac{Pr(y_m \mid x_m, F)}{Pr(y_m \mid \bar{F})}, \tag{4.32}$$

$$V_a = \frac{Pr(y_a \mid x_a, F)}{Pr(y_a \mid \bar{F})}. \tag{4.33}$$

4.2.5 A practical example

Suppose a burglary was committed. Entry was gained *via* a forced back door. The forensic scientist examining the scene recovered a right footwear mark on the linoleum floor near the point of entry. The print was lifted and submitted for laboratory analysis. The crime print showed general pattern markings typically found on a certain series of running shoes, say B, produced by manufacturer A. A few weeks later, a burglary was attempted at a local supermarket *via* a roof window. As the alarm went off, the offender left the scene. A nearby police patrol arrested a man fleeing the scene. He was found to wear a pair of running shoes of type B from manufacturer A.

Laboratory examination revealed that test prints made of the suspect's right shoe closely corresponded to the crime mark taken from the linoleum floor of the premises burgled a few weeks earlier. The size corresponded and agreements were found in general pattern and small cut marks. However, the suspect's shoe was slightly more worn and bore some cut marks that were not present in the mark recovered on the crime scene. How is this evidence to be evaluated using a Bayesian network?

Assume that Figure 4.4 provides an appropriate representation of the various issues to be addressed in the evaluation of this evidence. The key parameters are the following:

- The propositions in the light of which the evidence is to be considered are at the crime level, that is, the suspect is the offender, H_p, and some unknown person is the offender, H_d;
- There is uncertainty concerning the relevance of the crime mark with respect to the offence under consideration;
- There is uncertainty about whether the suspect, if he is the offender and the crime mark is relevant, wore the shoe at time of the offence;
- The evidence consists of features relating to the manufacture and to the use of the shoe(s) at the origin of the compared marks.

The numerical specification of these parameters requires an assessment of conditional and unconditional probabilities, which will now be discussed in more detail.

- Node H: $Pr(H_p)$ is the prior probability of guilt. As was mentioned in the previous examples, this parameter lies outside the competence of the forensic expert. Imagine that other available evidences linking the suspect to the first burglary are weak, so that an evaluator (investigator or court of law) would hold a low prior belief, for example, $Pr(H_p) = 0.01$.

- Node G: $Pr(G)$ is the relevance term, previously denoted as r. Although the fact that the crime mark was found near the point of entry suggests that the mark is relevant, a rather conservative value is assumed. Consider a probability of 0.5, for example.

- Node F: The term w was used to denote the probability that the crime mark was made by the suspect's shoe, given that the mark is relevant and the suspect is the offender. Clearly, this probability depends on the number of pairs of shoes in the possession of the suspect and the relative frequency of their use. In the case at hand, the suspect was wearing the pair of running shoes at the time of arrest. As this pair of shoes bore marks of wear, it is reasonable to assume that there is some chance that the suspect would have worn the pair of shoes if he were the offender. Consider a probability of 0.5 for $Pr(F \mid G, H_p)$.

- Node y_m: There are two probabilities of interest here. Firstly, consider $Pr(y_m \mid x_m, F)$, that is, the probability of the manufacturing features observed on the crime mark, given that it was made by the suspect's shoe and assuming knowledge about the manufacturing features of the shoe. A probability of 1 or close to 1 may be appropriate for this term. Secondly, consider $Pr(y_m \mid x_m, \bar{F})$, that is the probability of the observed manufactured features given that the mark had been left by some other running shoe of series B from manufacturer A. This probability may be estimated on the basis of information on sales and distribution provided by the manufacturer or supplier. One may also refer to statistics collected within a forensic laboratory. Assume the assignment of a probability of 0.01. Here, knowledge about x_m is clearly irrelevant, so it is permissible to set $Pr(y_m \mid x_m, \bar{F}) = Pr(y_m \mid \bar{F})$. If the suspect's shoe has left the crime mark, but the shoe's features of manufacture $\bar{x}_m$ are different from those described as x_m, then the probability of the mark bearing manufacturing features described as y_m can be considered impossible: $Pr(y_m \mid \bar{x}_m, F) = 0$.

- Node y_a: Again, consider two probabilities: $Pr(y_a \mid x_a, F)$ and $Pr(y_a \mid x_a, \bar{F})$. For the assessment of the former probability, it is necessary to answer a question of the type 'What is the probability of the crime mark's acquired features if they had been left by the suspect's shoe'?. Recall that the suspect's shoes exhibited slightly more wear and acquired features than the crime marks, so a value less than one is indicated, say 0.2 for the purpose of illustration. For assessment of the latter probability, it is necessary to answer a question of the type 'What is the probability of the crime mark's acquired features if they had been left by some other shoe of series B from manufacturer A'?. Assume that the expert considers parameters such as degree of wear as well as shape and the relative position of cut marks, on the basis of which an estimate of, say, 1/1000 is made. In analogy to the node y_m, x_a is irrelevant for y_a in cases where $\bar{F}$ is true: $Pr(y_a \mid x_a, \bar{F}) = Pr(y_a \mid \bar{x}_a, \bar{F}) = Pr(y_a \mid \bar{F})$. The remaining probability, $Pr(y_a \mid \bar{x}_a, F)$, can be estimated to be 0 (in analogy to the parameter $Pr(y_m \mid \bar{x}_m, F)$).

- Nodes x_m and x_a: $Pr(x_m)$ and $Pr(x_a)$ are the prior probabilities of the manufacturing and acquired features of the suspect's shoes. As both nodes x_m and x_a will usually be instantiated while using the Bayesian network, the probabilities assigned to these nodes are not crucial. Assume, for illustration, $Pr(x_m) = 0.01$ and $Pr(x_a) = 0.001$.

The remaining probabilities needed to complete the Bayesian network are either the logical complements of the parameters just discussed, or already defined through various assumptions made in Section 4.2.3. A summary of all the probabilities can be found in Tables 4.1, 4.2 and 4.3. A Bayesian network initialised with these values is shown in Figure 4.5(i).

An evaluation of the likelihood ratio may now be made. Using (4.31) and the various probabilities specified in the previous paragraph, one obtains:

$$V = 0.5 \times 0.5 \times (1/0.01) \times (0.2/0.001) + (1 - 0.5 \times 0.5) \simeq 5000\,. \tag{4.34}$$

A likelihood ratio of the same magnitude is obtained if one instantiates the nodes y_m, y_a, x_m and x_a. This situation is shown in Figure 4.5(ii), where the posterior probabilities for H_p have changed from 0.01 to 0.98 (rounded to two decimals).

Some readers may consider the various epistemic probabilities used throughout this example to be rather arbitrary. However, the numerical values adopted are not of primary importance. Indeed, any exact value appears hard to justify in practice. What is more important is the relative magnitude of the different parameters, notably those used in (4.31). Given that the variables r and w will usually have values less than 1 (*see* Tables 4.1 and 4.2), the reader may realise that to obtain a likelihood ratio greater than 1, the values of the parameters V_m and V_a play an essential role. Consider V_m for example. It is not a question as to whether the component probabilities of V_m, that is, $Pr(y_m \mid x_m, F)$ and $Pr(y_m \mid \bar{F})$, are exactly 1 and 0.001 respectively. What matters more is the understanding that the more

Table 4.1 Unconditional probabilities assigned to the nodes H, G, x_m, and x_a. Nodes x_m and x_a are associated with marks left by the manufacturing (x_m) and acquired (x_a) features of the shoe X. Node G has two states, G, the footwear mark was left by the offender and $\bar{G}$, the footwear mark was left by someone other than the offender. Node H has two states, H_p, the suspect is the offender and H_d, some unknown person is the offender.

H:	H_p	0.01
	H_d	0.99
G:	G	$r = 0.5$
	$\bar{G}$	$1 - r = 0.5$
x_m	x_m	0.01
	$\bar{x}_m$	0.99
x_a	x_a	0.001
	$\bar{x}_a$	0.999

Table 4.2 Conditional probabilities for the node F. Values given are $Pr(F \mid G, H)$. Node F has two states, F, the mark was left by shoe X and $\bar{F}$, the mark was left by some unknown shoe. Node G has two states, G, the footwear mark was left by the offender and $\bar{G}$, the footwear mark was left by someone other than the offender. Node H has two states, H_p, the suspect is the offender and H_d, some unknown person is the offender.

	H:	H_p		H_d	
	G:	G	$\bar{G}$	G	$\bar{G}$
F:	F	$w = 0.5$	0	0	0
	$\bar{F}$	$1 - w = 0.5$	1	1	1

Table 4.3 Conditional probabilities assigned to the nodes y_m and y_a. Node F has two states, F, the mark was left by shoe X and $\bar{F}$, the mark was left by some unknown shoe. Nodes x_m and x_a are associated with marks left by the manufacturing (x_m) and acquired (x_a) features of the shoe X. Node y_m and y_a are observations made on the crime mark, deemed to be associated with the manufacturing (y_m) and acquired (y_a) features of the shoe at the origin of the crime mark.

F:	F		$\bar{F}$			F		$\bar{F}$	
x_m:	x_m	$\bar{x}_m$	x_m	$\bar{x}_m$		x_a	$\bar{x}_a$	x_a	$\bar{x}_a$
y_m: y_m	1	0	0.01	0.01	y_a: y_a	0.2	0	0.001	0.001
$\bar{y}_m$	0	1	0.99	0.99	$\bar{y}_a$	0.8	1	0.999	0.999

the values of the parameters $Pr(y_m \mid x_m, F)$ and $Pr(y_m \mid \bar{F})$ diverge, the more is ratio V_m will increase. In a more formal notation, one may require that component probabilities satisfying $Pr(y_m \mid x_m, F) > Pr(y_m \mid \bar{F})$ and $Pr(y_a \mid x_a, F) > Pr(y_a \mid \bar{F})$, that is, $V_m > 1$ and $V_a > 1$, lead to an increase in the value of the likelihood ratio.

Such purely qualitative considerations are particularly well suited for addressing situations involving unique sets of circumstances. In forensic casework, qualitative reasoning can usefully supplement quantitative Bayesian networks. A more detailed discussion of this is given in Chapter 9.

4.2.6 An extension to firearm evidence

The two-component approach to footwear marks evidence as discussed in the previous Sections may analogously be applied to other kinds of evidence where observations can be divided into manufacturing and acquired features. Firearm evidence is a good example of this.

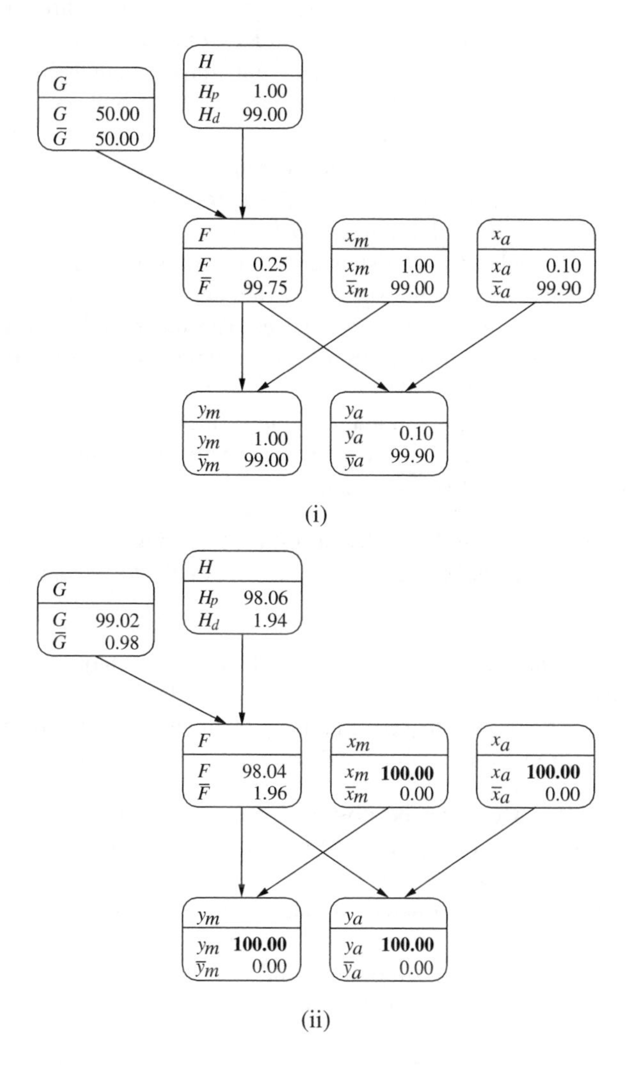

Figure 4.5 (i) Initialised Bayesian network for evaluating footwear marks evidence; (ii) state of the Bayesian network after instantiating the nodes y_m, y_a, x_m and x_a. Node F has two states, F, the mark was left by shoe X and $\bar{F}$, the mark was left by some unknown shoe. Nodes x_m and x_a are associated with the marks left by the manufacturing (x_m) and acquired (x_a) features of the shoe X. Node y_m and y_a are observations deemed to be associated with the manufacturing (y_m) and acquired (y_a) marks made on the crime mark. Node G has two states, G, the footwear mark was left by the offender and $\bar{G}$, the footwear mark was left by someone other than the offender. Node H has two states, H_p, the suspect is the offender and H_d, some unknown person is the offender.

When it is believed that a particular bullet found on the scene of a crime has been fired by a particular weapon found in the possession of a suspect, a firearm examiner may be asked to perform comparisons between the bullet from the scene and test-fired bullets from the suspect's weapon. The results of such comparative examinations may be used to support a conclusion with respect to the propositions that the bullet was or was not fired from the barrel of the suspect's weapon.

Expressed in very general terms, a firearm examiner's work will cover two main stages. Usually, features of manufacture will be examined first. These include parameters such as a bullet's calibre, number, width, angle and twist of grooves. When these parameters correspond to the observable characteristics on the test-fired bullets, further examinations may be undertaken. Thus, in a second step, the examiner will compare microscopic marks originating from marks on the inside of the barrel. The inner surface of a barrel holds a unique set of characteristics originating from, for example, its manufacture, extent of use, cleaning habits and storing conditions.

Although close correspondences in microscopic surface patterns may be observed when two bullets have been fired through the same barrel, there are a number of factors that may seriously compromise comparative examinations, notably:

- *The condition of the crime bullet*: As a result of striking different kinds of surfaces, bullets recovered on the scene of a crime are likely to be deformed or partially missing.

- *The condition of the suspect weapon*: Considerable time may have elapsed between the crime and the test firing of a suspect's firearm. Meanwhile, the suspect's weapon may have been fired and cleaned a number of times. In addition, the weapon may have been stored in unfavourable conditions, for example, humid conditions.

Consequently, differences may be observed between the bullet in question and the test-fired bullets even though they were fired from the same barrel.

This is surely an incomplete summary of all the subtleties that firearm examinations may entail. However, it appears sufficient to illustrate that practising forensic scientists face a difficult evaluative task. They are required to account for different sets of observations, each of which being described by similarities and differences. In addition, scientists may be asked to consider the findings in the light of a unique set of circumstantial information I.

The standpoint towards evaluating firearm evidence as outlined hereafter is on the basis of a general framework, with the aim of incorporating both features of manufacture and acquired features. To this end, consider a hypothetical shooting incident during which some private property has been severely damaged. On the scene, a bullet was extracted from the wooden frame of a window. No other elements, such as cartridges or other bullets, were found on the scene. There is no immediate suspect, but a few months later, a suspect is apprehended on the basis of information completely unrelated to the firearm evidence. This suspect is found to be in possession of a firearm. Laboratory analysis revealed that the manufacturing features of the suspect's weapon leave characteristic marks that 'correspond' to those present on the bullet recovered from the scene. However, only a few correspondences were found in acquired features. Is there a Bayesian network that can guide scientists in the evaluation of such evidence?

In analogy to Sections 4.2.1 to 4.2.5, the following propositions are defined:

- H_p: the suspect fired the incriminating bullet;
- H_d: some unknown person fired the incriminating bullet.

Notice that in the current scenario, there is no uncertainty about the relevance of the incriminating bullet. The bullet is necessarily relevant for the damage it caused. Thus, no relevance node, earlier named G, is defined.

However, a source node F is defined, which acts as a child variable for H:

- F: the bullet was fired by the suspect's weapon;
- $\bar{F}$: the bullet was fired by some unknown weapon.

In addition, there are nodes y_m and y_a that are used to describe those observable marks on a bullet which originated from the manufacturing and the acquired features of the barrel through which it was fired. The nodes x_m and x_a relate to the same features as observed on the test-fired bullets.

As far as the dependencies among these variables are concerned, a network structure as shown in Figure 4.4 appears appropriate (the sole difference being, however, that no node G is retained). The required node probabilities may be assessed as follows:

- Node H: This is the probability of the suspect being the shooter prior to the consideration of the firearm evidence. This probability may be arrived at according to the degree of there being other available evidence linking the suspect to the shooting incident. As was mentioned in earlier chapters, the assessment of prior probabilities for these propositions generally lies outside the scientist's area of competence. In addition, an assessment of the likelihood ratio does not require such prior probabilities to be specified. The reason for specifying these probabilities here is of a purely technical nature, they are needed to run the model. For the purpose of illustration, let $Pr(H_p) = Pr(H_d) = 0.5$.
- Node F: $Pr(F \mid H_p)$, abbreviated by w, denotes the probability that the suspect's weapon has been used to fire the incriminating bullet, given that the suspect is the offender. This probability depends, on the one hand, on the number of weapons in the possession of the suspect or the number of weapons to which the suspect had access at the time the crime happened, and, on the other hand, his eventual preferences in using one or other of these weapons. Imagine that no other weapon has been found in the possession of the suspect and that there is information that the suspect did not have access to a weapon other than the one being seized. The parameter w could then be taken to be 0.99, for example. The probability that the suspect's weapon was used for firing the incriminating bullet, given the suspect is not the offender, $Pr(F \mid H_d)$, depends in part to the number of persons that potentially had access to the suspect's weapon during the time the crime happened. For the scenario considered here, assume that it does not appear conceivable that someone other than the suspect could have used the suspect's weapon: $Pr(F \mid H_d) = 0$.
- Node y_m: Given the characteristics of manufacture of the suspect's firearm and given that it has fired the bullet found on the scene, one will almost certainly have the corresponding characteristics of manufacture on this bullet. For convenience, assume a

value of 1 for $Pr(y_m \mid x_m, F)$. If the bullet has not been fired by the suspect's weapon, then its characteristics (node x_m) are irrelevant for estimating the probability of y_m. Arguably, it is permissible to set $Pr(y_m \mid, x_m, \bar{F}) = Pr(y_m \mid \bar{x}_m, \bar{F}) = P(y_m \mid \bar{F})$. In the second situation, the bullet has been fired by another weapon and the probability of y_m is thus determined by the frequency of this configuration of manufacturing features in a relevant population. Assume there is a laboratory database available allowing the estimation of a value of 0.05 for $Pr(y_m \mid \bar{F})$. If the bullet was fired by the suspect's weapon, whose manufacturing characteristics were described by $\bar{x}_m$, one may consider it impossible to have correspondences with the manufacturing features of the bullet, described by y_m: arguably, $Pr(y_m \mid \bar{x}_m, F)$ is set to 0.

- Node y_a: Given that the suspect's firearm was seized a few months after the shooting incident, the firearm may have been exposed to a number of constraints, such as shooting, cleaning and different kinds of storage conditions. Consequently, there is some chance that the firearm will leave a pattern of individual marks that may be considerably different to the kind of marks the firearm produced a few months ago. Therefore, one may assume that the probability of observing a few correspondences in acquired features is as likely if the bullet were shot by the suspect's weapon as if it were shot by some other firearm: for example, $Pr(y_a \mid x_a, F) = Pr(y_a \mid \bar{F}) = 0.1$. Again, consideration of x_a can be omitted if $\bar{F}$ is true. The remaining parameter, $Pr(y_a \mid \bar{x}_a, F)$, can, as was done for the node x_m, be considered impossible.
- Nodes x_m and x_a: The prior probabilities assigned to these nodes are, in the most general setting, the frequencies of the manufacturing and acquired features in a relevant population. Again, different sources of information may be used to obtain these prior probabilities, for example, databases or expert knowledge. For the purpose of illustration, let $Pr(x_m)$ and $Pr(x_a)$ be 0.01 and 0.001 respectively. Notice, however, that the evaluation of the likelihood ratio is not necessarily influenced by the values that have actually been chosen. When evaluating the numerator of the likelihood ratio, the nodes x_m and x_a are set to 'known', that is, assumed to be certain, and, when evaluating the denominator, knowledge about the factual states of x_m and x_a may be irrelevant. The latter is notably the case when H_p implies $\bar{F}$ with certainty and the relation $Pr(y \mid x, \bar{F}) = Pr(y \mid \bar{x}, \bar{F})$ holds.

In the case at hand, the evidence consists of similarities and differences in the marks present on both, the crime bullet and the test-fired bullets from the suspect's weapon. When instantiating the respective evidence nodes in Figure 4.6, that is, y_m, y_a and x_m, x_a, the following may be considered:

- The weight of evidence can be expressed by a likelihood ratio V. Under the stated assumptions, (4.31) reduces to $V = wV_mV_a + (1 - w)$. Using the probabilities defined above,

$$V = 0.99(1/0.05)(0.01/0.01) + (1 - 0.99) = 19.81. \tag{4.35}$$

A likelihood ratio of the same magnitude is obtained with the Bayesian network. From the posterior probabilities of the node H (Figure 4.6), one has $95.195/4.805 \simeq 19.81$.

- The probability of the individual features y_a being present on the crime bullet is as likely if it were fired by the suspect's weapon (F), characterised by the acquired

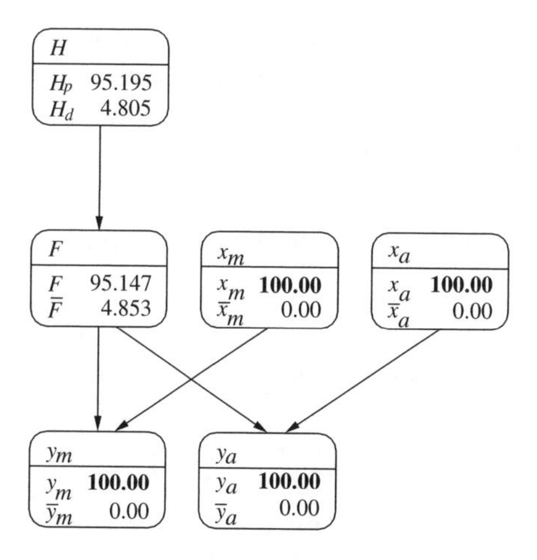

Figure 4.6 Bayesian network for evaluating firearm evidence. The evidence nodes y_m, y_a, x_m and x_a are instantiated. Node F has two states, F, the suspect's weapon has been used to fire the incriminated bullet and $\bar{F}$, the incriminated bullet has been fired from some other weapon. Nodes x_m and x_a are associated with the observable marks on test-fired bullets that originated from manufacturing (x_m) and acquired (x_a) features of the barrel through which the bullets were fired. Nodes y_m and y_a are associated with the observable marks on the incriminated bullet that originated from manufacturing (y_m) and acquired (y_a) features of the barrel through which the bullet was fired. Node H has two states, H_p, the suspect fired the incriminating bullet and H_d, some unknown person fired the incriminating bullet.

features described as (x_a), as if it were fired by some other weapon ($\bar{F}$): $Pr(y_a \mid x_a, F) = Pr(y_a \mid \bar{F})$. In such a situation, $V_a = 1$ and the likelihood ratio is solely confined to the parameters w and V_m, so that $V = wV_m + (1 - w)$.

Once again note that precise numerical values are not a necessary requirement for using the described Bayesian network. On the contrary, valuable insight can often be gained if the user evaluates different values for certain parameters of interest in order to see how they affect the outcomes. Such operations, also know as sensitivity analyses, will be discussed in more detail in Chapter 9.

The Bayesian network discussed in this section allows the joint evaluation of evidence at different levels of detail, including manufacturing and acquired features, together with uncertainties in relation to w, the probability that the suspect is the shooter and that he used the seized firearm. Moreover, a meaningful evaluation of evidence can be achieved even though part of the evidence may be neutral. In this section, neutral evidence was encountered in connection with y_a, the acquired features present on the crime bullet. Nevertheless, the overall evidential support for the node H is positive (*i.e.*, greater than one); this is so because of the relative rarity, represented by $Pr(y_m \mid \bar{F})$, of the manufacturing features. This probability determines, together with $Pr(y_m \mid x_m, F)$, the value of V_m. In particular, as $Pr(y_m \mid x_m, F) > Pr(y_m \mid \bar{F})$, the parameter V_m is greater than one and this tends to increase the likelihood ratio V (4.35).

In summary, the proposed Bayesian network approach provides a pictorial representation of how to reason from evidence to selected propositions of interest. Through the use of a Bayesian network, a deeper understanding may be gained of how, and at which level, selected sources of uncertainty intervene.

4.2.7 A note on the evaluation of the likelihood ratio

When evaluating the Bayesian networks for footwear and firearm evidence (Sections 4.2.5 and 4.2.6), the likelihood ratio was obtained *via* the posterior probabilities of the node H. Notice that the ratio of the posterior probabilities of the node H equals the likelihood ratio of the evidence only if the prior probabilities assumed for H are equal. The posterior probabilities of the node H were obtained by instantiating the evidence nodes of the numerically specified Bayesian network (*see* Figures 4.5 and 4.6).

However, forensic scientists will usually avoid addressing posterior probabilities of the major propositions of interest. They prefer to consider the probability of the evidence, given the pair of propositions of interest. Thus, an alternative way to evaluate the likelihood ratio would be to make different instantiations at the node H (*i.e.*, to H_p and H_d), and then observing the respective probabilities of the evidence. This is a feasible strategy whenever there is one evidence node (*see*, for example, the Bayesian network discussed in Section 4.1. However, the application of this procedure may not readily be clear if there is more than one evidence node. As an example, take the Bayesian networks for evaluating mark evidence. Following the instantiation of the node H, changes will occur in the values of the two evidence nodes, that is, y_m and y_a. How can the overall likelihood ratio be determined?

A possible approach would be to adopt a further node, which in some way brings together the available evidence. Let this node be denoted as Y and have entering arcs from both y_m and y_a. The node Y is binary, with states y and $\bar{y}$, which are described as follows:

- y: a combination of characteristics described as y_m and y_a is present on the crime mark;
- $\bar{y}$: a combination of characteristics different from y_m and y_a is present on the crime mark.

Notice that the crime mark may either be a footwear mark or a bullet. As the definition of the node Y is a logical combination of the two nodes y_m and y_a, the probability table associated with this node contains values equating to 0 and 1 (Table 4.4):

1. If both y_m and y_a are true, then y is true with probability 1.
2. If either or both of the nodes y_m or y_a are false, then y is false with probability 1.

As an example of the use of such an extended Bayesian network, reconsider the scenario involving footwear mark evidence discussed in Section 4.2.5. To this end, consider the Bayesian network shown in Figure 4.7. The numerical specifications of the network shown in this Figure are as defined in Section 4.2.5. The only difference is that a node Y has been added according to the definition given here.

Figure 4.7 represents the evaluation of the prosecution's case. Here, the probability of the evidence, y_m and y_a, or y for short, is evaluated given H_p, and x_m and x_a, the features

Table 4.4 Probabilities assigned to the node Y. Node Y has two states, y, a combination of characteristics described as y_m and y_a is present on the crime mark, and $\bar{y}$, a combination of characteristics different from y_m and y_a is present on the crime mark. States $\bar{y}_m$ and $\bar{y}_a$ denote manufacture marks and acquired marks, respectively, different from y_m and y_a.

	y_m:	y_m		$\bar{y}_m$	
	y_a:	y_a	$\bar{y}_a$	y_a	$\bar{y}_a$
Y:	y	1	0	0	0
	$\bar{y}$	0	1	1	1

of the control object. The nodes H, x_m and x_a are instantiated and a probability of about 5% is indicated for y.

An evaluation of the defence case is presented in Figure 4.8. Here, the node H is instantiated to H_d. Notice that no instantiations have been made in the nodes x_m and x_a. Features present on the control object do not affect consideration of the evidence y when assuming the alternative hypothesis to be true. The proposed Bayesian network illustrates this clearly. One may have any desired instantiation of the nodes x_m and x_a, as long as the node H is in the state H_d, the probability of y remains unchanged. In the case evaluated here, the probability indicated for y given H_d is 0.001%.

A so-called 'top-down' analysis has been performed. The numerator and denominator of the likelihood ratio have been evaluated separately according to the following formula:

$$V = \frac{Pr(y \mid x_m, x_a, H_p)}{Pr(y \mid H_d)}. \tag{4.36}$$

From the analysis outlined previously, a ratio of $5\%/0.001\% = 5000$ is obtained. This result is in agreement with the value obtained in Section 4.2.5, where the likelihood ratio was derived in a 'bottom-up' strategy, that is, by the posterior probabilities of the node H.

4.3 Scenarios with more than one stain

4.3.1 Two stains, one offender

The Bayesian networks discussed so far in this chapter, where the evidence is evaluated under propositions at the crime level, have been restricted to scenarios in which there is only one item of scientific evidence. This has enabled concentration on concepts such as *relevance* and *innocent acquisition* (Section 4.1). In addition, attention has been drawn to different levels of detail, or components, that may be considered when evaluating evidence (Section 4.2).

Consider the discussion of cases involving a single offender. Imagine the scenario developed by Stoney (1994), where two bloodstains, of profiles Γ_1 and of Γ_2, are recovered at the scene of a crime. It is not known which, if either, of the two stains is relevant. A suspect is found who is of profile Γ_1, so there is a match (or *correspondence*) between the suspect and one of the stains.

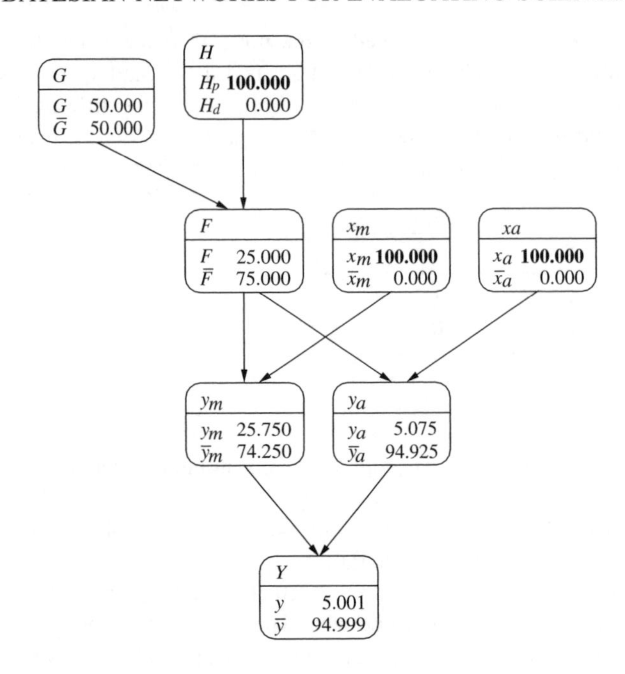

Figure 4.7 Extended Bayesian network for evaluating footwear mark evidence: evaluation of the probability of the evidence given the proposition H_p and knowledge about the features of the control object. The nodes H, x_m and x_a are instantiated. Node H has two states, the suspect is the offender (H_p) and some unknown person is the offender (H_d). Nodes x_m and x_a are associated with the observable marks of manufacturing (x_m) and acquired (x_a) features of control material (the suspect's shoe). States $\bar{x}_m$ and $\bar{x}_a$ denote manufacture marks and acquired marks, respectively, different from x_m and x_a. Node Y has two states, y, a combination of characteristics described as y_m and y_a is present on recovered mark, and $\bar{y}$, a combination of characteristics different from y_m and y_a is present on recovered mark. States $\bar{y}_m$ and $\bar{y}_a$ denote marks originating from manufacturing and acquired features, respectively, different from y_m and y_a. Node F has two states, the crime mark has been made by the suspect's shoe X and has not been made by the suspect's shoe. Node G has two states, the offender is associated with the mark and the mark is associated with someone other than the offender.

A formal approach to this scenario is on the basis of the definition of the following probabilities:

$$Pr(\text{The stain of profile } \Gamma_1 \text{ is from the offender}) = r_1,$$

$$Pr(\text{The stain of profile } \Gamma_2 \text{ is from the offender}) = r_2,$$

$$Pr(\text{Neither of the stain is from the offender}) = 1 - r_1 - r_2.$$

If H_p is true (*i.e.*, the suspect is the offender), there are three components to this probability:

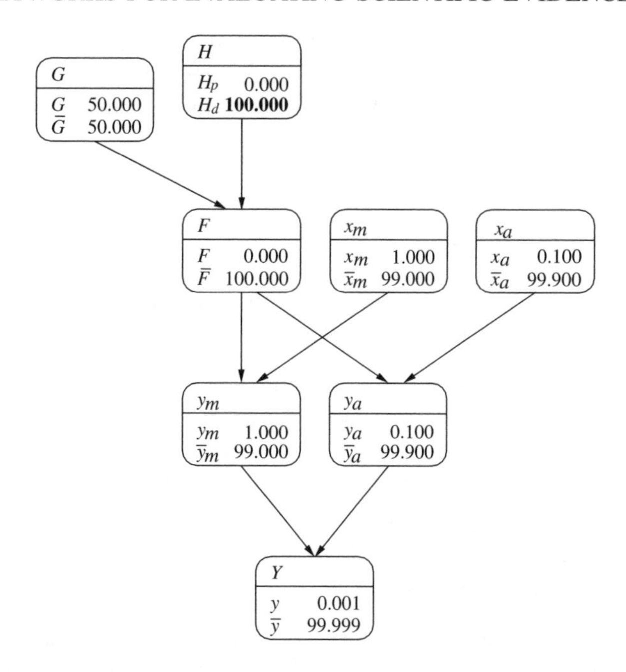

Figure 4.8 Extended Bayesian network for evaluating footwear mark evidence: evaluation of the probability of the evidence given the alternative proposition H_d. The node H is instantiated. Node H has two states, the suspect is the offender (H_p) and some unknown person is the offender (H_d). Nodes x_m and x_a are associated with the observable marks of manufacturing (x_m) and acquired (x_a) features of control material (the suspect's shoe). States $\bar{x}_m$ and $\bar{x}_a$ denote manufacture marks and acquired marks, respectively, different from x_m and x_a. Node Y has two states, y, a combination of characteristics described as y_m and y_a is present on the recovered mark, and $\bar{y}$, a combination of characteristics different from y_m and y_a is present on the recovered mark. States $\bar{y}_m$ and $\bar{y}_a$ denote manufacture marks and acquired marks, respectively, different from y_m and y_a. Node F has two states, the crime mark has been made by the suspect's shoe X and has not been made by the suspect's shoe. Node G has two states, the offender is associated with the mark and the mark is associated with someone other than the offender.

- Stain of profile Γ_1 is from the offender. There is a match with the profile of the suspect with probability r_1 (*i.e.*, $1 \times r_1$).
- Stain of profile Γ_2 is from the offender. There is no match with the profile of the suspect. This event has probability zero (*i.e.*, $0 \times r_2$) since the suspect is assumed to be the offender and only one offender is assumed.
- Neither stain is from the offender. This event has probability $(1 - r_1 - r_2)$, and if it is true, there is a probability γ_1 of a match between the suspect's profile (Γ_1) and the crime stain of the same profile. The probability of the combination of these events is $(1 - r_1 - r_2)\gamma_1$.

These three components are mutually exclusive and the probability in the numerator of the likelihood ratio is the sum of these three probabilities, namely, $r_1 + (1 - r_1 - r_2)\gamma_1$.

If H_d is true (*i.e.*, the suspect is not the offender), the probability of a correspondence is as before, namely, γ_1.

The likelihood ratio is then

$$V = \frac{Pr(\Gamma_1, \Gamma_2 \mid H_p)}{Pr(\Gamma_1, \Gamma_2 \mid H_d)} = \frac{r_1 + (1 - r_1 - r_2)\gamma_1}{\gamma_1}. \quad (4.37)$$

A graphical structure for the 'two stain–one offender' scenario may be obtained in two steps. First, consider the case at the source level, where two propositions are of interest:

- Node F: This node represents the uncertainty about which of the two stains, if either, has been left by the suspect. The possible states of this variable are f_0, neither of the two stains comes from the suspect, f_1, the stain of profile Γ_1 comes from the suspect, and f_2, the stain of profile Γ_2 comes from the suspect.
- Node E: E is the evidence node, defined as a 'correspondence in Γ_1 between the profile of the suspect and the profile of one of the crime stains'. This variable may either be true (state e_1) or false (state e_2).

Clearly, F is unobservable and its nature is such as to affect the states of the variable E. The node F is thus chosen as a parent. The second step in deriving a network structure consists of constructing an argument from the propositions at the source level concerning the origin of the stain, that is, node F, to the propositions, H_p and H_d, at the crime level. This move from source level to crime level may be achieved with the use of the following two nodes.

- Node H: This node represents propositions at the crime level, 'the suspect is the offender (H_p)' and 'the suspect is not the offender (H_d)'.
- Node G: This node expresses the relevance of the recovered stains. Three states may be defined: g_0, neither of the two stains comes from the offender, g_1, the stain of profile Γ_1 comes from the offender, g_2, the stain of profile Γ_2 comes from the offender. Notice that g_0, g_1 and g_2 are mutually exclusive states. This implies that no more than one, if either, of the two stains may be relevant.

The proposition as to which, if either, of the two stains has been left by the suspect (node F) depends on whether the suspect is the offender (node H) and which of the two stains has been left by the offender (relevance, node G). These three nodes are thus combined in a converging connection such that $G \rightarrow F \leftarrow H$.

As may be seen, these considerations lead to an overall network structure that is analogous to the network discussed in Section 4.1 (Figure 4.1), applicable to one-trace transfer cases assuming there is to be one offender. The quantitative part of the proposed Bayesian network consists of the following assessments.

- Node H: The probabilities assigned to H_p and H_d are the prior probabilities of guilt and innocence respectively.
- Node G: The definition of this node is an expression of the relevance of the pair of recovered stains. In analogous notation with previous considerations of the concept

of relevance (*e.g.*, Section 4.1), the probability of stain 1 being left by the offender is denoted as r_1 and the probability of stain 2 being left by the offender is denoted as r_2. Consequently, there is a probability of $1 - r_1 - r_2$ that neither of the stains came from the offender.

- Node F: If the suspect is the offender (H_p) and a crime stain is relevant (*i.e.*, it has been left by the offender), then it has been left by the suspect since the suspect and the offender are one. Formally, $Pr(f_i \mid g_j, H_p) = 1$ if $i = j = 0, 1, 2$. Conversely, for all $i \neq j$ $(i, j = 0, 1, 2)$ $Pr(f_i \mid g_j, H_p) = 0$. If the suspect is not the offender (H_d), then it may be assumed that none of the stains, irrespective of their relevance, comes from the suspect: $Pr(f_0 \mid g_j, H_d) = 1$ for all j.
- Node E: If the suspect has left the stain of profile Γ_1, then certainly there is a match in Γ_1: $Pr(e_1 \mid f_1) = 1$. There cannot be a match in Γ_1 if the stain of type Γ_2 comes from the suspect: $Pr(e_1 \mid f_2) = 0$. The probability of a correspondence in Γ_1, given the suspect left neither of the two stains, equals the random match probability of Γ_1: $Pr(e_1 \mid f_0) = \gamma_1$.

A summary of the probabilities assigned to the nodes G, F and E is given in Tables 4.5, 4.6 and 4.7.

For illustration, consider a hypothetical case in which the assumption is made that either of the two stains is as likely to be left by the offender as neither of the two so that $Pr(r_0) = Pr(r_1) = Pr(r_2) = 1/3$. Take the random match probability γ_1 to be 0.05. Figure 4.9(i) shows a fully specified Bayesian network assuming equal prior probabilities for the ultimate probandum H.

Figures 4.9 (ii) and (iii) provide an illustration of the evaluation of the probability of the evidence, given the prosecution's and defence's cases respectively. First, consider the value of the numerator of the likelihood ratio, that is, $Pr(e_1 \mid H_p)$. From (4.37) and the values defined here, one obtains the following: $r_1 + (1 - r_1 - r_2)\gamma_1 = 1/3 + 1/3 \times 0.05 = 0.35$. As may be seen, this is the value obtained in node E (Figure 4.9(ii)).

Recall that, given H_p, there are three components to the probability of the evidence. When using the described Bayesian network in an appropriate computerised environment, such component probabilities may be easy to evaluate. One needs to set the node H to H_p and then make various instantiations at the node G:

Table 4.5 Unconditional probabilities assigned to the node G. This node expresses the relevance of the recovered stains. Three states are defined: neither stain is from the offender (g_0), the stain of profile Γ_1 is from the offender (g_1), the stain of profile Γ_2 is from the offender (g_2).

G:	g_0	$1 - r_1 - r_2$
	g_1	r_1
	g_2	r_2

Table 4.6 Conditional probabilities assigned to the node F. Values given are $Pr(F \mid G, H)$. The possible states of this variable are f_0, neither of the two stains comes from the suspect, f_1, the stain of profile Γ_1 comes from the suspect, and f_2, the stain of profile Γ_2 comes from the suspect. Node H has two states, H_p, the suspect is the offender, and H_d, the suspect is not the offender. Node G has three states: g_0, neither stain is from the offender, g_1, the stain of profile Γ_1 comes from the offender, and g_2, the stain of profile Γ_2 comes from the offender.

	H:	H_p			H_d		
	G:	g_0	g_1	g_2	g_0	g_1	g_2
F:	f_0	1	0	0	1	1	1
	f_1	0	1	0	0	0	0
	f_2	0	0	1	0	0	0

Table 4.7 Conditional probabilities assigned to the node E. Values given are $Pr(E \mid F)$. Node E is the evidence node, defined as a 'correspondence in Γ_1'. This variable may either be true (state e_1) or false (state e_2). Node F represents the uncertainty about which one of the two stains, if either, has been left by the suspect. It has three states: f_0, neither of the two stains comes from the suspect; f_1, the stain of profile Γ_1 comes from the suspect; and f_2 the stain of profile Γ_2 comes from the suspect.

	F:	f_0	f_1	f_2
E:	e_1	γ_1	1	0
	e_2	$1-\gamma_1$	0	1

- If G is set to g_1, that is, $r_1 = 1$, then the node E would correctly display the value of the component probability $(r_1 \times 1)$, that is, 1.
- If G is set to g_2, that is, $r_2 = 1$, the node E would show the component probability $(r_2 \times 0)$, that is, 0.
- If G is set to g_0, that is, $1 - r_1 - r_2 = 1$, the node E would indicate the value for the component probability $(1 - r_1 - r_2)\gamma_1$, that is, 0.05.

Notice that the states of the Bayesian network, as applicable to these operations, are not shown in Figure 4.9. Readers are invited to proceed themselves to arrive at these evaluations.

Next, consider the value of the denominator of the likelihood ratio. If H_d is true, the probability of a correspondence is γ_1 (4.37). This situation is shown in Figure 4.9(iii): the node E displays 0.05, a value earlier defined for γ_1. Note that under H_d, the probability of the evidence will be given by γ_1 irrespective of instantiations made at the node G.

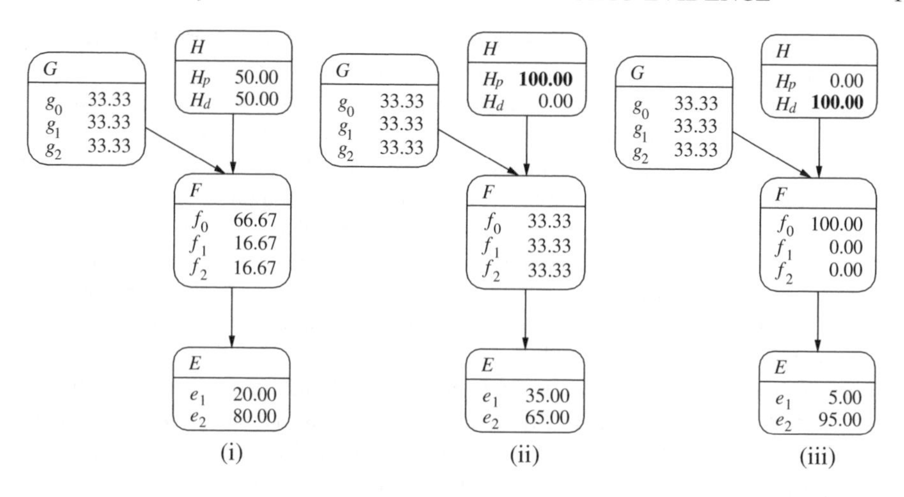

Figure 4.9 Evaluation of the Bayesian network describing the 'two stain–one offender' scenario: (i) initial state; (ii) the prosecution's case (H_p is true); (ii) the defence case (H_d is true). Node E is the evidence node, defined as a 'correspondence in Γ_1'. This variable may either be true (state e_1) or false (state e_2). Node F has three states, f_0: neither of the two stains comes from the suspect, f_1: the stain of profile Γ_1 comes from the suspect, f_2: the stain of profile Γ_2 comes from the suspect. Node G has three states: neither stain is from the offender (g_0), the stain of profile Γ_1 is from the offender (g_1), the stain of profile Γ_2 is from the offender (g_2). Node H has two states, H_p: the suspect is the offender, H_d: the suspect is not the offender.

In summary, separate evaluations of $Pr(e_1 \mid H_p)$ and $Pr(e_1 \mid H_d)$ have led to a likelihood ratio of $0.35/0.05 = 7$. The same value may be obtained when the Bayesian network is evaluated *via* the posterior probabilities of the proposition H (a situation not represented in Figure 4.9): instantiating E to e_1 will change the equal prior probabilities assigned to H_p and H_d to 0.875 and 0.125 respectively, the ratio of which is 7.

Certain special cases can be distinguished. As r_1 and r_2 tend to zero, which implies that neither stain is relevant, then the likelihood ratio tends to 1. The proposed Bayesian network may provide a clear illustration of this. When the node G is instantiated to g_0, knowing the variable E to be in state e_1 or e_2 would no longer affect the probability of H. A likelihood ratio of 1 provides no support for either proposition, a result in this case which is entirely consistent with the information that neither stain is relevant.

Analogously, one may verify that when assuming $r_1 = r_2 = 0.5$, $V = 1/2\gamma_1$. For $r_1 = 1$, $V = 1/\gamma_1$. As $r_2 \rightarrow 1$, then $r_1 \rightarrow 0$ and $V \rightarrow 0$. All of these are perfectly reasonable results.

4.3.2 Two stains, no putative source

There is another frequently encountered situation in which scientists are confronted with more than one stain. Two or more traces may be recovered, for example, at different locations where, at temporally distinct instances, crimes have been committed. At an early stage of an investigation, it may well be that no potential source is available for comparison. However, analyses may be performed on the crime samples with the aim of evaluating possible linkages between cases. Such scenarios currently play an important role in a domain known as 'forensic intelligence' (Ribaux and Margot 2003).

Imagine two separate offences. Naked, dead bodies have been found at two distinct locations. In both cases, evidence has been collected from the bodies. Let E_1 and E_2 denote items of evidence collected in the first and second case respectively. The sets of evidence could consist of, for example, textile fibres, hairs, semen or saliva.

The scenario is approached here at a rather general and disaggregated level. For simplicity, it will be assumed that the compared characteristics are discrete. In the case of textile fibres, this could be a particular combination of fibre type and fibre colour, for example, red wool. A development involving continuous data together with an extension to crime level propositions can be found in Taroni et al. (2005b).

Suppose that the two items of evidence are analysed and both are found to be of type A, some sort of discrete physical attribute, for example. For the current level of discussion, it will not be necessary to specify A in further detail. Addressing the case at the source level, propositions of interest can be defined as follows:

- H_1: the items of evidence E_1 and E_2 come from the same source;
- H_2: E_1 and E_2 come from different sources.

These propositions do not specifically relate to a prosecution's or defence's standpoint, so simple numerical subscripts are used for the variable H. Notice that issues such as the relevance of the evidential material are not considered here.

A likelihood ratio V may be developed in order to evaluate how well the available evidence allows one to discriminate between H_1 and H_2. Formally, one can write

$$V = \frac{Pr(E_1 = A, E_2 = A \mid H_1)}{Pr(E_1 = A, E_2 = A \mid H_2)} . \tag{4.38}$$

Consider the numerator first. What is the probability of both items of evidence E_1 and E_2 being of type A, given that they come from the same source? Under H_1, there is a single source, and thus the probability of E_1 and E_2 both being of type A depends on the probability of the single (common) source being of type A. In fact, the common source of the two items of evidence could either be of type A or $\bar{A}$, abbreviated by $H_{1,1}$ and $H_{1,2}$, respectively. With E denoting the conjunction of $E_1 = A$ and $E_2 = A$, the numerator becomes

$$Pr(E \mid H_1) = \underbrace{Pr(E \mid H_{1,1})}_{1}\,\underbrace{Pr(H_{1,1})}_{\gamma_A} + \underbrace{Pr(E \mid H_{1,2})}_{0}\,\underbrace{Pr(H_{1,2})}_{\gamma_{\bar{A}}}$$

where γ_A denotes the probability by which a source from a relevant population would be found to be of type A. Note that $Pr(E \mid H_{1,1}) = 1$ is an expression of the assumption that the methodology used by the scientist is expected to report a match if there is, in fact, a match.

Then, consider the denominator. If H_2 is true, there may be different possibilities for there being two *different* sources: both sources are of type A ($H_{2,1}$), both sources are of type $\bar{A}$ ($H_{2,2}$), or one source is of type A and the other is of type $\bar{A}$ ($H_{2,3}$). The denominator can be re-written as

$$Pr(E \mid H_2) = \sum_{i=1}^{3} Pr(E \mid H_{2,i})Pr(H_{2,i}) .$$

The conjunction E can only be true when the distinct sources of E_1 and E_2 are both of type A. Consequently, only the first product of this equation remains non-zero:

$$\begin{aligned} Pr(E \mid H_2) = & \underbrace{Pr(E \mid H_{2,1})}_{1} \underbrace{Pr(H_{2,1})}_{\gamma_A^2} + \\ & \underbrace{Pr(E \mid H_{2,2})}_{0} \underbrace{Pr(H_{2,2})}_{\gamma_{\bar{A}}^2} + \\ & \underbrace{Pr(E \mid H_{2,3})}_{0} \underbrace{Pr(H_{2,3})}_{2\gamma_A \gamma_{\bar{A}}} . \end{aligned}$$

The likelihood ratio then becomes

$$V = \frac{Pr(E_1 = A, E_2 = A \mid H_1)}{Pr(E_1 = A, E_2 = A \mid H_2)} = \frac{\gamma_A}{1 \times \gamma_A^2} = \frac{1}{\gamma_A} . \quad (4.39)$$

In the numerator and denominator, γ_A is a suitable statistic drawn from the same population. Notice that the validity of this assumption needs to be reviewed whenever other kinds of evidences are considered. See, for example, the differences in definition of relative frequencies γ and γ' presented in Section 3.3.3.

A serial connection, $H \rightarrow H_i \rightarrow E$ (with H_i representing the sub-hypotheses of H), appropriately translates (4.39) into a Bayesian network. However, a more detailed Bayesian network may be constructed. The stepwise construction of such a Bayesian network is explained below.

Start by considering the first item of evidence (collected in the first case). Let E_1 denote the outcomes of the observations made on this piece of evidence. The variable E_1 assumes the binary states A or $\bar{A}$, where $\bar{A}$ covers all potential outcomes other than A. Information on the observed attributes of the first item of evidence is used to draw an inference about the characteristics of the source from which it comes. Let S_{E_1} denote the characteristics of the source of the first item of evidence. This variable is also binary and has the following states:

- S_{E_1}: the source of the first item of evidence is of type A;
- $\bar{S}_{E_1}$: the source of the first item of evidence is of type $\bar{A}$.

Logically, the observed characteristics of the evidence directly depend on the attributes of the source, so, graphically, a direct edge may be drawn from S_{E_1} to E_1.

The item of evidence recovered in the second case may be evaluated analogously. A variable E_2 is used to describe the observations made on the second item of evidence. A variable S_{E_2} describes the characteristics of the source. These two variables are related so that the following network fragment is obtained: $S_{E_2} \rightarrow E_2$.

Next, a variable H is defined, representing the proposition according to which E_1 and E_2 come from the same source (H_1), or come from different sources (H_2). H needs to be combined with the network fragments $S_{E_1} \rightarrow E_1$ and $S_{E_2} \rightarrow E_2$ in some meaningful way. One possibility would be to condition S_{E_2} on both H and S_{E_1}, as shown in Figure 4.10. This allows one to assess the following probabilities (Table 4.8):

- If the two items of evidence come from the same source and the source of the first item of evidence is of type A, then certainly the source of the second item of evidence is of type A: $Pr(S_{E_2} = A \mid H_1, S_{E_1} = A) = 1$.

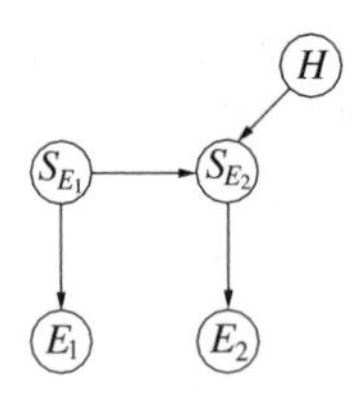

Figure 4.10 A Bayesian network for a scenario in which two stains are found at two different scenes. Nodes E_1 and E_2 denote the outcomes of the observations made on the evidence at the first and second crime scenes, respectively, with two states A and $\bar{A}$ for each node. Nodes S_{E_1} and S_{E_2} have two states, the source of the first or second item of evidence is of type A or of type $\bar{A}$. Node H represents the proposition according to which E_1 and E_2 come from the same source (H_1) or come from different sources (H_2). [Reproduced with permission of ASTM International.]

Table 4.8 Conditional probabilities applicable to the node S_{E_2} in Figure 4.10. Values given are $Pr(S_{E_2} \mid S_{E_1}, H)$. Nodes S_{E_1} and S_{E_2} have two states, the source of the first or second item of evidence is of type A or of type $\bar{A}$. Node H represents the proposition according to which E_1 and E_2 come from the same source (H_1) or come from different sources (H_2). The frequency of A in a relevant population is γ_A. [Reproduced with permission of ASTM International.]

	H :	H_1		H_2	
	S_{E_1} :	A	$\bar{A}$	A	$\bar{A}$
S_{E_2} :	A	1	0	γ_A	γ_A
	$\bar{A}$	0	1	$1-\gamma_A$	$1-\gamma_A$

- If the two items of evidence come from the same source and the source of the first item of evidence is of type $\bar{A}$, then certainly the source of the second item of evidence is also of type $\bar{A}$: $Pr(S_{E_2} = \bar{A} \mid H_1, S_{E_1} = \bar{A}) = 1$.
- If the two pieces of evidence come from different sources, then the probability of the source of the first piece of evidence being of type A is given just by the frequency of that characteristic in a relevant population, denoted as γ_A. Notice that given H_2, S_{E_2} is assumed to be independent from S_{E_1}: $Pr(S_{E_2} = A \mid H_2, S_{E_1} = A) = Pr(S_{E_2} = A \mid H_2, S_{E_1} = \bar{A}) = \gamma_A$.

The conditional probabilities associated with the evidence nodes E_1 and E_2 reflect the degree to which the characteristics observed by the scientist are indicative of the characteristics of the sources from which the items of evidence originate. Generally, scientists tend to prefer methods for which $Pr(E_1 = A \mid S_{E_1} = A) \rightarrow 1$ and $Pr(E_1 = A \mid S_{E_1} = \bar{A}) \rightarrow 0$.

Notice that the proposed Bayesian network is somewhat more detailed than the formula (4.39) developed earlier in this section. The major difference is that a distinction has been drawn between the observations made on the items of evidence (nodes E_1 and E_2) and the (true) characteristics of their sources (nodes S_{E_1} and S_{E_2}). The latter are in fact unobserved variables not explicitly represented in the derivation of the likelihood ratio formula. Moreover, two distinct variables are used to represent the items of evidence.

An alternative network structure is shown in Figure 4.11. A summary of the node definitions is given in Table 4.9. In this model, the observed characteristics – fibre type and colour – of each item of evidence have been represented using distinct nodes. The characteristics of a potential common source of the two items of evidence are modelled by the nodes CS_t and CS_c, where the subscripts t and c refer to fibre type and colour, respectively. The term 'potential common source' is used here in order to express the idea that uncertainties about the variables CS_t and CS_c will be allowed to affect the factual state of the nodes $S_{i,\{t,c\}}$ only if the two items of evidence, do in fact, come from a common

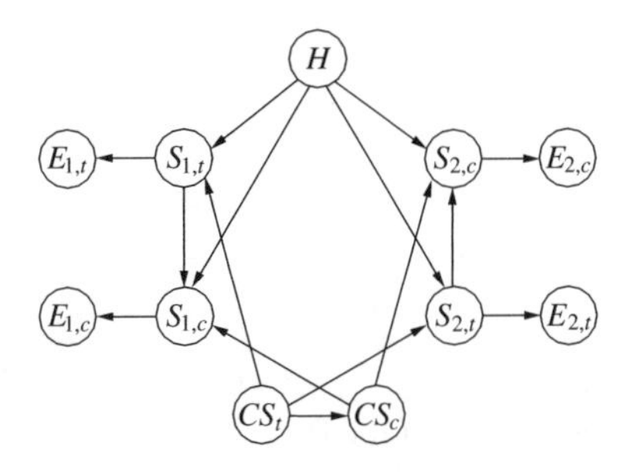

Figure 4.11 An alternative network structure for a scenario in which two stains are found on two different scenes. A summary of the node definitions is given in Table 4.9. [Reproduced with permission of ASTM International.]

Table 4.9 Definitions of the nodes present in the network shown in Figure 4.11. Note that $i = 1$ and $i = 2$ denote the first and the second item of evidence, respectively. [Reproduced with permission of ASTM International.]

Node	Definition	States	
H	The items of evidence come from the same source	H_1:	yes
		H_2:	no
$E_{i,t}$	Observed fibre type of the item of evidence i	$E_{i,t}$:	wool
		$\bar{E}_{i,t}$:	other
$E_{i,c}$	Observed colour of the item of evidence i	$E_{i,c}$:	red
		$\bar{E}_{i,c}$:	other
$S_{i,t}$	Fibre type of the source of the item of evidence i	$S_{i,t}$:	wool
		$\bar{S}_{i,t}$:	other
$S_{i,c}$	Fibre colour of the source of the item of evidence i	$S_{i,c}$:	red
		$\bar{S}_{i,c}$:	other
CS_t	Fibre type of a potential common source	CS_t:	wool
		$\bar{C}S_t$:	other
CS_c	Fibre colour of a potential common source	CS_c:	red
		$\bar{C}S_c$:	other

Table 4.10 Probability table of the nodes $S_{i,t}$ in Figure 4.11 for $i = 1, 2$. Values given are $Pr(S_{i,t} \mid H, CS_t)$. Node H has two states, H_1, the items of evidence come from the same source and H_2, the items of evidence do not come from the same source. Node CS_t has two states, the fibre type of a common source is wool or is of some fibre other than wool. Node $S_{i,t}$ has two states, the fibre type of the source of the item of evidence i is wool or is of some fibre other than wool. Note that when H_2 is true, the state of node CS_t is not relevant for the state of $S_{i,t}$. The frequency of wool in a relevant population is γ_w. [Reproduced with permission of ASTM International.]

	H:	H_1		H_2	
	CS_t:	*wool*	*other*	*wool*	*other*
$S_{i,t}$:	*wool*	1	0	γ_w	γ_w
	other	0	1	$1 - \gamma_w$	$1 - \gamma_w$

source, that is, H_1 is true. Note that the nodes referring to fibre colour are conditioned on the fibre type, this should allow one to adjust the frequency of fibre colour for a given fibre type.

The probability tables associated with the observation nodes $E_{i,\{t,c\}}$ can account for the relative performance of the examiner. For example, when it is thought that the examiner will always recognise a fibre type as wool when it is indeed wool, and never declare a fibre type as wool when in fact it is not, then the following assessments are appropriate: $Pr(E_{i,t} \mid S_{i,t}) = 1$ and $Pr(E_{i,t} \mid \bar{S}_{i,t}) = 0$, for $i = 1, 2$. The argument applies analogously to the nodes $E_{i,c}$ as far as the observation of the fibre's colour is concerned.

The probability tables associated with the source nodes $S_{i,\{t,c\}}$, either contain logical probabilities or frequencies. Consider, for example, the probabilities assigned to $S_{i,t}$ given H: if the two items of evidence come from a common source (H_1 being true), and, that common source consists of wool ($Pr(CS_t = wool) = 1$), then, logically, $Pr(S_{i,t} = wool \mid H, CS_t = wool) = 1$ with $i = 1, 2$ denoting items of evidence 1 and 2, respectively. Analogously $Pr(S_{i,t} = other \mid H, CS_t = other) = 1$. However, if the two items of evidence do not have a common source (H_2 being true), then the factual state of CS_t is not relevant when assessing the nodes $S_{i,t}$. In the latter case, the probability of the source of an item of evidence being wool is given just by the frequency γ_w of wool in a relevant population, for both states of CS_t. Thus, $Pr(S_{i,t} = \text{wool} \mid H_2) = \gamma_w$. Table 4.10 provides a summary of the probabilities assigned to the nodes $S_{i,t}$.

The probability tables of the nodes $S_{i,c}$ are completed in a similar way (*see* Table 4.11). When H_1 is true, that is, the two items of evidence come from a common source, then the probability of $S_{i,c}$ being red, for example, depends on the probability of the common source being red, thus $Pr(S_{i,c} = red \mid H, CS_c = red) = 1$ and $Pr(S_{i,c} = red \mid H, CS_c = other) = 0$. Note that when H_1 is true, the truth or otherwise of the $S_{i,t}$ does not affect the $S_{i,c}$. The contrary holds when the two items of evidence do not come from a common source, that is, H_2 being true. Then, the probability of the source of an item of evidence being red, for example, is dependent solely on the fibre type. The variable γ_r denotes the

Table 4.11 Probability table of the nodes $S_{i,c}$ in Figure 4.11 for $i = 1, 2$. Values given are $Pr(S_{i,c} \mid H, S_{i,t}, CS_c)$. Node H has two states, H_1, the items of evidence come from the same source and H_2, the items of evidence do not come from the same source. Node $S_{i,t}$ has two states, the fibre type of the source of the item of evidence i is woollen or is of some type other than wool. Node CS_c has two states, the colour of a potential common fibre source is red or is of some other colour than red. Node $S_{i,c}$ has two states, the fibre colour of the source of the item of evidence i is red or is of some colour other than red. The frequency of red colour amongst sources of fibres that are woollen is γ_r and amongst sources of fibres other than wool is γ_r'. Note that when H_2 is true, the state of node CS_c is not relevant for the state of $S_{i,c}$, though the state of $S_{i,t}$ is relevant. [Reproduced with permission of ASTM International.]

			$S_{i,c}$	
H	$S_{i,t}$	CS_c	*red*	*other*
H_1	*wool*	*red*	1	0
		other	0	1
	other	*red*	1	0
		other	0	1
H_2	*wool*	*red*	γ_r	$1 - \gamma_r$
		other	γ_r	$1 - \gamma_r$
	other	*red*	γ_r'	$1 - \gamma_r'$
		other	γ_r'	$1 - \gamma_r'$

frequency of red colour among sources of fibres that are woollen. The variable γ_r' denotes the frequency of red colour among sources of fibres other than wool. Note that the directed edge between the nodes $S_{i,t}$ and $S_{i,c}$ is justified as long as the parameters γ_r and γ_r' assume different values.

5

DNA evidence

Interest in the probabilistic evaluation of DNA evidence has grown considerably during the last decade. Topics such as the assessment of mixtures, consideration of error rates and effects of database selection have increased the interest in forensic statistics. This chapter discusses these topics and also includes discussion of issues concerning relatedness testing and sub-population structures.

5.1 DNA likelihood ratio

Imagine a crime has been committed. Examination of the scene has revealed the presence of a single bloodstain. From the position where the stain was found, as well as its condition (*e.g.*, freshness), investigators believe the stain to be relevant to the case. The crime sample is submitted to a forensic laboratory, where some kind of DNA typing technique is applied. Imagine further that, on the basis of information completely unrelated to the blood evidence, a suspect is apprehended. A blood sample provided by the suspect is submitted for DNA profiling. For a situation in which the suspect is found through a search in a database, refer to Section 5.10.

If the suspect and the crime sample were found to be of the same type, Γ_1 for instance, what is the value to be assigned to this correspondence? Let E denote the correspondence between the suspect and the crime sample and imagine the calculation of the likelihood ratio with I representing the available background information and H_p and H_d a pair of propositions at the source-level, that is:

- H_p: the suspect is the source of the stain,
- H_d: another person, unrelated to the suspect, is the source; that is, the suspect is not the source of the stain.

The likelihood ratio considers the probability of the evidence given each of these propositions as follows,

$$V = \frac{Pr(E \mid H_p, I)}{Pr(E \mid H_d, I)}. \tag{5.1}$$

Bayesian Networks and Probabilistic Inference in Forensic Science F. Taroni, C. Aitken, P. Garbolino and A. Biedermann

The variable E may be considered as a combination of two distinct components, E_s and E_c, denoting the DNA profile of the suspect and the crime sample, respectively. Following the arguments set out in Section 3.3.1, the likelihood ratio can be expressed as

$$V = \frac{Pr(E_c \mid E_s, H_p, I)}{Pr(E_c \mid E_s, H_d, I)} . \quad (5.2)$$

Assume that the DNA typing system is sufficiently reliable that two samples from the same person will be found to match and that there are no false-negatives. Take the case in which both the profile of the suspect and the crime sample are of type Γ_1, say. Then, assuming the suspect is the source of the crime stain and knowing that he is of type Γ_1, the probability of the recovered sample being of type Γ_1 is 1: $Pr(E_c = \Gamma_1 \mid E_s = \Gamma_1, H_p, I) = 1$.

When the alternative proposition is assumed to be true (H_d), the suspect and the donor of the stain are two different persons. Assuming that the DNA profiles of the suspect and the true donor of the stain are independent, one is allowed to write $Pr(E_c = \Gamma_1 \mid E_s = \Gamma_1, H_d, I) = Pr(E_c = \Gamma_1 \mid H_d, I)$. This parameter is generally estimated using the so-called *profile probability*, the probability that an unknown person would have a Γ_1 profile in a relevant population.

Usually, a DNA profile consists of pairs of alleles at several loci. An individual is said to be *homozygous* at a particular locus if the genotype consists of two indistinguishable alleles, say $A_i A_i$ for example. If a genotype consists of two different alleles, for example, $(A_i A_j, i \neq j)$, the individual is considered to be *heterozygous*. Individual allele probabilities may be used to estimate genotype frequencies, denoted P_{ij}. Assuming Hardy-Weinberg equilibrium,

$$\begin{aligned} P_{ij} &= 2\gamma_i\gamma_j \ (i \neq j), \\ &= \gamma_i^2 \ (i = j). \end{aligned} \quad (5.3)$$

Notice that these considerations are a widely accepted simplification. In reality, the evidential value of a match between the profile of the crime sample and that of a suspect needs to take into account the fact that there is a person (the suspect) who has already been seen to have that profile (type Γ_1). So, the probability of interest is $Pr(E_c = \Gamma_1 \mid E_s = \Gamma_1, H_d, I)$ and this can be quite different from $Pr(E_c = \Gamma_1 \mid H_d, I)$ (Weir 2000). Stated otherwise, knowledge of the suspect's genotype should affect one's uncertainty about the offender's genotype in cases where, for example, the offender is a close relative of the suspect or both the suspect and the offender share ancestry due to membership in the same population.

In fact, observation of a gene in a sub-population increases the chance of observing another of the same type. Hence, within a sub-population, DNA profiles with matching allele types are more common than suggested by the independence assumption, even when two individuals are not directly related.

The conditional probability (a *random-match probability*) incorporates the effect of population structure or other dependencies between individuals, such as that imposed by family relationships (Balding and Nichols 1994). This aspect is presented in Section 5.12.

5.2 Network approaches to the DNA likelihood ratio

Consider the variable E as a global assignment for the event that the crime and suspect sample correspond with respect to their DNA profile. A basic two-node network fragment allowing knowledge about E to be used for drawing an inference to the propositions at the source level has been discussed in Section 4.1. If the latter propositions are represented by H, the corresponding network fragment is $H \rightarrow E$. Besides the prior probabilities that need to be assessed for the node H, two further probabilities are essential for the numerical specification of the network fragment: $Pr(E \mid H_p, I)$ and $Pr(E \mid H_d, I)$. It has been assumed that if the suspect were the source of the crime sample, then the suspect's blood would certainly match: $Pr(E \mid H_p, I) = 1$. If the crime stain came from a person other than the suspect, the probability of the evidence may be estimated using the profile probability: $Pr(E \mid H_d, I) = \gamma$. As noted above, these two parameters represent the numerator and denominator, respectively, of the likelihood ratio (5.1).

One may also consider a less coarse level of detail when evaluating DNA evidence. Dawid et al. (2002) proposed more fine-grained network fragments focusing on individual genes and genotypes. An example of this is shown in Figure 5.1. Here the node gt, representing a genotype, is modelled as a logical combination of the alleles inherited from the mother and father respectively. These parentally inherited genes are represented by the nodes mg and pg, to be read 'maternal gene' and 'paternal gene', respectively.

A gene, A for instance, can take one of several different forms, also called alleles. Suppose there are n alleles at gene A. These alleles may be denoted $A_1, A_2, \ldots, A_n$. In the Bayesian network shown in Figure 5.1 the states A_1, A_2 and A_x are assumed for the nodes mg and pg where the third state, A_x, is an aggregation of all unobserved alleles $A_3, \ldots, A_n$. The possible states of the genotype node gt may then be defined as A_1A_1, A_1A_2, A_1A_x, A_2A_2, A_2A_x, and A_xA_x.

On the basis of these notions, imagine the construction of a Bayesian network for evaluating the DNA likelihood ratio. However, instead of reducing the evidence to E, as was assumed in the so-called 'match approach' outlined at the beginning of this section, consider the typing results on the suspect and the crime sample as distinct components: $E = (E_s, E_c)$. Recall from Section 5.1 that the variables E_s and E_c were used to denote the *profile* of the suspect and the crime sample, respectively. For ease of argument, the definition of the evidence is now restricted to the typing results of a *single locus* or *marker*. Thus, let sgt and tgt, read as 'suspect genotype' and 'trace genotype' respectively, denote the allelic configuration at a certain locus, where the trace genotype is the genotype of the crime sample.

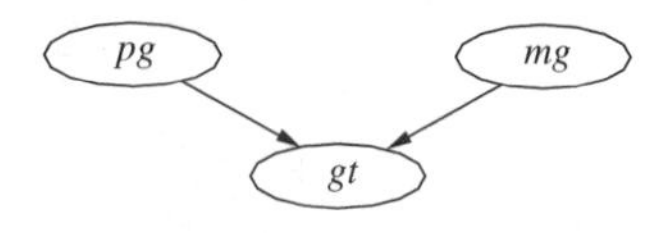

Figure 5.1 Representation of a genotype, node gt, with pg and mg denoting the alleles inherited from the father, pg, and the mother, mg, respectively.

Modelling the genotype of the suspect and crime stain using distinct nodes suggests that two network fragments of the kind shown in Figure 5.1 are necessary. One such network fragment consists of the nodes 'suspect genotype' (sgt), 'suspect paternal gene' (spg) and 'suspect maternal gene' (smg). The other contains the nodes 'trace genotype' (tgt), 'trace paternal gene' (tpg) and 'trace maternal gene' (tmg).

Next, a logical combination is needed between these two network fragments and the propositions at the source level. This may be achieved by conditioning the paternal genes of the crime stain by both the paternal genes of the suspect and the propositions at the source level. This is an expression of the idea that if the suspect were known to be the source of the crime stain, then his allelic composition provides relevant information for estimating the allelic configuration of the crime stain. Graphically, the following connections are adopted: $H \rightarrow tpg$, $H \rightarrow tmg$, $spg \rightarrow tpg$ and $smg \rightarrow tmg$. The resulting Bayesian network is shown in Figure 5.2.

If the states of the variables are as defined above, then a numerical specification may be adopted as follows:

- Nodes spg and smg: the probability tables associated with these nodes contain unconditional probabilities given by the population frequencies of the alleles. The frequencies of the alleles A_1 and A_2 are denoted by γ_1 and γ_2. The probability of an allele being different from A_1 and A_2 (*i.e.*, one of $A_3, \ldots, A_n$ or A_x) is then $1 - \gamma_1 - \gamma_2 = \gamma_x$.
- Nodes sgt and tgt: the probability tables of these nodes can logically be completed using values equating certainty and impossibility. For example, if $spg = A_1$ and $smg = A_2$ then certainly $sgt = A_1A_2$. Stated otherwise, all the states of a genotype node not strictly containing the alleles that are given by actual configuration of the respective parental gene nodes, must be false.
- Nodes tpg and tmg: if the crime stain comes from the suspect ($Pr(H = H_p) = 1$), then the state of a trace parental gene must equate the state of the respective suspect parental gene: for $i = 1, 2, x$, $Pr(tpg = A_i \mid spg = A_i, H_p) = 1$ and $Pr(tmg = A_i \mid smg = A_i, H_p) = 1$. If the crime stain does not come from the suspect, then the probability of a crime stain parental gene being A_1, A_2 or A_x is just given by the population frequencies of these alleles. Probabilities $Pr(tpg = A_i \mid spg = A_i, H_d) = Pr(tpg = A_i \mid smg = A_i, H_d) = \gamma_i,\ i = 1, 2, x$.

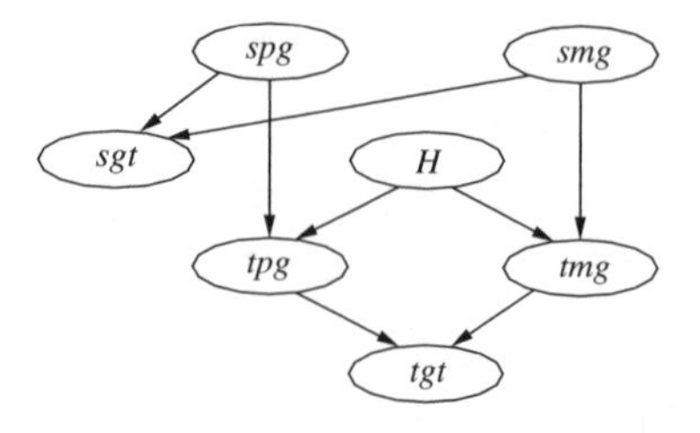

Figure 5.2 Bayesian network for evaluating DNA typing results obtained from a crime stain and a sample provided by a suspect. H: the suspect is the source of the crime stain; sgt and tgt: genotype of the suspect and the crime stain, respectively; spg and smg: suspect paternal and suspect maternal gene; tpg and tmg: trace paternal and trace maternal gene.

A summary of these assignments can be found in the Tables 5.1, 5.2 and 5.3.

For illustration, imagine DNA typing results for a single marker. The suspect's blood and the crime stain are both found to be of type A_1A_2. Consider a Bayesian network as shown in Figure 5.2 and let the states of the parental gene nodes consist of, respectively, the states A_1 A_2, and A_x. Let γ_1 and γ_2 be the frequencies of the alleles A_1 and A_2 respectively. A value of 0.05 is assumed for both γ_1 and γ_2.

Table 5.1 Unconditional probabilities applicable to the nodes modelling the suspect's parental genes, that is, *spg* and *smg*.

spg, smg:	A_1	γ_1
	A_2	γ_2
	A_x	$\gamma_x = 1 - \gamma_1 - \gamma_2$

Table 5.2 Conditional probabilities assigned to the nodes 'trace paternal gene' (*tpg*) and 'trace maternal gene' (*tmg*). Factor H has two states, H_p, the suspect is the source of the crime sample and H_d, the suspect is not the source of the crime sample. The nodes, suspect paternal gene, *spg*, and suspect maternal gene, *smg*, have three states, alleles A_1, A_2 and A_x ($A_x = A_3, \ldots, A_n$).

	H:	H_p			H_d		
	$spg(smg)$:	A_1	A_2	A_x	A_1	A_2	A_x
$tpg(tmg)$:	A_1	1	0	0	γ_1	γ_1	γ_1
	A_2	0	1	0	γ_2	γ_2	γ_2
	A_x	0	0	1	γ_x	γ_x	γ_x

Table 5.3 Conditional probabilities applicable for a genotype node, that is, *sgt* or *tgt*. The nodes, suspect (trace) paternal gene, *spg* (*tpg*), and suspect (trace) maternal gene, *smg* (*tmg*), have three states, alleles A_1, A_2 and A_x ($A_x = A_3, \ldots, A_n$).

	$spg(tpg)$:	A_1			A_2			A_x		
	$smg(tmg)$:	A_1	A_2	A_x	A_1	A_2	A_x	A_1	A_2	A_x
$sgt(tgt)$:	A_1A_1	1	0	0	0	0	0	0	0	0
	A_1A_2	0	1	0	1	0	0	0	0	0
	A_1A_x	0	0	1	0	0	0	1	0	0
	A_2A_2	0	0	0	0	1	0	0	0	0
	A_2A_x	0	0	0	0	0	1	0	1	0
	A_xA_x	0	0	0	0	0	0	0	0	1

For evaluation of the numerator of the likelihood ratio, one needs to consider the probability of the crime stain being of type A_1A_2 given the prosecution's case is true and given the suspect's blood is of type A_1A_2. To this end, the node H is set to H_p and the node sgt is set to A_1A_2. In the node tgt one can now read the probability of the trace being of type A_1A_2 given the stated conditions (Figure 5.3(i)): $Pr(tgt = A_1A_2 \mid sgt = A_1, A_2, H_p) = 1$.

The denominator of the likelihood ratio is a consideration of the probability that the stain is of type A_1A_2 assuming the defence case is true. An evaluation of this scenario is shown in Figure 5.3(ii): the node H is instantiated to H_d and the effect of this propagated to the evidence node tgt, which displays $Pr(tgt = A_1A_2 \mid H_d) = 0.005$ (0.50%). Notice that when assuming H_d to be true, any information regarding the suspect's genotype would have no effect on the probability of the crime stain's genotype.

As a result of this analysis, one obtains a likelihood ratio of $1/0.005 = 200$. This result is in agreement with what may be obtained using the so-called 'match approach': here the Bayesian network consists of $H \rightarrow E$ and the likelihood ratio is given by $1/2\gamma_1\gamma_2$.

Notice also that a 'top-down' analysis has been performed, that is, alternatively instantiating H to H_p and H_d. By doing so, it is not necessary to elicit prior probabilities for the node H, since they do not enter the probability calculations. This view is consistent with the requirement that forensic scientists should solely focus on the ratio of the probabilities of the evidence, *given* the pair of competing propositions considered; that is, the likelihood ratio.

5.3 Missing suspect

One may legitimately ask whether there is a need for such sophisticated models as discussed in the previous section, when the same result may be obtained by basic algebraic calculus. However, as pointed out by Dawid et al. (2002), the aim of using a probabilistic network approach is to extend the considerations to scenarios where one needs to account for genetic information of further individuals. This may be necessary if, for example, biological material of one or more target individuals cannot be obtained, but samples of one or more close relatives are available. Clearly, a purely arithmetic solution to such problems becomes increasingly difficult.

Imagine a bloodstain of some relevance is found on the scene of a crime. The principal suspect is unavailable for profiling. However, the suspect is known to have a brother and this individual is willing to provide a blood sample for analysis. How can knowledge of the brother's genotype be used to infer something about the suspect's genotype? What inferences can be drawn to the propositions at the source or the crime level? These are two of the questions for which meaningful answers may be found through the use of Bayesian networks.

As a starting point, reconsider the Bayesian network shown in Figure 5.2. This model accounts for genotypic information available from the suspect and from the crime stain. Now, if one were to extend the considerations to the genotype of the suspect's brother, a further network fragment as shown in Figure 5.1, representing the genotype of an individual, must be incorporated in the existing network.

Let the brother's genotype be modelled by three nodes arranged in a converging connection: $bpg \rightarrow bgt \leftarrow bmg$. The nodes are defined as follows: brother genotype (bgt), brother paternal gene (bpg) and brother maternal gene (bmg). How is this network fragment

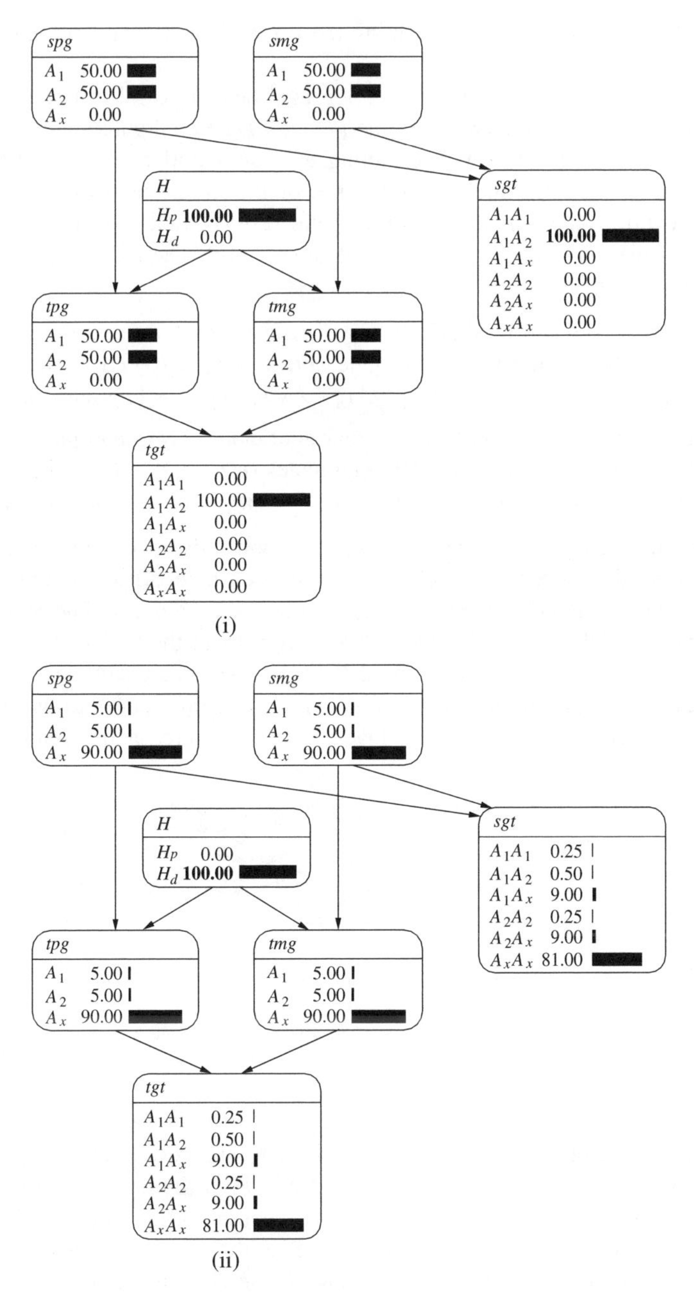

Figure 5.3 Evaluation of the likelihood ratio in a case where the evidence consists of a crime stain and a suspect with same genotype, that is, A_1A_2: (i) the prosecution's case, and (ii) the defence case. Node H has two states, H_p, the suspect is the source of the crime sample, and H_d, the suspect is not the source of the crime sample. Nodes spg, smg, tpg and tmg, the paternal and maternal genes for the suspect and trace samples have three states A_1, A_2 and A_x. Nodes sgt and tgt, the suspect and trace genotypes, have six states, A_1A_1, A_1A_2, A_1A_x, A_2A_2, A_2A_x, A_xA_x. (A_x is an aggregation of all unobserved alleles $A_3, \ldots, A_n$).

logically combined with the network relating the suspect's genotype with that of the crime stain?

Notice that the suspect and the brother are assumed to be full brothers, that is, they share the same mother and father. Then, the parental gene configurations of the latter two individuals are solely relevant for determining the configuration of the suspect's and the brother's parental genes. Let the parental genes of the mother and the father be defined using the following nodes: mother paternal gene (mpg), mother maternal gene (mmg), father paternal gene (fpg) and father maternal gene (fmg). The following connections can now be adopted:

- The suspect's and the brother's paternal gene nodes (spg and bpg) are both conditioned on their father's parental gene nodes (fpg and fmg). This results in the following set of arcs: $fpg \rightarrow spg$, $fmg \rightarrow spg$, $fpg \rightarrow bpg$ and $fmg \rightarrow bpg$.
- The suspect's and the brother's maternal gene nodes (smg and bmg) are both conditioned on their mother's parental gene nodes (mpg and mmg). The set of adopted arcs thus is as follows: $mpg \rightarrow smg$, $mmg \rightarrow smg$, $mpg \rightarrow bmg$ and $mmg \rightarrow bmg$.

These structural assumptions are represented graphically in Figure 5.4. Note that the node representing the suspect's genotype, sgt, has been maintained although it is desired to evaluate a scenario in which the suspect is assumed to be missing. The reason for this is that useful insight might be gained in how knowledge about the brother's genotype effects the probability distribution of the node modelling the suspect's genotype.

The Bayesian network discussed here contains further differences to the one discussed in Section 5.2. Notably, an extension is made from the source node, denoted F here, to

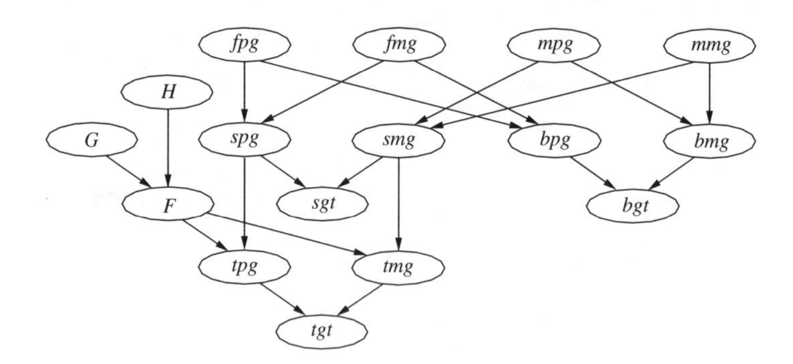

Figure 5.4 Bayesian network for evaluating DNA typing results when genotypic information from the suspect's brother is available. Node H is the crime level node with two states, H_p, the suspect is the offender, H_d, the suspect is not the offender. Node F is the source-level node, with two states, F, the suspect is the source of the crime sample, $\bar{F}$, the suspect is not the source of the crime sample. Node G is the relevance node, with two states, G, the crime sample was left by the offender, and $\bar{G}$, the crime sample was not left by the offender. Nodes $fpg, fmg, mpg, mmg, spg, smg, bpg, bmg, tpg$ and tmg are gene nodes, with three states, A_1, A_2 and A_x. Nodes sgt, bgt and tgt are genotypic nodes, with six states, A_1A_1, A_1A_2, A_1A_x, A_2A_2, A_2A_x, A_xA_x. Note that f denotes father, p paternal, m mother (if in first place) or maternal (if in second place), s suspect, b brother and t trace (if in first place).

Table 5.4 Conditional probabilities assigned to the node *spg*, the suspect's paternal gene. Nodes *fpg* and *fmg* are the paternal and maternal genes of the suspect's father. In each case, the gene can take one of three forms, A_1, A_2 and A_x.

	fpg:	A_1			A_2			A_x		
	fmg:	A_1	A_2	A_x	A_1	A_2	A_x	A_1	A_2	A_x
spg:	A_1	1	0.5	0.5	0.5	0	0	0.5	0	0
	A_2	0	0.5	0	0.5	1	0.5	0	0.5	0
	A_x	0	0	0.5	0	0	0.5	0.5	0.5	1

the propositions at the crime level, denoted H. This transition is made analogously to the procedure described in Section 4.1: a node G, defined as 'the stain comes from the offender', is adopted in order to account for the uncertainty in relation to the relevance of the recovered stain.

The node probability tables are essentially the same as discussed before, except for those of the nodes representing the suspect's and the brother's parental genes. These nodes do not contain unconditional probabilities as was the case for the network shown in Figure 5.2. In the Bayesian network discussed here, the parental gene nodes of the suspect and the brother are conditioned on the parental genes of the mother or the father. For illustration, the probability table of the node *spg* is now explained in more detail:

- $Pr(spg = A_i \mid fpg = fmg = A_i) = 1$ for $i = 1, 2, x$;
- for $i, j = 1, 2, x$ and $i \neq j$ $Pr(spg = A_i \mid fpg = A_i, fmg = A_j) = Pr(spg = A_j \mid fpg = A_i, fmg = A_j) = 0.5$.

A summary of the probability table of the node *spg* is given in Table 5.4. This probability table is applicable analogously for the nodes *smg*, *bpg*, *bmg*.

Note that the current Bayesian network can readily be adopted to analyse other scenarios, analysis of which would generally be very tedious if approached in a purely arithmetic way. For example, when biological material is available from the suspect's parents (mother, father or both), a Bayesian network as shown in Figure 5.5 could be used. At this point, numerical examples will not be considered. Instead, the discussion will continue in Section 5.4 with the study of further modifications.

5.4 Analysis when the alternative proposition is that a sibling of the suspect left the stain

In Section 5.2 Bayesian networks were discussed for the evaluation of one-trace scenarios in which a sample of the suspect's blood was found to be of the same type as that of a crime stain. Under the assumptions stated there, the value of the likelihood was the inverse of the profile probability of the observed characteristics. For example, the likelihood ratio for a correspondence in the genotype A_1A_2 with the propositions chosen at the source level, is $1/2\gamma_1\gamma_2$, where γ_1 and γ_2 denote, respectively, the individual probabilities of the alleles A_1 and A_2.

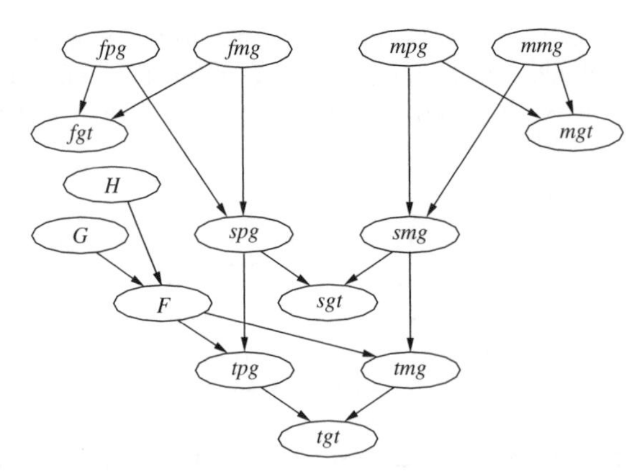

Figure 5.5 Bayesian network for evaluating DNA typing results when genotypic information from the suspect's parents is available. Node H is the crime level node with two states, H_p, the suspect is the offender, H_d, the suspect is not the offender. Node F is the source node, with two states, F, the suspect is the source of the crime sample, $\bar{F}$, the suspect is not the source of the crime sample. Node G is the relevance node, with two states, G, the crime sample was left by the offender, and $\bar{G}$, the crime sample was not left by the offender. Nodes $fpg, fmg, mpg, mmg, spg, smg, tpg$ and tmg are gene nodes, with three states, A_1, A_2 and A_x. Nodes fgt, mgt, sgt and tgt are genotypic nodes, with six states, $A_1A_1, A_1A_2, A_1A_x, A_2A_2, A_2A_x, A_xA_x$. Note that f denotes father, p paternal, m mother (if in first place) or maternal (if in second place), s suspect, and t trace (if in first place).

A modification to this scenario was discussed in Section 5.3 where it was assumed that the individual suspected of leaving the stain was unavailable. For such a scenario, a Bayesian network was constructed and discussed (*see* Figure 5.4), allowing an inference of the suspect's genotype to be made when genotypic information of a sibling of the suspect, for example, the suspect's brother, was available.

In this section yet another scenario will be considered. Imagine a case in which the alternative proposition put forward by the defence is not, as was assumed in the previous scenarios, that another person left the stain, but that a sibling of the suspect, for example, a brother, is the source of the stain. While unrelated individuals are very unlikely to share the same alleles at a certain locus, brothers will share zero, one or two identical alleles. Thus, if the brother is not available[1] for DNA typing, the inferential problem consists of inferring the genotypic probabilities of the brother, given knowledge of the suspect's genotype.

An approach for evaluating such scenarios has been described by Evett (1992). This approach considers a pair of propositions at the source level, which will be denoted F here:

- F: the suspect left the crime stain;
- $\bar{F}$: a brother of the suspect left the crime stain.

There is a crime stain and a suspect, both were found to be of genotype A_1A_2. The numerator of the likelihood ratio considers the probability of this correspondence given that the

[1]For example, a brother may be missing or may refuse to co-operate.

suspect is the source of the stain. Following the development discussed in Section 5.1, this probability can be set to 1. It is less obvious how the denominator may be evaluated, but one may gain a reasonable idea of the order of magnitude of the denominator, if one imagines a case where the alleles A_1 and A_2 are both very rare. Then it is very likely that one of the suspect's parents has a genotype of the kind A_1A_x and the other parent has a genotype of the kind A_2A_x. Again, A_x is notation for the set of alleles $A_3, \ldots, A_n$.

Next, assume that the alleles A_1 and A_2 are so rare that parental configurations such as A_1A_1/A_2A_x, A_1A_2/A_2A_x and so on may be ignored. If the parental genotype configuration is A_1A_x/A_2A_x, then the probability that a brother of the suspect would be of type A_1A_2 is $1/4$. Consequently, one obtains a likelihood ratio of 4. Notice however, that the likelihood ratio is in fact slightly less than 4, since the probability of parental genotype configuration being A_1A_x/A_2A_x is not exactly 1.

Two principal issues must be considered in the construction of a Bayesian network for this scenario. First, an argument must be constructed allowing for the revision of the probability of the brother's genotype, given knowledge about the suspect's genotype. Basically, this is equivalent to the problem discussed in Section 5.3, where knowledge about a brother's genotype was used to infer something about the suspect's genotype. The corresponding graphical approach (Figure 5.4) may serve as a starting point here.

Secondly, an appropriate graphical structure should enable evaluation of the following two probabilities:

- $Pr(tgt \mid sgt, F)$, that is, the probability of the crime stain's genotype (tgt), given knowledge about the suspect's genotype (sgt) and given that the suspect is the source of the crime stain (F).
- $Pr(tgt \mid sgt, \bar{F})$, that is, the probability of the crime stain's genotype (tgt), given knowledge about the suspect's genotype (sgt) and given that a brother of the suspect is the source of the crime stain ($\bar{F}$).

A graphical structure satisfying these requirements is shown in Figure 5.6. The major differences between this network and the model previously discussed, that is, Figure 5.4, are as follows:

- The crime stain's parental gene nodes (nodes tpg and tmg) are explicitly conditioned on the parental gene nodes of the brother (nodes bpg and bmg). This conditioning is needed for evaluation of the probability of the genotype of the crime stain given the alternative proposition, that is, a brother of the suspect is the source of the stain.
- The propositions addressed are at the source level, represented by the node F. No extension is made to the crime level. However, a node B is adopted as a parental node for F. The node B is binary with states 'yes/no', and is used to represent uncertainty which may exist in cases where it may not be known whether the suspect has in fact a brother.
- The genotype nodes of the parents are explicitly represented: mgt and fgt denote the mother's and father's genotype, respectively. Although the scenario under consideration assumes the parents to be unavailable, these nodes will later be used to evaluate specific cases, notably the aforementioned development by Evett (1992).

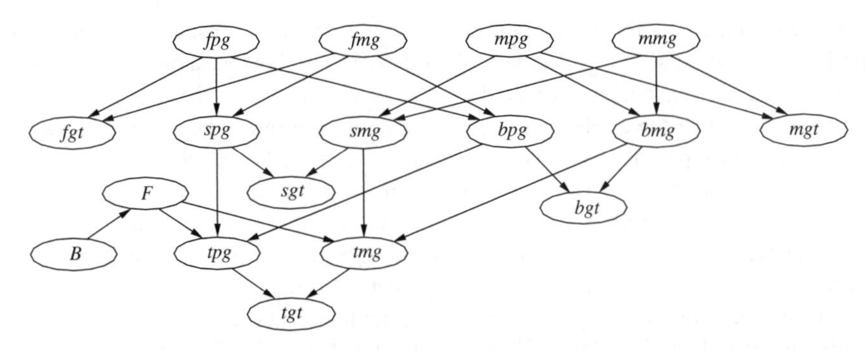

Figure 5.6 Bayesian network for evaluating DNA typing results in a one-stain scenario when the alternative proposition at the source level is that the stain was left by the suspect's brother. Node F is the source node, with two states, F, the suspect is the source of the crime sample, $\bar{F}$, a brother of the suspect is the source of the crime sample. a node B is adopted as a parental node for F. Node B has two states: B, the suspect has a brother, and $\bar{B}$, the suspect does not have a brother. Nodes $fpg, fmg, mpg, mmg, spg, smg, bpg, bmg, tpg$ and tmg are gene nodes, with three states, A_1, A_2 and A_x. Nodes fgt, mgt, sgt, bgt and tgt are genotypic nodes, with six states, A_1A_1, A_1A_2, A_1A_x, A_2A_2, A_2A_x, A_xA_x. Note that f denotes father, p paternal, m mother (if in first place) or maternal (if in second place), s suspect, b brother and t trace (if in first place).

For these structural relationships, the following node probabilities are adopted:

- Nodes tpg and tmg: if the suspect is the source of the crime stain, then the configuration of the parental gene nodes of the crime stain equals that of the suspect: for $i = 1, 2, x$, $Pr(tpg = A_i \mid spg = A_i, F) = 1$ and $Pr(tmg = A_i \mid smg = A_i, F) = 1$. Note that the brother's actual parental gene configuration is irrelevant under F, the suspect being the source of the crime stain. Table 5.5 provides a summary of these assignments. Notice that this table applies analogously to the node tmg also. If a brother of the suspect is the source of the crime stain, then it is the parental gene configuration of this individual that determines the actual state of the parental

Table 5.5 Conditional probabilities assigned to the node tpg, the paternal gene of the crime sample, assuming F to be true, that is, the suspect is the source of the crime stain. Nodes spg and bpg, are paternal gene nodes for the suspect and the brother, with three states, A_1, A_2 and A_x. The relationship is obtained from Figure 5.6.

	F:	F								
	spg:	A_1			A_2			A_x		
	bpg:	A_1	A_2	A_x	A_1	A_2	A_x	A_1	A_2	A_x
tpg:	A_1	1	1	1	0	0	0	0	0	0
	A_2	0	0	0	1	1	1	0	0	0
	A_x	0	0	0	0	0	0	1	1	1

Table 5.6 Conditional probabilities assigned to the node tpg, the paternal gene of the crime sample, assuming $\bar{F}$ to be true, that is, a brother of the suspect is the source of the crime stain. Nodes spg and bpg, are paternal gene nodes for the suspect and the brother, with three states, A_1, A_2 and A_x. The relationship is obtained from Figure 5.6.

	F :	$\bar{F}$								
	spg:	A_x			A_2			A_x		
	bpg:	A_1	A_2	A_x	A_1	A_2	A_x	A_1	A_2	A_x
tpg:	A_1	1	0	0	1	0	0	1	0	0
	A_2	0	1	0	0	1	0	0	1	0
	A_x	0	0	1	0	0	1	0	0	1

gene configuration of the crime stain: for $i = 1, 2, x$, $Pr(tpg = A_i \mid bpg = A_i, \bar{F}) = 1$ and $Pr(tmg = A_i \mid bmg = A_i, \bar{F}) = 1$. Thus, $\bar{F}$ being true, it is the suspect's parental gene configuration that is irrelevant here. Table 5.6 provides a summary of the conditional probabilities applicable for the nodes tpg and tmg given the brother is the source of the crime stain.

- Node F: the probability that the suspect or a brother left the crime stain depends on whether the suspect has in fact a brother. So, if the suspect has in fact a brother, that is, B is true, equal probabilities will be assigned here for $Pr(F \mid B)$ and $Pr(\bar{F} \mid B)$. Notice, that in practice, these values will be assessed in the light of the circumstantial information I. If the suspect has no brother, that is, $\bar{B}$ is true, then certainly the suspect must have left the stain, so $Pr(F \mid \bar{B}) = 1$ and $Pr(\bar{F} \mid \bar{B}) = 0$.
- Nodes fgt and mgt: these are genotype nodes according to the definition provided in Section 5.2. Thus, conditional probabilities as defined in Table 5.3 are applicable.

For illustration, consider again a case where both a crime stain and a suspect were found to be of type A_1A_2. As in the examples discussed in the previous paragraphs, the respective allele probabilities γ_1 and γ_2 are assumed to be 0.05. Figure 5.7 represents this scenario. As the suspect is assumed to have a brother (node B is instantiated), there are equal prior probabilities for both possible outcomes of the variable F, that is, the propositions at the source level. So the ratio of the posterior probabilities displayed in the node F, given the instantiation $tgt = A_1A_2$ and $sgt = A_1A_2$, will equal the likelihood ratio. This situation is shown in Figure 5.7: from the node F it may be found that $78.35/21.65 = 3.62$. A likelihood ratio inferior to 4 has thus been obtained, in agreement with the result suggested by Evett (1992).

The Bayesian network in Figure 5.7 may be used to verify further implications of the approach of Evett (1992), notably the upper and lower limit of the likelihood ratio.

Consider the upper limit first. The rarer the alleles A_1 and A_2, the more likely it becomes that one of the two parents is of type A_1A_x and the other parent is of type A_2A_x. Assuming the parental genotype configuration to be in fact A_1A_x/A_2A_x, then, as mentioned earlier in this section, a likelihood ratio of 4 is obtained. This result may be obtained by use of the Bayesian network shown in Figure 5.7: if, in addition to the instantiations already

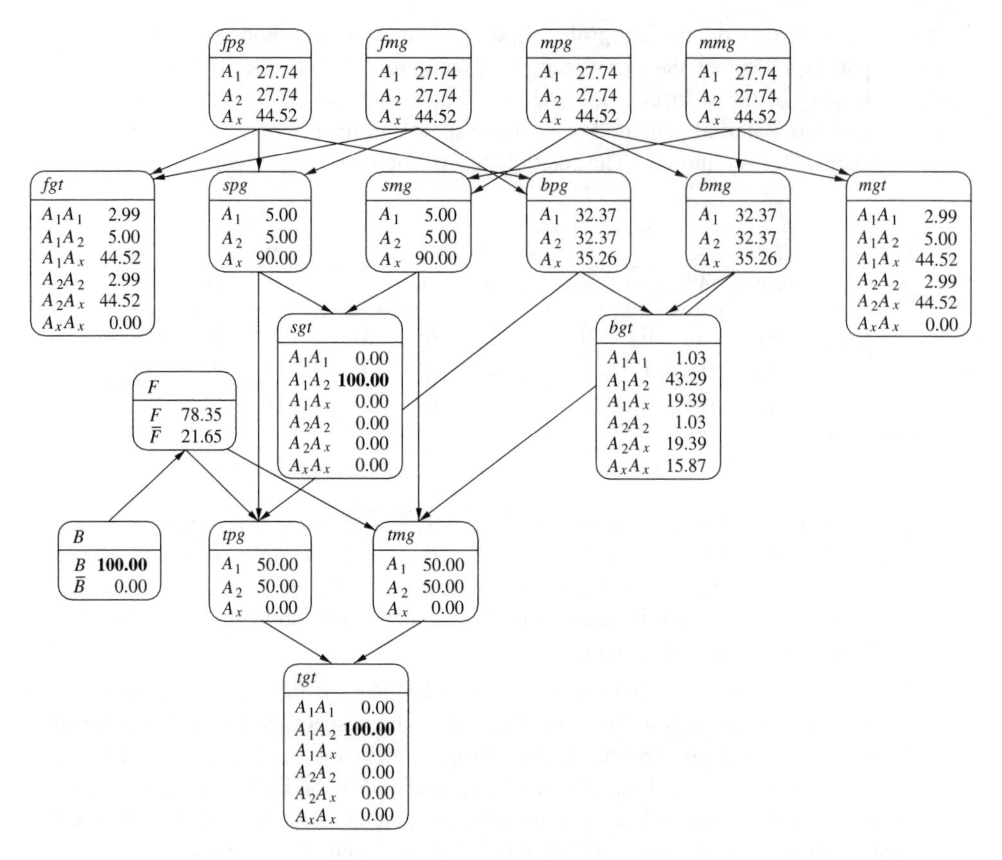

Figure 5.7 Bayesian network for evaluating DNA typing results in a one-stain scenario when the alternative proposition at the source level is that the stain was left by the suspect's brother. The nodes B, tgt and sgt are instantiated, meaning that the suspect is assumed to have a brother and that information is available about the suspect's and the crime stain's genotype. Node F is the source node, with two states, F, the suspect is the source of the crime sample, $\bar{F}$, a brother of the suspect is the source of the crime sample. Node B has two states: B, the suspect has a brother, and $\bar{B}$, the suspect does not have a brother. Nodes $fpg, fmg, mpg, mmg, spg, smg, bpg, bmg, tpg$ and tmg are gene nodes, with three states, A_1, A_2 and A_x. Nodes fgt, mgt, sgt, bgt and tgt are genotypic nodes, with six states, A_1A_1, A_1A_2, A_1A_x, A_2A_2, A_2A_x, A_xA_x. Note that f denotes father, p paternal, m mother (if in first place) or maternal (if in second place), s suspect, b brother and t trace (if in first place).

made, one sets the node fgt to A_1A_x and the node mgt to A_2A_x (or vice-versa), then the node F would display the values 80 and 20 respectively. This corresponds to a likelihood ratio of 4. Notice also that for a parental genotype configuration A_1A_x/A_2A_x, the Bayesian network correctly displays a probability of 1/4 for the brother's genotype being of type A_1A_2.

Now consider the lower limit of the likelihood ratio. Following Evett (1992), the probability of a full sibling's genotype ($fsgt$) being, for example, A_1A_2 may be obtained *via* the following formula:

$$Pr(fsgt = A_1A_2 \mid sgt = A_1A_2) = \frac{1}{Pr(A_1A_2)} \sum_i Pr^2(fsgt = A_1A_2 \mid \phi_i)Pr(\phi_i). \quad (5.4)$$

The variable ϕ_i denotes the genotypic configuration of the parents. Notice that with the two alleles A_1 and A_2 being very frequent, the genotypic configuration of the parents is less likely to contain the allele A_x. For illustration, imagine an extreme case where both A_1 and A_2 have a frequency of 0.5. The possible parental configurations can then be reduced to: A_1A_1/A_2A_2, A_1A_1/A_1A_2, A_1A_2/A_2A_2, and A_1A_2/A_1A_2. Evaluation of (5.4) according to these parameters yields $5/8$, the inverse of which, 1.6, represents the lower limit of the likelihood ratio.

If the Bayesian network discussed above is to be used to verify this result, the unconditional probabilities of the nodes fpg, fmg, mpg and mmg need to be changed so that $Pr(A_1) = Pr(A_2) = 0.5$ and $Pr(A_x) = 0$. Then three queries may be processed. Only the last of these is represented graphically here, due to limitations of space. It is hoped, however, that these queries demonstrate well the flexibility of Bayesian networks and their wide range of possibilities for the evaluation of scenarios.

- Firstly, an evaluation of the probability that a brother of the suspect is of the same type as the suspect, that is, A_1A_2. This is operated by instantiating the suspect genotype node (sgt) to A_1A_2. The Bayesian network would propagate this information and update the probability distribution of the brother's genotype node (bgt). In the case at hand, the probability of node bgt being in state A_1A_2 would increase from 0.5 to 0.625: this is just what was obtained *via* (5.4), that is, $5/8$.
- Secondly, an evaluation of the denominator of the likelihood ratio may be obtained if, in addition to the node sgt, the node representing the proposition at the source level, F, is instantiated to $\bar{F}$. Then the Bayesian network will display the value 0.625 for the trace genotype node (tgt) being in state A_1A_2. The inverse of this represents the likelihood ratio, that is, 1.6.
- An alternative evaluation of the likelihood ratio may be obtained when considering the posterior probabilities of the propositions at the source level (node F). After initialising the Bayesian network, the genotype nodes of the suspect and the crime stain (nodes sgt and tgt) are both instantiated to A_1A_2. In addition, the node B is set to 'true'. Then, the node F displays the posterior probabilities of the propositions at the source level, given the information that both the suspect and the crime stain are of the same genotype, that is, A_1A_2. The value of these two probabilities are 0.6154 and 0.3846. As may be seen, their ratio is 1.6. Figure 5.8 provides a graphical representation of this evaluation.

5.5 Interpretation with more than two propositions

Consider again a situation in which the evidence E consists of a DNA profile of a stain of body fluid found at a crime scene and of the DNA profile from a suspect, which matches

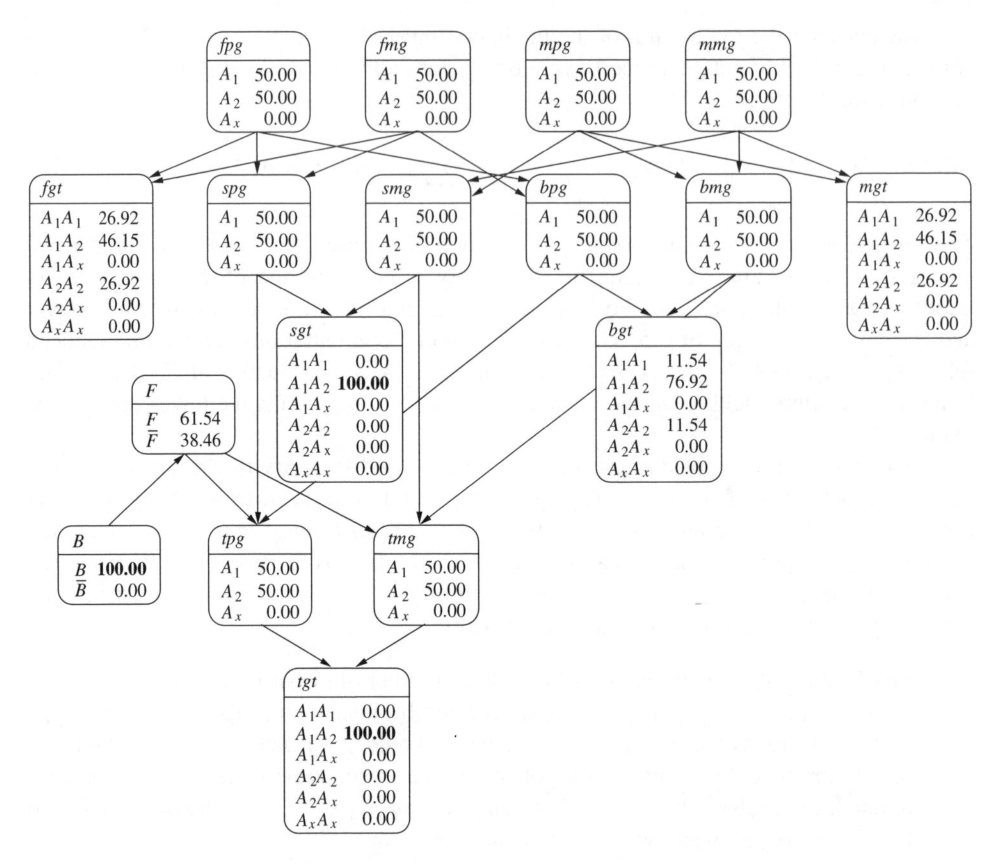

Figure 5.8 Bayesian network for evaluating the lower limit of the likelihood ratio in a one-stain scenario involving DNA typing results. The alternative proposition at the source level is that the stain was left by a brother of the suspect. The allele frequencies for A_1 and A_2 have been set to 0.5 (50%). The nodes B, tgt and sgt are instantiated, meaning that the suspect is assumed to have a brother and that information is available about the suspect's and the crime stain's genotype. Node F is the source node, with two states, F, the suspect is the source of the crime sample, $\bar{F}$, a brother of the suspect is the source of the crime sample. Node B has two states: B, the suspect has a brother, and $\bar{B}$, the suspect does not have a brother. Nodes $fpg, fmg, mpg, mmg, spg, smg, bpg, bmg, tpg$ and tmg are gene nodes, with three states, A_1, A_2 and A_x. Nodes fgt, mgt, sgt, bgt and tgt are genotypic nodes, with six states, A_1A_1, A_1A_2, A_1A_x, A_2A_2, A_2A_x, A_xA_x. Note that f denotes father, p paternal, m mother (if in first place) or maternal (if in second place), s suspect, b brother and t trace (if in first place).

in some sense the crime stain. Until now, such scenarios have been evaluated with respect to *pairs* of propositions H, for example, the suspect is the source of the crime stain (H_p), and, a random member of the population left the crime stain (H_d).

However, it may well be that the relevant population contains close relatives of the suspect. It is important to account for this possibility, as recently reiterated by Buckleton and

Triggs (2005), since close relatives are far more likely to match than other members of the population. The defence may thus require a forensic scientist to incorporate in their analysis more than one proposition. An approach for such a situation, using posterior probabilities, has been discussed by Evett (1992). The following three explanations are considered:

- H_p: the suspect left the crime stain;
- H_{d1}: a random member of the population left the crime stain;
- H_{d2}: a brother of the suspect left the crime stain.

Let θ_0, θ_1 and θ_2 denote the prior probabilities for H_p, H_{d1} and H_{d2}, respectively, so that $\theta_0 + \theta_1 + \theta_2 = 1$. Assume that $Pr(E \mid H_p) = 1$. Denote $Pr(E \mid H_{d1})$ by ϕ_1 and $Pr(E \mid H_{d2})$ by ϕ_2 (notice that ϕ here has a different definition than the one used in the previous section). It is further assumed that H_d, the complement of H_p, is the conjunction of H_{d1} and H_{d2}. The posterior probability of H_p given E can now be written as follows:

$$Pr(H_p \mid E) = \frac{Pr(E \mid H_p)\theta_0}{Pr(E \mid H_p)\theta_0 + Pr(E \mid H_{d1})\theta_1 + Pr(E \mid H_{d2})\theta_2}$$

$$= \frac{\theta_0}{\theta_0 + \phi_1\theta_1 + \phi_2\theta_2} . \tag{5.5}$$

Similarly, the posterior probability of H_d, given E, is given by:

$$Pr(H_d \mid E) = \frac{\phi_1\theta_1 + \phi_2\theta_2}{\theta_0 + \phi_1\theta_1 + \phi_2\theta_2} . \tag{5.6}$$

Hence, the posterior odds in favour of H_p are

$$\frac{\theta_0}{\phi_1\theta_1 + \phi_2\theta_2} . \tag{5.7}$$

Notice that an assessment of the evidence through (5.7) requires prior probabilities to be specified for the explanations put forward by the prosecution and the defence. Notice also that such a specification is not necessary when evaluating the weight of the evidence *via* the likelihood ratio, where prior odds do not enter into consideration.

A feasible way to specify prior probabilities for (5.7) is to consider what may be called a 'pool' of possible suspects. Let the size of this suspect population, including the defendant, be N individuals. Let n denote the number of brothers of the suspect; these brothers are also included in the suspect population. Thus, there are $N - n$ unrelated individuals among the N members of the suspect population. If the other (non-DNA) evidence does not allow any distinction to be drawn between these N individuals, then it can be argued that the prior probability for each and every member of the suspect population is $1/N$. Consequently, $\theta_0 = 1/N$, $\theta_1 = (N - n)/N$ and $\theta_2 = (n - 1)/N$.

Using these figures and assuming $N >> n$, (5.7) yields posterior odds for H_p approximately equal to

$$\frac{1}{\phi_1 N + \phi_2(n - 1)} . \tag{5.8}$$

How can the difficulty of more than two propositions be approached by means of a Bayesian network? Basically, it is possible to reconsider the Bayesian network described in

Section 5.4 (Figure 5.6). This network may be modified in order to account for the specific properties of the problem considered here.

The first modification concerns the definition of the source node F, where F has the same states as H immediately above. The change from H to F is to ensure consistency between Figures 5.8 and 5.9. Instead of having binary states, three states are necessary here:

- F_p: the suspect is the source of the crime stain;
- F_{d1}: a random member of the population is the source of the crime stain;
- F_{d2}: a brother of the suspect left the crime stain.

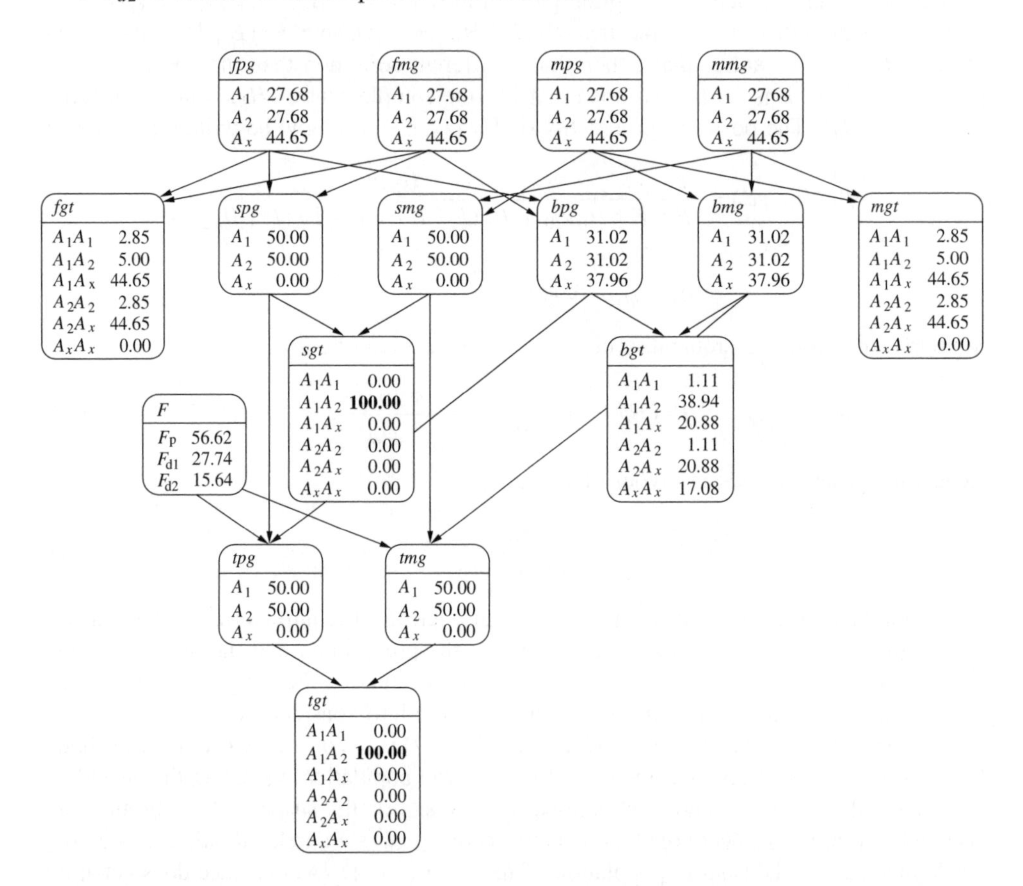

Figure 5.9 Bayesian network for evaluating DNA typing results considering more than one alternative proposition. The allele frequencies for A_1 and A_2 have been set to 0.05. Node F is the source node, with three states, F_p, the suspect is the source of the crime sample, F_{d1}, a random member of the population is the source of the crime stain, F_{d2}, a brother of the suspect left the crime stain. Nodes $fpg, fmg, mpg, mmg, spg, smg, bpg, bmg, tpg$ and tmg are gene nodes, with three states, A_1, A_2 and A_x. Nodes fgt, mgt, sgt, bgt and tgt are genotypic nodes, with six states, A_1A_1, A_1A_2, A_1A_x, A_2A_2, A_2A_x, A_xA_x. Note that f denotes father, p paternal, m mother (if in first place) or maternal (if in second place), s suspect, b brother and t trace (if in first place).

As a consequence of this, more probabilities must be specified for the probability tables of the trace parental gene nodes (tpg and tmg). In particular, it is necessary to assess the probability of each trace parental gene node given that a random member of the population left the crime stain. Thus, in addition to the Tables 5.5 and 5.6 already defined in Section 5.4, probabilities must be added that comply with the following:

$$Pr(tpg = A_i \mid F_{d1}) = \gamma_i, \quad i = 1, 2, x. \tag{5.9}$$

Analogous probabilities apply for the node tmg. Notice that (5.9) holds irrespective of the actual state of the parental gene nodes of the suspect and the brother.

A second modification of Figure 5.6 concerns the node B, which is omitted here for the time being. Consequently, there will be unconditional probabilities that need to be specified for the node F. A possibility for assigning these priors has been outlined above.

Next, consider a numerical example of this Bayesian network. In order to facilitate comparisons with results obtained in earlier Sections, allele probabilities of 0.05 (more generally, γ_1 and γ_2) for both A_1 and A_2, are assumed. The pool of potential suspects is set to one hundred individuals, including the suspect and one brother. Then, following Evett (1992), the prior probabilities obtained for the node F are $Pr(F_p) = 0.01$, $Pr(F_{d1}) = 0.98$, $Pr(F_{d2}) = 0.01$. The remaining probabilities are assessed as outlined above.

The evidence consists of a match between the profile of the crime stain and the profile of the suspect, both of which are found to be of type A_1A_2. This evidence is communicated to the model by instantiating the nodes tgt and sgt. The corresponding state of the Bayesian network is shown in Figure 5.9. The allele frequencies A_1 and A_2 have been set equal to 0.05. These frequencies are not shown in Figure 5.9 because it displays a propagated network and not an initialised Bayesian network. A Bayesian network in a propagated state no longer shows the initial frequencies for the parental gene nodes. Node F indicates the posterior probabilities of the various competing hypotheses at the source level. For example, the probability that the suspect is the source of the crime stain, given the evidence, is 0.5662.

The same result can be obtained from (5.5):

$$\begin{aligned} Pr(H_d \mid E) &= \frac{\theta_0}{\theta_0 + \phi_1\theta_1 + \phi_2\theta_2} \\ &= \frac{0.01}{0.01 + 0.98 \times 0.005 + 0.01 \times 0.2763} \approx 0.5662\ . \end{aligned} \tag{5.10}$$

The value of ϕ_1, *i.e.* 0.005, is given by $2\gamma_1\gamma_2$. Notice that it may also be read off the initialised Bayesian network: before any evidence is entered, all genotype nodes (sgt, bgt, tgt, fgt, mgt) display 0.005 for the state A_1A_2.

There is a second parameter in the denominator that may require further explanation: the probability of the evidence given that a brother of the suspect is the source of the crime stain, that is, ϕ_2. This parameter depends directly on the probability of a brother sharing the same characteristics as the matching suspect. In order to express the idea that a brother is more likely to match than an unrelated member of the suspect population, scenarios proposed in the literature have, for example, assumed a value of 10^{-2} for a brother versus 10^{-6} for each of the (unrelated) individuals of the suspect population (Balding 2000).

Such a subjective estimate can be avoided when working with a Bayesian network. Here, ϕ_2 can be inferred directly from available knowledge about the suspect's genotype.

Notably, the probability of the crime stain being of type A_1A_2, given that a brother of the suspect is the source of the stain (F instantiated to F_{d2}) and given that the suspect is of type A_1A_2, may be processed as a query. The value obtained is 0.2763.

5.6 Evaluation of evidence with more than two propositions

In the previous section, the interpretation of DNA evidence has been extended to more than two propositions. This has led to consideration of posterior odds in favour of one of the specified propositions. However, when assessing scientific evidence, working with likelihood ratios is preferable. As likelihood ratios are used to compare propositions in pairs, some meaningful procedure is required to combine propositions, when more than two of them need to be considered.

An approach for comparing more than two propositions has been described by Aitken and Taroni (2004). Consider a number n of competing exclusive propositions $H_1, \ldots, H_n$. Let E denote the evidence to be evaluated under each of the n propositions, and consider $Pr(E \mid H_i)$, with $i = 1, \ldots, n$. If the prior probabilities of each of the propositions H_i are available, the ratio of the probability of E given each of the pair of competing propositions H_1 and $\bar{H}_1 = (H_2, \ldots, H_n)$ can be evaluated as follows:

$$\frac{Pr(E \mid H_1)}{Pr(E \mid \bar{H}_1)} = \frac{Pr(E \mid H_1)\{\sum_{i=2}^{n} Pr(H_i)\}}{\sum_{i=2}^{n} Pr(E \mid H_i)Pr(H_i)} . \tag{5.11}$$

If one were to apply this approach to the scenario discussed in Section 5.5, the source-level propositions

- F_p: the suspect is the source of the crime stain;
- F_{d1}: a random member of the population is the source of the crime stain;
- F_{d2}: a brother of the suspect left the crime stain.

could be combined to

- H_p: the suspect is the source of the crime stain;
- H_d: the suspect is not the source of the crime stain, that is, either a random member of the suspect population or a brother of the suspect is the source of the crime stain.

With E denoting the results of the DNA typing analysis, the following form of the likelihood ratio can thus be obtained:

$$\frac{Pr(E \mid H_p)}{Pr(E \mid H_d)} = \frac{Pr(E \mid F_p)\{Pr(F_{d1}) + Pr(F_{d2})\}}{Pr(E \mid F_{d1})Pr(F_{d1}) + Pr(E \mid F_{d2})Pr(F_{d2})} . \tag{5.12}$$

Using the values defined in Section 5.5, (5.12) yields:

$$\frac{Pr(E \mid H_p)}{Pr(E \mid H_d)} = \frac{1 \times \{0.98 + 0.01\}}{0.005 \times 0.98 + 0.2763 \times 0.01} \approx 129 . \tag{5.13}$$

How can the Bayesian network described in Section 5.5 (Figure 5.9) be modified in order to account for the approach considered here? Basically, a node, say H, that regroups the various competing propositions, needs to be adopted. As H will be used to evaluate a likelihood ratio, let H have the binary states H_p and H_d as described above. Considering H_p to be true whenever the suspect is in fact the source of the crime stain (F_p), and considering H_d to be true only if either a random member of the suspect population (F_{d1}) or a brother of the suspect (F_{d2}) is the source of the crime stain, the conditioning $F \rightarrow H$ can be adopted. The node probabilities assigned to H are described in Table 5.7.

Notice that from this it follows that the prior probability of H_p equals the prior probability assigned to F_p, whereas the prior probability of H_d is the sum of the prior probabilities assigned to F_{d1} and F_{d2}.

Figure 5.10(i) shows an evaluation of the numerator of the likelihood ratio. The suspect is known to be of type A_1A_2, the node sgt is thus instantiated to A_1A_2. Under the

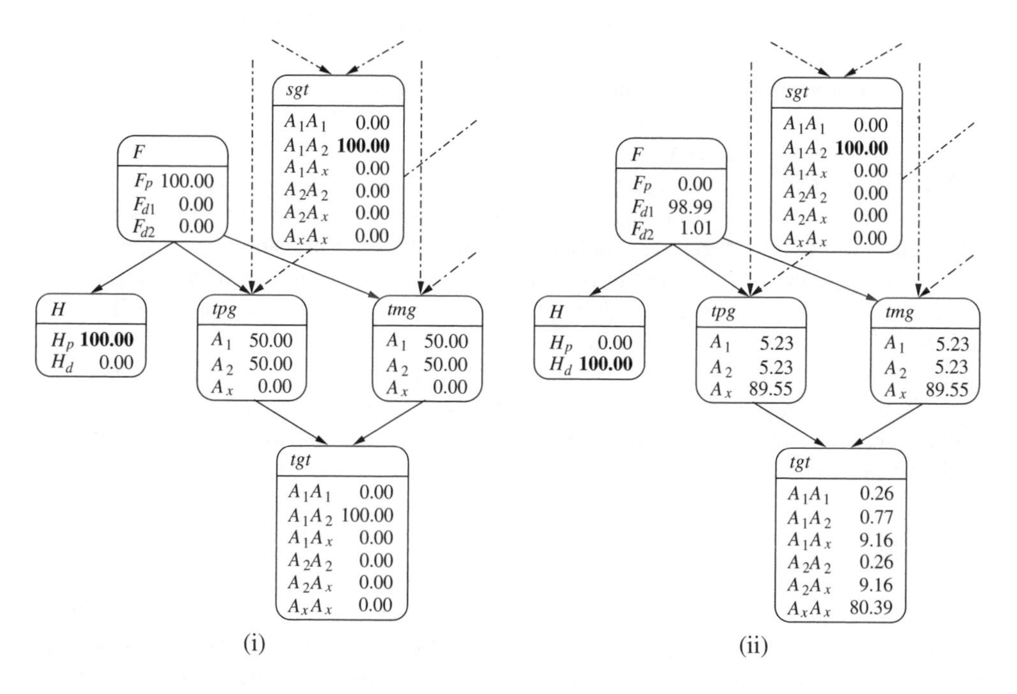

Figure 5.10 Partial representations of the Bayesian network shown in Figure 5.9, including an extension to a node H, regrouping the various source-level propositions of the node F such that $H_p = F_p$, the suspect is the source of the crime stain, and H_d is the union of F_{d1} and F_{d2}, either a random member of the suspect population or a brother of the suspect is the source of the crime stain. Evaluation of the numerator (i) and the denominator (ii) of the likelihood ratio follow from instantiation of H_p in (i) and of H_d in (ii). Instantiation has also been made at sgt. The allele frequencies for A_1 and A_2 have been set to 0.05. Node F is the source node, with three states, F_p, the suspect is the source of the crime sample, F_{d1}, a random member of the population is the source of the crime stain, F_{d2}, a brother of the suspect left the crime stain. Nodes tpg and tmg are gene nodes, with three states, A_1, A_2 and A_x. Nodes sgt and tgt are genotypic nodes, with six states, A_1A_1, A_1A_2, A_1A_x, A_2A_2, A_2A_x, A_xA_x. Note that p denotes paternal, m denotes maternal, s denotes suspect and t denotes trace (if in first place).

Table 5.7 Conditional probabilities assigned to the node H (*see* also Figure 5.10) such that $H_p = F_p$, the suspect is the source of the crime stain, and H_d is the union of F_{d1} and F_{d2}, either a random member of the suspect population or a brother of the suspect is the source of the crime stain.

	F :	F_p	F_{d1}	F_{d2}
H :	H_p	1	0	0
	H_d	0	1	1

assumption that the suspect is the source of the crime stain, H_p, the probability of the crime stain being A_1A_2 can be found to be 1 (node tgt).

An evaluation of the denominator of the likelihood ratio is provided by Figure 5.10(ii). Here, the node H is instantiated to H_d and the node tgt indicates the probability of the crime stain being of type A_1A_2, that is, 0.0077. The likelihood ratio is thus approximately 129 (*i.e.*, 1/0.0077) and is in agreement with the result obtained *via* (5.13).

Notice that the Bayesian network described in this section is useable for addressing a wide range of different scenarios simply by changing the prior probabilities assigned to the node F. For example, if it is not an issue that a brother of the suspect may be a potential source of the crime stain, then the prior probability assigned to F_{d2} is 0 and the Bayesian network yields a likelihood ratio equal to $1/2\gamma_1\gamma_2$, where γ_1 and γ_2 are the frequencies of the alleles A_1 and A_2 respectively. On the other hand, if the suspect and a brother are the only possible sources of the crime stain, then F_{d1} would be set to 0 and the Bayesian network would yield the same results as described in Section 5.4: the likelihood ratio falls slightly below 4.

5.7 Partial matches

So far, Bayesian networks have been constructed and discussed for situations in which there is a match between the suspect's genotype (sgt) and that of a crime stain (tgt). The primary aim of the proposed models was to assess the strength of the link between the suspect and the crime stain by means of a likelihood ratio. As was seen in Section 5.2, for a fairly general one-stain one-offender case, the likelihood ratio reduces, under certain assumptions, to the inverse of the profile frequency. Several variations of this scenario have also been investigated. In Section 5.4, for example, changes in the value of the likelihood ratio have been studied for scenarios in which the alternative proposition was that a sibling of the suspect was the source of the crime stain. In Section 5.6, the population of potential sources of a crime stain was allowed to cover individuals related and unrelated to the suspect.

In this section, scenarios will be addressed in which there is no full match between the suspect's characteristics and those of a crime stain. In such situations, the suspect is usually considered as excluded as the donor of the crime stain. However, as pointed out by Sjerps and Kloosterman (1999), there can be situations in which the two non-matching DNA profiles suggest that a close relative of the suspect might match the crime stain. This may be the case, for example, when the two non-matching DNA profiles share several very rare alleles. One could also imagine database searches focusing on individuals that have corresponding alleles in several loci.

Consider a case involving a crime stain and a non-matching reference sample provided by a suspect. For the time being, the discussion is restricted to a single locus with possible alleles A_1, A_2 and A_x, where the latter is a variable that covers all possible outcomes other than A_1 and A_2. Let the genotype of the crime stain (tgt) be A_1A_1, for instance. Assume further that the suspect's genotype (sgt) is found to be A_1A_2. The pair of propositions under which these observations are evaluated is defined as follows:

- H_p: the crime stain comes from the suspect's brother;
- H_d: the crime stain comes from an unrelated individual.

The likelihood ratio is

$$V = \frac{Pr(sgt = A_1A_2, tgt = A_1A_1 \mid H_p)}{Pr(sgt = A_1A_2, tgt = A_1A_1 \mid H_d)} . \tag{5.14}$$

This can be rewritten as:

$$V = \frac{Pr(tgt = A_1A_1 \mid H_p)}{Pr(tgt = A_1A_1 \mid H_d)} \times \frac{Pr(sgt = A_1A_2 \mid tgt = A_1A_1, H_p)}{Pr(sgt = A_1A_2 \mid tgt = A_1A_1, H_d)} . \tag{5.15}$$

In the absence of knowledge about the genotype of the suspect and that of an eventual brother, the probability of the trace genotype being of type A_1A_1 is the same given H_p and H_d. Thus, the first term on the right-hand side of (5.14) is 1. If the crime stain has in fact been left by the suspect's brother and the trace genotype is A_1A_1, then the genotype of the suspect's brother (bgt) must be of type A_1A_1. Therefore, the numerator of the second term on the right-hand side of (5.15) can be expressed, more shortly, as $Pr(sgt = A_1A_2 \mid bgt = A_1A_1)$. The denominator can be reduced to $Pr(sgt = A_1A_2)$ if an absence of correlation is assumed between the DNA profile of the donor of the crime stain and that of the suspect. The likelihood ratio thus can be approximated by (Sjerps and Kloosterman 1999):

$$V \approx \frac{Pr(sgt = A_1A_2 \mid bgt = A_1A_1)}{Pr(sgt = A_1A_2)}. \tag{5.16}$$

This result can be tracked in several different ways. One possibility would be to reconsider the Bayesian network studied in Section 5.5 (Figure 5.9). In order for this model to be used here, the probability of the state F_p, defined as 'the suspect is the source of the crime stain', is set to zero. The remaining states of the node F, namely, F_{d1} ('a random member of the population is the source of the crime stain') and F_{d2} ('a brother of the suspect left the crime stain'), reflect the propositions that are of interest in the current scenario. Allele frequencies γ_{A_1} and γ_{A_2} are once again set to 0.05. Assuming equal prior probabilities for F_{d1} and F_{d2}, the following can be considered:

- If the node tgt is set to A_1A_1 and the node sgt to A_1A_2, then the posterior probabilities for F_{d1} and F_{d2} are 0.16 and 0.84 respectively. This corresponds to a likelihood ratio of 5.25 in support of a brother of the suspect being the donor of the crime stain.

- The numerator and denominator of (5.16) can also be evaluated separately. If bgt is instantiated to A_1A_1, then the probability of sgt being A_1A_2 is 0.02625. Initially, the probability of sgt being of type A_1A_2 is given by $2\gamma_{A_1}\gamma_{A_2} = 0.005$. The quotient of these two values is

$$0.02625/0.005 = 5.25.$$

An alternative Bayesian network for evaluating a partial match scenario is shown in Figure 5.11, where consideration is also given to the possibility that a son of the suspect is the offender. The principal features are the following:

- A node H is used to evaluate the probability of a brother (H_{p1}), a son (H_{p2}) or an unrelated individual (H_d) being the source of the crime stain.

- Knowledge about the suspect's genotype (sgt) is used to adjust the genotypic frequencies of the suspect's brother (bgt) and son ($son - gt$).

Consider again a scenario in which only two hypotheses are retained: either a brother of the suspect or an unrelated individual are the source of the crime stain. The probability of H_{p2} is thus set to zero. Assigning equal prior probabilities to H_{p1} and H_d and using allele frequencies of $\gamma_{A_1} = \gamma_{A_2} = 0.05$, the Bayesian network yields the same results as those mentioned above:

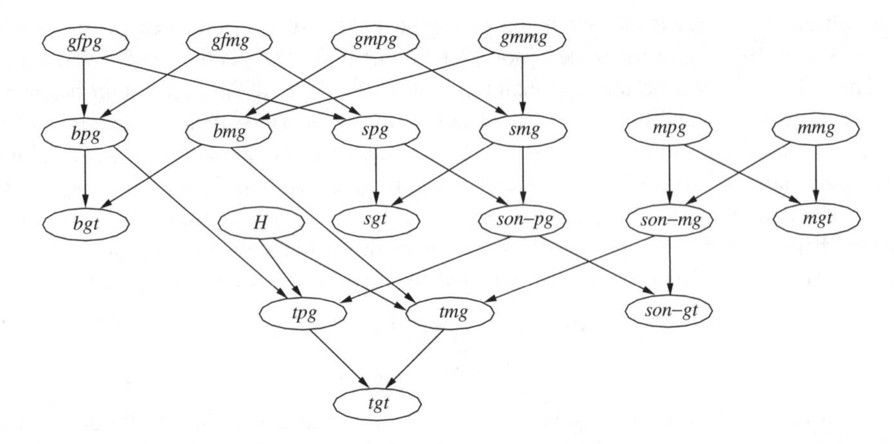

Figure 5.11 Bayesian network for evaluating scenarios in which the suspect's genotype does not match that of a crime stain. Close relatives of a non-matching suspect, that is, a brother or a son, are modelled as potential donors of the crime stain. Nodes are $gfpg$, $gfmg$, grandfather's paternal and maternal gene, $gmpg$, $gmmg$, grandmother's paternal and maternal gene, bpg, bmg, bgt, brother's paternal and maternal gene and genotype, spg, smg, sgt, suspect's paternal and maternal gene and genotype, mpg, mmg, mgt, mother's paternal and maternal gene and genotype, $son - pg$, $son - mg$, $son - gt$ refer to a son of the suspect for paternal gene, maternal gene and genotype, tpg, tmg, tgt, trace paternal and maternal genes and genotype. Node H has three states, H_{p1}, a brother of the suspect is the source of the crime stain, H_{p2}, a son of the suspect is the source of the crime stain, H_d, an unrelated individual is the source of the crime stain.

Table 5.8 Posterior probabilities of the major propositions, given genotypic information on the suspect, *sgt*, the crime stain, *tgt* and the son's mother, *mgt* ('*n.a.*' stands for 'genotypic information not available'). Node H has three states, corresponding to the three propositions, a brother, a son or an unrelated individual is the source of the crime stain.

	sgt:	A_1A_2	A_1A_2	A_1A_2	A_1A_2
	tgt:	A_1A_1	A_1A_1	A_1A_1	A_1A_1
	mgt:	*n.a.*	A_1A_1	A_1A_2	A_2A_2
H :	*brother*	0.323	0.025	0.050	0.840
	son	0.615	0.970	0.941	0.000
	unrelated	0.062	0.005	0.009	0.160

- If sgt is instantiated to A_1A_2 and tgt to A_1A_1, then the posterior probabilities for H_{p1} and H_d are 0.84 and 0.16 respectively. This corresponds to a likelihood ratio of 5.25.
- Evaluating the numerator gives $Pr(sgt = A_1A_2 \mid bgt = A_1A_1)$ as 0.02625, whereas the denominator, $Pr(sgt = A_1A_2)$, is given as 0.005.

Whenever a non-zero prior probability is used for H_{p2}, a son of the non-matching suspect being the donor of the crime stain, the network can be used to discriminate between three potential donors, a situation not considered by (5.16).

Another possibility for the use of the proposed network is the assessment of the potential of genotypic information of other close relatives to achieve further discrimination between the main propositions. For the purpose of illustration, consider again a scenario where the propositions of interest involve a brother, a son and a person unrelated to the partial matching suspect. One of the questions of interest may now be whether the analysis of the genotype of the son's mother, for example, could add relevant information that leads to changes in the posterior probabilities of the three main propositions.

Assume again a locus described by alleles A_1, A_2 and A_x with $\gamma_1 = \gamma_2 = 0.05$. From Table 5.8 it can be seen that there are situations in which the posterior probabilities of the major propositions can change given information on the genotype of the son's mother.

5.8 Mixtures

The previous sections have dealt with scenarios involving evidential material assumed to originate from a single individual. Here the analysis is extended to cases where the crime stain contains material from more than one contributor. The occurrence of mixed samples may be more probable in some cases than in others, according to particular circumstances. For example, in cases of rape and other physical assaults, evidence recovered on a victim may contain material from both the victim and assailant(s).

Recall that for the kind of genetic markers considered here, an individual has at most two alleles. So, the presence of more than two alleles is usually taken as a clear indication that the sample contains biological material from more than one individual.

The inferential complexity of evaluating such scenarios consists of regrouping the various competing propositions that may be formulated and the number of combinations of genotypes that enter into consideration.

The approach discussed here will assume, as was done in the previous sections, independence of an individual's alleles both within and across markers, that is, absence of any sub-population effects. In addition, all contributors to the mixed stain are considered to be unrelated to each other. Therefore, when assessing the denominator of the likelihood ratio, the probability of the evidence is not conditioned on the already observed genotypes, for example, those of the suspect(s), (Harbison and Buckleton 1998). Information pertaining to peak areas and heights (*see*, for example, Evett et al. 1998a or Buckleton et al. 2004) is not taken into account.

5.8.1 A three-allele mixture scenario

Consider an alleged rape. A vaginal swab is taken and submitted for laboratory analysis. Consider the DNA typing results for a hypothetical marker for which three alleles have been found, for example, A_1, A_2 and A_3. Samples are also available from the victim and a suspect. The victim was found to be heterozygous A_1A_3 and the suspect is homozygous A_2.

Assume that, according to information provided by the victim, the number of contributors can be restricted to two individuals, the victim herself and the assailant. Assume further that the propositions put forward by the prosecution and the defence are as follows:

- H_p: the crime stain contains DNA from the victim and the suspect;
- H_d: the crime stain contains DNA from the victim and an unknown individual.

Denote the genotype of the victim and the suspect by, respectively, vgt and sgt. The profile of the mixed crime sample will be abbreviated csp. The likelihood ratio can then be formulated as follows:

$$\begin{aligned} V &= \frac{Pr(csp, vgt, sgt \mid H_p, I)}{Pr(csp, vgt, sgt \mid H_d, I)} \\ &= \frac{Pr(csp \mid vgt, sgt, H_p, I)}{Pr(csp \mid vgt, sgt, H_d, I)} \times \frac{Pr(vgt, sgt \mid H_p, I)}{Pr(vgt, sgt \mid H_d, I)} . \end{aligned} \tag{5.17}$$

Considering the victim's and the suspect's genotype to be independent from whether the suspect is or is not a contributor to the mixed stain, (5.17) can be reduced to:

$$V = \frac{Pr(csp \mid vgt, sgt, H_p, I)}{Pr(csp \mid vgt, sgt, H_d, I)} . \tag{5.18}$$

Proposition H_d assumes the victim and a person other than the suspect, that is, the true offender, are contributors to the mixed stain. Assuming independence between the genotypes of the suspect and the true offender, sgt can be omitted in the denominator. The likelihood ratio is now

$$V = \frac{Pr(csp \mid vgt, sgt, H_p, I)}{Pr(csp \mid vgt, H_d, I)} . \tag{5.19}$$

If H_p is true, $vgt = A_1A_3$ and $sgt = A_2$ then the mixed sample will contain the three alleles A_1, A_2 and A_3. A probability of 1 can then be assigned to the numerator.

For the denominator one needs to consider that the second contributor, an unknown person, may possess one of the following genotypes: A_1A_2, A_2A_2 or A_2A_3, but not A_2A_x for $x \neq 1, 2, 3$ as A_x would then appear in the crime stain profile. Denote these genotypes U_i with

$$U_1 = A_1A_2,\ U_2 = A_2A_2,\ U_3 = A_2A_3\ .$$

Extending the conversation to U and assuming independence between the genotypes of the victim and suspect, the denominator can be written:

$$Pr(csp \mid vgt, H_d, I) = \sum_i Pr(csp \mid vgt, U_i, H_d, I)Pr(U_i \mid H_d, I)\ . \tag{5.20}$$

Each of the genotypes U_i, combined with that of the victim, that is, A_1A_3, would produce an allelic configuration as observed on the crime stain. Thus, again ignoring peak heights, $Pr(csp \mid vgt, U_i, H_d, I) = 1$ for $i = 1, 2, 3$. Consequently, the denominator reduces to:

$$Pr(csp \mid vgt, H_d, I) = \sum_i Pr(U_i \mid H_d, I)\ . \tag{5.21}$$

Assuming that the possible genotypes of the unknown individual do not depend on H_d, the probabilities of the U_i are given by the products of the respective allele frequencies.

Let γ_1, γ_2 and γ_3 denote the frequencies of the alleles A_1, A_2 and A_3, respectively. The likelihood ratio can then be written

$$V = \frac{1}{2\gamma_1\gamma_2 + \gamma_2^2 + 2\gamma_2\gamma_3}\ . \tag{5.22}$$

5.8.2 A Bayesian network

When examining the mixture scenario in terms of a Bayesian network, the variables representing the genotype of the suspect (sgt) and the victim (vgt) need to be represented in some way. Also, a certain number of intermediate variables are necessary in order to provide a logical connection to a variable representing the observed crime stain profile (csp). This can be achieved in much the same way as was done before for single stain scenarios.

Notice that the probabilistic approach to mixed stains as described in the previous Section (Section 5.8.1) assumes there to be exactly two individuals contributing to the mixture, that is, the victim and the suspect or an unknown individual; the suspect's profile is known, but not considered in the development under H_d. The mixture can thus be thought of in terms of a combination of two distinct stains, for each of which a specific individual is considered as a potential source.

This point of view can be translated into a Bayesian network by choosing two submodels that are analogous to the graph shown in Figure 5.2 (Section 5.2). These submodels are used to evaluate whether the victim and the suspect are the respective sources of the two stains that the mixture contains. The source nodes are denoted $T1 = v?$ and $T2 = s?$. The two stains are unobserved variables and their genotypes are denoted $T1gt$ and $T2gt$. Further details are given in Figure 5.12.

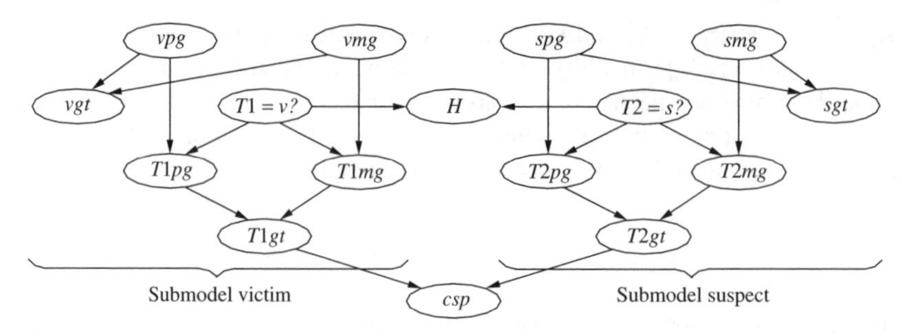

Figure 5.12 Bayesian network for evaluating a scenario where the profile of the crime stain (csp) possibly contains more than two alleles. There are four states (sources) for H, $s\&v$, suspect and victim, $v\&U$, victim and unknown individual, $s\&U$, suspect and unknown individual, $2U$, two unknown individuals. Nodes $T1 = v?$ and $T2 = s?$ take values *yes* or *no* according as, $T1 = v?$, the victim has contributed to the crime stain or, $T2 = s?$, the suspect has contributed to the crime stain. Nodes vpg, vmg, spg and smg denote victim's or suspect's paternal or maternal genes, with vgt and sgt denoting the victim's or suspect's genotypes. Parameters $T1$ and $T2$ denote the two stains which contribute to the common stain profile csp. Nodes $T1pg$, $T2pg$, $T1mg$ and $T2mg$ denote their paternal and maternal genes, with $T1gt$ and $T2gt$ denoting their genotypes. [Reproduced with permission of Academic Press, Ltd.]

The two submodels can then be related to each other by adopting two nodes, H and csp, both of which are logical conjunctions of nodes contained in the submodels. The node H represents a collection of propositions as to how pairs of individuals may contribute to the mixture, that is, the suspect and the victim (s & v), the victim and an unknown individual (v & U), the suspect and an unknown individual (s & U) or two unknown individuals ($2U$). The node probability table associated with H may be completed logically with values 0 and 1 (Table 5.9).

The definition of the node csp, denoting the profile of the mixed stain, is strongly dependant on the level of detail of the model's allele and genotype nodes. In order to keep the sizes of the probability tables moderate, and for the ease of argument, consider a hypothetical marker with only three alleles A_1, A_2 and A_3 and their respective frequencies being γ_1, γ_2 and γ_3. These then are the states of the victim's and the suspect's parental gene nodes (vpg, vmg, spg and smg) as well as the trace parental gene nodes ($T1pg$, $T1mg$, $T2pg$ and $T2mg$). Consequently, there are six states indicated for each of the genotype nodes vgt, sgt, $T1gt$ and $T2gt$: A_1A_1, A_1A_2, A_1A_3, A_2A_2, A_2A_3, A_3A_3. The probability tables of the genotype nodes are completed analogously to the description provided in Section 5.2 (Table 5.3).

Given the states of the genotype nodes as stated above, there are various possible combinations of alleles that a mixture may contain. Restrict the number of states of the crime stain profile node (csp), for the sake of simplicity, to $A_1A_2A_3$ and $\overline{A_1A_2A_3}$. The latter state is a global assignment to all combinations of alleles other than $A_1A_2A_3$. Despite this reduction, the probability table associated to the node csp is still too large to be given here in full detail. However, one may formulate more generally the way in which the table must

Table 5.9 Conditional probabilities assigned to the node H, the contributors to the mixed stain. There are four states (sources) for H, $s\&v$, suspect and victim, $v\&U$, victim and unknown individual, $s\&U$, suspect and unknown individual, $2U$, two unknown individuals. Factors $T1 = v?$ and $T2 = s?$ take values *yes* or *no* according as, $T1 = v?$, the victim has contributed to the crime stain or, $T2 = s?$, the suspect has contributed to the crime stain. [Adapted from Mortera et al. (2003) reproduced with permission from Academic Press.]

	$T1 = v?$:	*yes*		*no*	
	$T2 = s?$:	*yes*	*no*	*yes*	*no*
H :	$s \& v$	1	0	0	0
	$v \& U$	0	1	0	0
	$s \& U$	0	0	1	0
	$2U$	0	0	0	1

be completed. Notably, a probability of 1 is assigned to $Pr(csp = A_1A_2A_3 \mid T1gt, T2gt)$ whenever

- both $T1gt$ and $T2gt$ are heterogeneous and have not more than one allele in common, or
- one of the stains is heterogeneous and the other is homogeneous and the two genotypes have no allele in common.

Otherwise, a zero probability is assigned. Notice that the probability assigned to a state $\overline{A_1A_2A_3}$ of the node csp, given a particular conditioning by $T1gt$ and $T2gt$, is just the complement of the probability assigned to state $A_1A_2A_3$.

Consider a numerical example of the Bayesian network as presented in Figure 5.12. Imagine a scenario as discussed in Section 5.8.1 involving a victim being heterogenous A_1A_3, a suspect being homogenous A_2A_2 and mixed crime stain with alleles $A_1A_2A_3$. The allele frequencies will be chosen as follows: $\gamma_1 = 0.2$, $\gamma_2 = 0.3$ and $\gamma_3 = 0.5$. Equal prior probabilities will be assigned to the source nodes $T1 = v?$ and $T2 = s?$. Notice that the way in which the latter probabilities are chosen is irrelevant for the analysis here. Various instantiations will be made at the node H, and the changes in the evidence node csp recorded. From the particular choice of the (logical) conditional probabilities assigned to the node H one has that, given an instantiation made at the node H, that the source nodes $T1 = v?$ and $T2 = s?$ become either true or false. It can then be seen that the assignment of prior probabilities to these variables is merely a technical matter necessary to run the model; any pair of prior probabilities summing up to 1 may be used.

An evaluation of the numerator of the likelihood ratio may be obtained through the following instantiations: $vgt = A_1A_3$, $sgt = A_2A_2$ and $H = v \& s$. As may be expected, the effect of these instantiations is that the probability of the crime stain being of type $A_1A_2A_3$ is 1 (node csp). The alternative proposition is that the victim and an unknown individual have contributed to the mixed stain. Thus for evaluating the denominator, the state of the node H must be changed to v & U. The change in the conditional H reduces the probability of the crime stain being of type $A_1A_2A_3$ to 0.51. The same results may be obtained via (5.22), the denominator of which yields $2 \times 0.2 \times 0.3 + 0.3^2 + 2 \times 0.3 \times 0.5 = 0.51$.

Notice also that given $H = v$ & U, the actual genotype of the suspect (node sgt) does not affect the probability of the crime stain profile (node csp), a result which is entirely in agreement with the stated assumptions.

In this section, a Bayesian network has been set up for the specific scenario involving a mixed stain with three alleles and assuming exactly two contributors. A Bayesian network with the same structure may also serve to evaluate scenarios involving, for example, a mixed stain covering four alleles, simply by extending the lists of possible alleles involved in the scenario.

The Bayesian network presented here has initially been described in Mortera (2003) and Mortera et al. (2003). In the latter paper the interested reader can find further extensions to issues including more than two contributors, missing individuals or silent alleles. Such technicalities have not been developed here in more detail since the major aim was to familiarise the reader with the idea that basic building blocks, or so-called network fragments, which have separately been set up and validated earlier, can logically be reused in order to approach further complications that the evaluation of DNA profiling results may entail, for example, the occurrence of mixed stains.

5.9 Relatedness testing

The application of graphical probabilistic models, notably Bayesian networks, to inference problems involving the results of DNA typing analysis represents a lively area of research. A particular application is relatedness testing as discussed by Dawid et al. (2002). These authors have shown how appropriate graphical structures for Bayesian networks can be derived from initial pedigree representations of forensic identification problems. In the next section, this will be considered through an example of a classic case of disputed paternity.

5.9.1 A disputed paternity

A certain male, denoted as the putative father pf, is supposed to be the father of a certain child c. Results are available of measurements taken on genetic markers on the mother m, the putative father pf and the child c. The parameter of interest is the likelihood ratio for the putative father pf being the true father tf, noted $tf = pf?$, given knowledge about the child's genotype cgt and the mother's genotype mgt.

The set-up of this scenario can be represented in terms of a basic paternity pedigree as shown in Figure 5.13(i) where squares represent males and circles females. In order to evaluate the scenario in terms of a Bayesian network, Dawid et al. (2002) proposed the use of certain submodels, as shown in Figure 5.14. The network fragment represented by Figure 5.14(i) describes a child's genotype as the result of a pair of parental genes. These parental gene nodes are unobserved variables as one may not know which of the two alleles of a child's genotype has been transmitted by the mother and which one has been transmitted

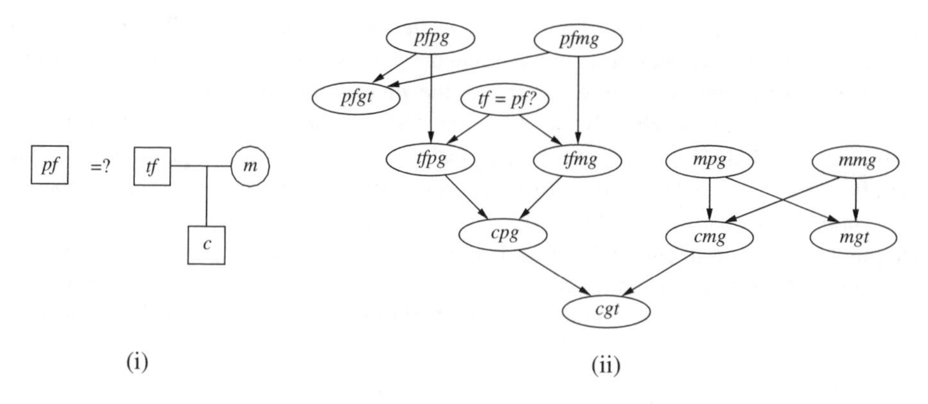

Figure 5.13 Different representations of a case of disputed paternity: (i) paternity pedigree, where squares represent males and circles females; pf denotes the putative father, tf denotes the true father, m denotes the mother and c denotes the child and (ii) a Bayesian network. For the Bayesian network, $tfpg, tfmg, mpg$ and mmg denote the paternal p and maternal m (in second place) genes of the true father tf and the mother m (in first place); cpg and cmg denote the child's paternal and maternal genes, respectively; $pfpg$ and $pfmg$ denote the putative father's paternal and maternal genes, respectively; $pfgt, mgt$ and cgt denote the genotypes of the putative father, the mother and the child, respectively, and $tf = pf?$ takes two values in answer: 'yes' or 'no' as to whether the true father is the putative father. [Reproduced with permission of Blackwell Publishers, Ltd.]

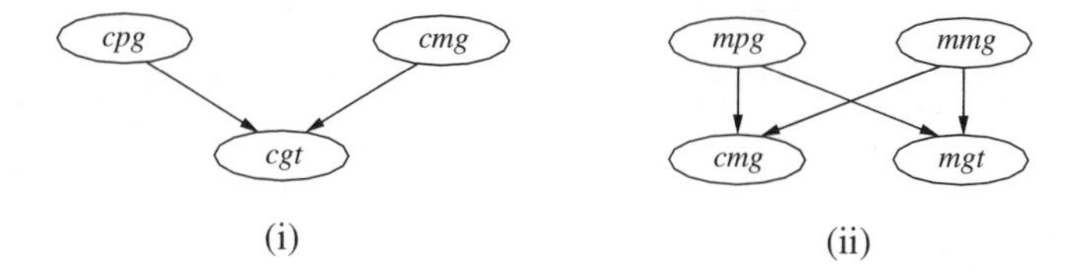

Figure 5.14 Basic submodels useable for (i) inferring a child's paternal and maternal genes (cpg and cmg), based on the child's genotype (cgt), and (ii) inferring a child's maternal gene (cmg), based on the mother's genotype (mgt), and the mother's paternal and maternal genes (mpg and mmg), respectively.

by the father. The submodel shown in Figure 5.14(ii) provides a means for evaluating the uncertainty in relation to which of two parental alleles is transmitted to the offspring.

If related to each other in an appropriate way, such network fragments allow available information on the genotypes of the child, the mother and the putative father to be used to infer whether the true father is the putative father. Dawid et al. (2002) have proposed a structure for a Bayesian network which is as shown in Figure 5.13(ii). If set up properly, this network permits one to obtain the same results as with the classic arithmetic calculus of Essen-Möller under Hardy-Weinberg assumptions of independence.

For illustration, consider a scenario involving a heterozygous child with genotype A_1A_2. The undisputed mother is homozygous A_1A_1 and the putative father is heterozygous A_1A_2.

The numerator of the likelihood ratio is a consideration of the probability of the child's genotype given the genotypes of the mother and the putative father and given the proposition that the putative father is the true father. As the mother will certainly transmit an allele A_1, the probability of the child being A_1A_2 depends on the probability of the true father transmitting the allele A_2, an event with a probability of 0.5. Under the proposition of non-paternity, some unrelated individual from the population is the true father. The probability of this individual contributing the allele A_2 is just given by the allele frequency of that allele, that is, γ_2. The likelihood ratio in favour of paternity thus is given by $1/2\gamma_2$.

Figure 5.15 provides a numerical evaluation of this scenario assuming the allele frequencies $\gamma_1 = \gamma_2 = 0.05$. As equal prior probabilities have been assumed for the target

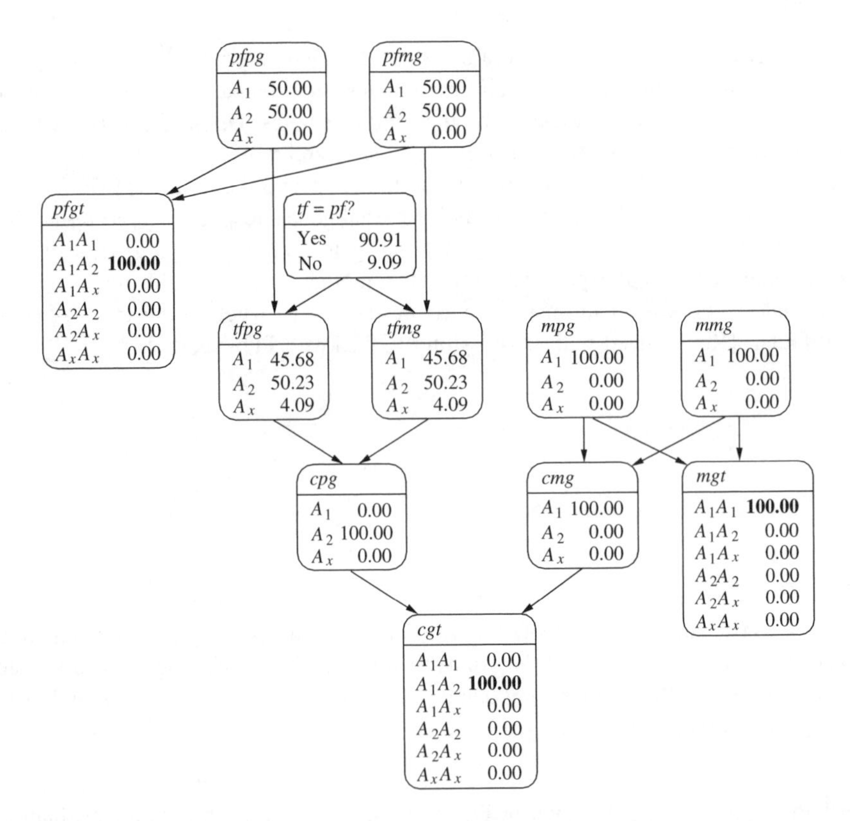

Figure 5.15 Evaluation of a case of disputed paternity. The mother is homozygous $mgt = (A_1A_1)$ whereas the putative father ($pfgt$) and the child (cgt) are both heterozygous A_1A_2. The probability of paternity is displayed in the node $tf = pf$?. Nodes $tfpg, tfmg, mpg$ and mmg denote the paternal p and maternal m (in second place) genes of the true father tf and the mother m (in first place); cpg and cmg denote the child's paternal and maternal genes, respectively; $pfpg$ and $pfmg$ denote the putative father's paternal and maternal genes, respectively; $pfgt, mgt$ and cgt denote the genotypes of the putative father, the mother and the child, respectively, and $tf = pf$? takes two values in answer: 'yes' or 'no' as to whether the true father is the putative father.

node $pf = tf?$, the ratio of its posterior probabilities after instantiating the nodes cgt, mgt and $pfgt$, equals the likelihood ratio. In the case at hand, both by use of the described Bayesian network and the formula developed above, a likelihood ratio of 10 is obtained.

The network represented in Figure 5.15 may also be used for calculating separately the parameters of the likelihood ratio. For evaluating the numerator, the node $tf = pf?$ is set to 'yes' whereas the genotype nodes of the mother and the putative father are set to A_1A_1 and A_1A_2 respectively. The probability of the child's genotype being A_1A_2 can then be read from the node cgt, which would display 0.5. In order to switch to an evaluation of the denominator it is sufficient to change the node $tf = pf?$ to 'no'. For the state A_1A_2, the node cgt would then display, as may be expected since the likelihood ratio is already known to be 10, a probability of 0.05, which is the frequency defined for A_2. Notice also that for the instantiation $tf = pf?$='no', the actual genotype of the putative father no longer affects the probability of the node cgt.

An extension that may be handled by the network represented in Figure 5.15 is the possibility of mutation. Mutation refers to those events that lead to a change in the sequence of the nucleotides of DNA. In relatedness testing, mutation may lead to a constellation in which a child carries an allele different from those carried by its parents. Dawid et al. (2002) proposed an approach for evaluating such seeming exclusions in paternity. In this approach, a difference is made between an individual's 'actual' parental gene and the parental gene 'originally' carried by the respective parent. Figure 5.16 summarises this extension pictorially for an individual's maternally inherited gene. The nodes *mamg* and *mapg* denote, respectively, the mother's 'actual' maternal and paternal gene. Notice that the procedure applies analogously to the paternal line. In Figure 5.16, the nodes *comg* and *camg* denote, respectively, the child's 'original' and 'actual' maternal gene. Mutation rates are considered when assessing probabilities for original genes mutating into actual genes. Further details are given in Dawid et al. (2002) and Dawid (2003).

5.9.2 An extended paternity scenario

In cases of disputed paternity, account has frequently to be taken of further individuals or of the fact that one or more of the principal individuals, such as the alleged father for

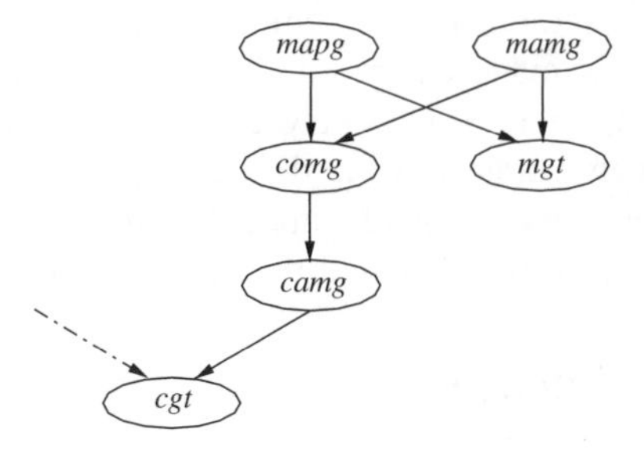

Figure 5.16 Maternal inheritance line in a paternity network with mutation. The nodes *mamg* and *mapg* denote, respectively, the mother's 'actual' maternal and paternal gene. The nodes *comg* and *camg* denote, respectively, the child's 'original' and 'actual' maternal gene. [Adapted from Dawid et al. (2002), reproduced with permission from Blackwell Publishers Ltd.]

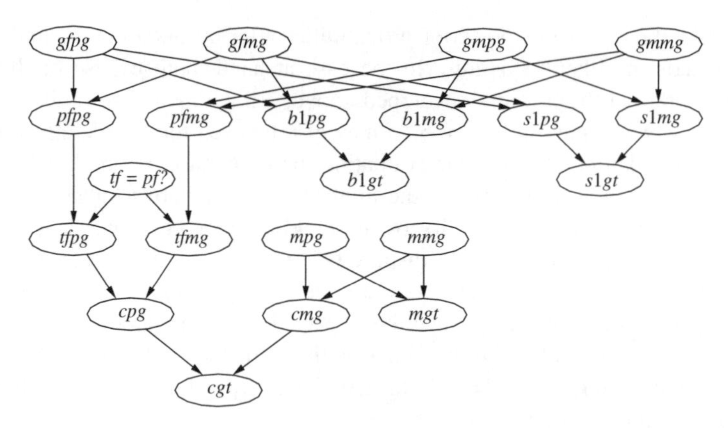

Figure 5.17 Evaluation of a case of disputed paternity where the putative father is not available, but a brother and a sister of the putative father are available. Nodes $gfpg, gfmg, gmpg, gmmg$ denote the grandfather (gf) and grandmother (gm) paternal p and maternal m genes. Nodes $pfpg, pfmg, b1pg, b1mg, s1pg$ and $s1mg$ denote the putative father, pf, the brother $b1$, and sister $s1$ paternal p and maternal m genes. Nodes $tfpg, tfmg, mpg, mmg, cpg$ and cmg denote the true father tf, mother m (in first place) and child c paternal p and maternal m (in second place) genes. Nodes $b1gt, s1gt, mgt$ and cgt denote the brother $b1$, sister $s1$, mother m and child c genotypes. Node $tf = pf$? takes two values: 'yes' if $tf = pf$, the true father is the putative father and 'no' if $tf \neq pf$, the true father is not the putative father. [Adapted from Dawid et al. (2002), reproduced with permission from Blackwell Publishers Ltd.]

example, are unavailable for DNA typing. It is here that the paternity network discussed in Section 5.9.1 provides a valuable basis, as it may be modified and extended as required.

As an example, consider the question whether a certain child is the genetical child of an alleged father. The case is such that the grandmother, the grandfather and the alleged father are deceased. The individuals investigated are a brother and a sister of the alleged father, the child and his mother.

Figure 5.17 provides a pictorial representation of this scenario. The extension to siblings of the alleged father is made using arguments expressed in the submodels shown in Figure 5.14. This extension is also largely analogous to the discussion provided in Section 5.3, where an inference was drawn to the suspect's genotype on the basis of knowledge about a brother's genotype (*see* also Figure 5.4). The probability tables are completed in much the same way as in Section 5.3, and thus are not explained in further detail here.

5.9.3 Y-chromosomal analysis

Forensically relevant polymorphisms may also be found on the male-specific Y chromosome, which now is commonly analysed by many forensic laboratories (Sinha 2003). Y chromosomal markers can be useful when male-specific information needs to be extracted from a sample. As the Y chromosome has no homologue, a male will only show one peak per locus. Y chromosomal analysis may thus be helpful, for example, to determine the number of male contributors to a mixed stain. Potential applications also include deficiency

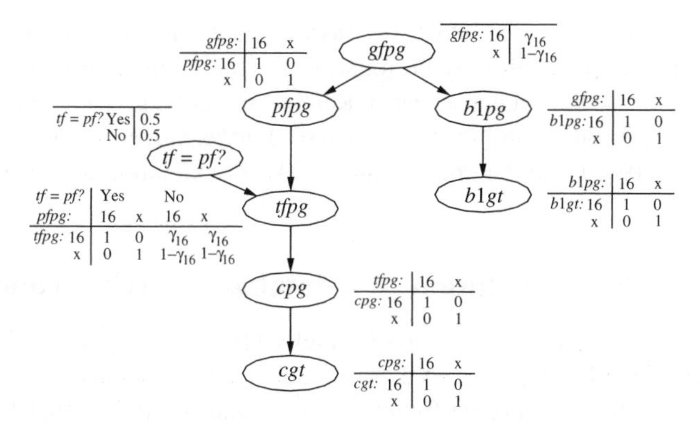

Figure 5.18 Evaluation of DNA profiling results obtained from the analysis of a hypothetical Y-chromosomal marker. The scenario is the same as discussed in Section 5.9.2. Node $gfpg$ denotes the grandfather paternal gene. Nodes $pfpg, tfpg, b1pg$ and cpg, denote the putative father, pf, true father, tf, brother, $b1$ and child c paternal p genes. Nodes $b1gt$ and cgt denote the brother $b1$ and child c genotypes. Node $tf = pf$? takes two values: 'yes' if $tf = pf$, the true father is the putative father and 'no' if $tf \neq pf$, the true father is not the putative father. The state 'x' stands for all outcomes other than 16. The respective conditional node probabilities are shown in tables beside each node, γ_{16} denoting the frequency of the sequence 16.

paternity testing involving male offspring, for example, where the father is not available and inferences are made on the basis of the alleged father's relatives.

As an example, reconsider the scenario discussed in Section 5.9.2 and assume that analyses have been extended to Y chromosomal markers. What is an appropriate structure for a Bayesian network when working with such markers? Basically, one can start with the Bayesian network shown in Figure 5.17 and adapt its structure as required. As the observed characteristic on the Y chromosome is strictly inherited from male predecessors, there is no need to retain nodes representing female parental genes such as cmg and $tfmg$. As a result, one can obtain a Bayesian network as shown in Figure 5.18. The network accounts for typing results obtained for a hypothetical marker and allows the evaluation of a scenario in which the child is found to have the sequence, say, 16 (short for a result of the kind 16-...). The state 'x' stands for all outcomes other than 16. The respective node probabilities are shown besides each node, γ_{16} denoting the frequency of the sequence 16.

From the relationships among the variables as shown in Figure 5.18 and the respective node probabilities one can find that the likelihood ratio in favour of paternity, based on a single marker, is $1/\gamma_{16}$ for a case in which both the child and a brother of the alleged father are found to have the sequence 16.

Notice that similar constructions of Bayesian networks may be used to evaluate profiling results obtained for the maternally inherited mitochondrial DNA.

5.10 Database search

Throughout this chapter it has been assumed that a suspect has been found on the basis of information which is completely unrelated to the crime stain.

A different situation is one in which the suspect has been selected through a search in a database. Different approaches may be found in the literature as to how the value of the evidence may be assessed in such a scenario. Reference will be given here to a probabilistic solution proposed by Balding and Donnelly (1996). Further information may also be found in Evett and Weir (1998), Aitken and Taroni (2004), Buckleton et al. (2004) and Balding (2005).

5.10.1 A probabilistic solution to a database search scenario

In the scenario considered by Balding and Donnelly (1996), a suspect has been found as a result of a search of the DNA profile of a crime stain against a database of N suspects. The suspect's profile was the only profile found to match; that is, all the other $(N - 1)$ profiles contained in the database did not match.

Close examination of this scenario suggests there to be two distinct pieces of information. Let E stand for the *match* which has been observed between the profile crime stain (E_c) and the profile of the suspect (E_s). Let D denote the proposition that the other $(N - 1)$ profiles in the database do not match. Notice that the evaluation of a likelihood ratio incorporating both pieces of information E and D is essentially a problem of combining evidence, a topic which will be discussed in more detail in Chapter 7. Part of this discussion is anticipated here in that a likelihood ratio is formulated for evaluating the combined effect of the two pieces of evidence E and D. It is written as

$$V = \frac{Pr(E, D \mid H_p, I)}{Pr(E, D \mid H_d, I)} . \tag{5.23}$$

Using the product rule, (5.23) can be rewritten as:

$$V = \frac{Pr(E \mid H_p, D, I)}{Pr(E \mid H_d, D, I)} \times \underbrace{\frac{Pr(D \mid H_p, I)}{Pr(D \mid H_d, I)}}_{database\ search\ LR} . \tag{5.24}$$

Following the analysis of Balding and Donnelly (1996), the first ratio on the right-hand side of (5.24) reduces approximately to $1/\gamma$, where γ is the random-match probability. The probability of a match may thus be assumed to be independent of the information that there has been no match among the $(N - 1)$ other individuals contained in the database.

More detailed considerations are required for evaluating the second ratio on the right-hand side of (5.24), a parameter that Evett and Weir (1998) have referred to as the *database search* likelihood ratio. Consider the numerator first. What is the probability that none of the $(N - 1)$ suspects match, given that the suspect is the source of the crime stain (H_p)? Notice that D is not conditioned on any information relating to the observed profile, so the probability of interest is concerned with the potential of the profiling system to differentiate between any two people chosen at random, a property also known as the *discriminating power*. Following a notation used by Evett and Weir (1998), a parameter $\psi_{(N-1)}$ is used to denote the probability that none of the $(N - 1)$ innocent individuals has a genotype matching that of any unspecified crime stain.

For evaluating the denominator of the database search likelihood ratio, one needs to consider that if the suspect is not the source of the crime stain, then necessarily someone else is. Thus, let A denote the proposition that the true source of the crime stain is among the $(N-1)$ other members of the database. The denominator is then

$$Pr(D \mid H_d) = Pr(D \mid A, H_d)Pr(A \mid H_d) + Pr(D \mid \bar{A}, H_d)Pr(\bar{A} \mid H_d) \tag{5.25}$$

If the true offender is among the $(N-1)$ other members of the database, that is, A is true, then the probability that there is no match among the $(N-1)$ profiles can be considered impossible. Accordingly, the product of the first pair of terms in (5.25) is zero. $Pr(D \mid \bar{A}, H_d)$ is the probability that the $(N-1)$ profiles do not match, given that the true source is not among these $(N-1)$ profiles and that the suspect is not the source of the crime stain. Again, this is an estimate that can be made based on $\psi_{(N-1)}$. If ϕ denotes the probability that the true source is among the $(N-1)$, then (5.25) reduces to $\psi_{(N-1)}(1-\phi)$ and the database search likelihood ratio becomes

$$\frac{Pr(D \mid H_p)}{Pr(D \mid H_d)} = \frac{\psi_{(N-1)}}{\psi_{(N-1)}(1-\phi)} = \frac{1}{1-\phi} \, . \tag{5.26}$$

Now, the greater the size of the database, the more likely it becomes that the true source is among the $(N-1)$ individuals, thus ϕ is increasing. Consequently, the denominator of (5.26) diminishes and the database search likelihood ratio increases. The primary result of the analysis thus is that, under the stated assumptions, the fact that the suspect has been selected through a search in a database has a tendency to increase the overall likelihood ratio (5.24).

5.10.2 A Bayesian network for a database search scenario

The probabilistic approach described in Section 5.10.1 considers the three binary variables H, E and D, which are defined as follows:

- H: the suspect (H_p) or some other person (H_d) is the source of the crime stain;
- E: the profile of the crime stain matches that of the suspect;
- D: the other $(N-1)$ profiles of the database do not match.

Notice that $\bar{E}$ and $\bar{D}$ denote the negations of the propositions E and D, respectively. For deriving an appropriate graphical structure relating these three variables it is necessary to inquire about the assumed dependencies among them. Consider thus (5.23) which states that the variables E and D are both conditioned on the variable H. A graphical representation of these assumptions is that H is chosen as a parental variable for, respectively, E and D so that a diverging connection with $E \leftarrow H \rightarrow D$ is obtained (Figure 5.19).

Notice that the Bayesian network shown in Figure 5.19 is an explicit representation of the assumption that, knowing H were true, the probability of a match between the suspect's profile and the profile of the crime stain is not influenced by the fact that there has been a search in a database. The conditional probabilities assigned to the variables E and D are presented in Table 5.10.

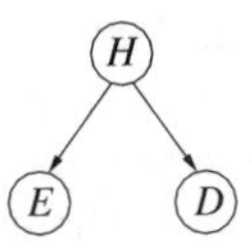

Figure 5.19 Bayesian network for a scenario in which the suspect has been found as a result of a database search. Node H has two states: the suspect (H_p) or some other person (H_d) is the source of the crime stain. Node E has two states: the profile of the crime stain matches or does not match that of the suspect. Node D has two states: all other $(N - 1)$ profiles of the database do not match or one and only one of the other $(N - 1)$ profiles of the database do match. [Reproduced with permission from Elsevier.]

Table 5.10 Probabilities assigned to the variables E and D, conditional on the state of H. Node H has two states: the suspect (H_p) or some other person (H_d) is the source of the crime stain. Node E has two states: the profile of the crime stain matches or does not match that of the suspect. Node D has two states: all other $(N - 1)$ profiles of the database do not match or one and only one of the other $(N - 1)$ profiles of the database do match. The random-match probability is γ. The probability that the true source of the crime stain is amongst the other $N - 1$ profiles is denoted ϕ. The term $\psi_{(N-1)}$ is used to denote the probability that none of the $(N - 1)$ innocent individuals has a genotype matching that of any unspecified crime stain.

	H:	H_p	H_d
E:	E	1	γ
	$\bar{E}$	0	$1 - \gamma$
D:	D	$\psi_{(N-1)}$	$(1 - \phi)\psi_{(N-1)}$
	$\bar{D}$	$1 - \psi_{(N-1)}$	$1 - [(1 - \phi)\psi_{(N-1)}]$

5.11 Error rates

When evaluating DNA evidence, forensic scientists consider whether their findings are what they would expect to observe given specified propositions. For example, if a crime stain is found to have the same profile as a sample provided by a suspect, then this would be something that one may expect to observe if the suspect were in fact the source of the crime stain. However, a forensic scientist would also be required to explain what chance there may be to observe the findings when some alternative propositions were true, for example, that the crime stain comes from an unknown person who is unrelated to the suspect.

If the suspect is assumed not to be the source of the crime stain, then necessarily someone else must be, so one needs to consider the probability that another person would also have a matching profile. This is the random-match probability mentioned several times already. It may be thought of as the probability of a coincidental match, a match of DNA profiles between two different people. This probability has been the main focus of attention for the evaluation of evidence in this chapter. However, there may be other relevant issues to consider when evaluating DNA evidence. One of these other issues is the potential for error.

One such source of error is a false-positive, which may be defined as an event that occurs when a laboratory erroneously reports a match between two samples that actually have different profiles. A false-positive may be because of error in the collection or handling of samples, misinterpretation of test results, or incorrect reporting of results (National Research Council 1992; Thompson 1995; Thompson and Ford 1991).

The effect that the potential of error through a false-positive may have on the value of a reported match has been evaluated in a Bayesian framework by Thompson et al. (2003). Their analysis shows that for the evaluation of DNA evidence, accurate information is not only necessary about the random-match probability but also on the rate of laboratory error.

5.11.1 A probabilistic approach to error rates

Imagine a single stain scenario as described in Section 5.1. Let H denote a pair of propositions at the source level:

- H_p: the specimen came from the suspect;
- H_d: the specimen did not come from the suspect.

The available evidence, denoted R, consists of the forensic scientist's report of a match between the DNA profile of the crime stain and that of a sample provided by the suspect.

In the traditional approach for considering the proposition H in the light of the evidence R, a likelihood ratio is developed as follows:

$$V = \frac{Pr(R \mid H_p, I)}{Pr(R \mid H_d, I)} . \tag{5.27}$$

In order to account for the potential for error, an intermediate variable M is introduced. The variable M designates a true match, and is differentiated from R, a reported match. As noted by Thompson et al. (2003), this distinction is an assumption of two possible underlying states of reality:

- M: the suspect and the specimen have matching DNA profiles;
- $\bar{M}$: the suspect and the specimen do not have matching DNA profiles.

Notice that it is impossible to know with certainty whether M or $\bar{M}$ is true because the only information available about M and $\bar{M}$ is the laboratory report, which might be mistaken. So both M and H are unobserved variables, but their truth-state may be revised based on new information, such as R.

Thus, when conditioning R on M and H, (5.27) can be extended to:

$$V = \frac{Pr(R \mid M, H_p)Pr(M \mid H_p) + Pr(R \mid \bar{M}, H_p)Pr(\bar{M} \mid H_p)}{Pr(R \mid M, H_d)Pr(M \mid H_d) + Pr(R \mid \bar{M}, H_d)Pr(\bar{M} \mid H_d)} . \tag{5.28}$$

Next, the following assumptions may reasonably be made:

- If there is really a match (M), then the probability that a match will be reported (R) is not affected by whether the match is coincidental: $Pr(R \mid M, H_p) = Pr(R \mid M, H_d) = Pr(R \mid M)$.
- If the suspect is the source of the crime stain, then there certainly is a match: $Pr(M \mid H_p) = 1$. Notice that the same assumption has already been made in Section 4.1, where no distinction was made between a true match and a reported match.
- A non-match $\bar{M}$ can only arise under H_d so that $Pr(R \mid \bar{M}, H_d)$ can be reduced to $Pr(R \mid \bar{M})$.

(5.28) can now be written as

$$V = \frac{Pr(R \mid H_p)}{Pr(R \mid H_d)} = \frac{Pr(R \mid M)}{Pr(R \mid M)Pr(M \mid H_p) + Pr(R \mid \bar{M})Pr(\bar{M} \mid H_d)}. \tag{5.29}$$

The conditional probabilities contained in (5.29) may be assessed as follows:

- $Pr(R \mid M)$: this is the probability that a laboratory will report a match if the suspect and the specimen have matching DNA profiles. Provided that samples are adequate in quality and quantity, a competent laboratory may be expected to be unlikely to fail a true match, so that values for $Pr(R \mid M)$ close to 1 appear appropriate. For present purposes, assume that $Pr(R \mid M) = 1$.
- $Pr(M \mid H_d)$: this is the probability of a coincidental match. For a comparison between single-source samples, $Pr(M \mid H_d)$ is the random-match probability, or, the frequency of the matching profile in a relevant reference population, incorporating sub-population effect (Section 5.12). In analogy to the notation used in Section 4.1, the random-match probability will be denoted γ here. Notice that M and $\bar{M}$ are mutually exclusive and exhaustive events, so that $Pr(\bar{M} \mid H_d)$ is the complement of γ, that is, $1 - \gamma$.
- $Pr(R \mid \bar{M})$: this term is the false-positive probability, denoted fpp.

Substituting the terms, the likelihood ratio given by (5.29) takes the following form:

$$V = \frac{1}{\gamma + [fpp \times (1 - \gamma)]}. \tag{5.30}$$

5.11.2 A Bayesian network for error rates

In order to set up an appropriate Bayesian network, the number and definition of nodes, as well as their dependencies, need to be considered carefully. From the probabilistic approach described in the previous section, three binary variables may be defined: H, the specimen comes from the suspect, M, the suspect and the specimen have matching DNA profiles, and R, the forensic scientist's report of a match between the suspect's profile and the profile of the sample.

These three variables then need to be combined in a way so that the dependence and independence properties implied by the structure correspond to one's perception of the problem. In deriving (5.28), it was assumed that $Pr(R \mid M, H_p) = Pr(R \mid M, H_d) = Pr(R \mid M)$, so that the joint probability of the three variables R, M and H could be written as:

$$Pr(R, M, H) = Pr(R \mid M) \times Pr(M \mid H) \times Pr(H) . \qquad (5.31)$$

The appropriate graph structure that correctly represents these dependency assumptions is serial: $H \rightarrow M \rightarrow R$ (Figure 5.20). This can be understood by considering the chain rule for Bayesian networks (Chapter 2), that yields (5.31) for the defined three variables R, M and H, and for which the structural relationships are such as specified in Figure 5.20. The probabilities associated to the nodes M and R are shown in Tables 5.11 and 5.12.

In the Bayesian networks for the DNA likelihood ratio earlier described in Section 5.2, it has been assumed that the evidence node represents the true state of affairs. For example, by instantiating the node E in the simple network $H \rightarrow E$, or, analogously, the nodes sgt and tgt in the network shown in Figure 5.2, one's observations are assumed faithfully to reflect the true state of these parameters. In practice, however, care should be taken to examine the conditions under which such an assumption is acceptable. This is equivalent to the distinction between a proposition and evidence for that proposition.

Figure 5.20 Bayesian network for the problem of error rates. There are three binary nodes: H, the specimen comes or does not come from the suspect, M, the suspect and the specimen have, or do not have, matching DNA profiles, and R, the forensic scientist reports a match or a non-match between the suspect's profile and the profile of the sample. [Adapted from Taroni et al. (2004), reproduced with permission from Elsevier.]

Table 5.11 Conditional probabilities assigned to the node M. Factor H has two states, H_p, the specimen comes from the suspect and H_d the specimen does not come from the suspect. Factor M has two states, M, the suspect and the specimen have matching DNA profiles and $\bar{M}$, the suspect and the specimen do not have matching DNA profiles.

	H:	H_p	H_d
M:	M	1	γ
	$\bar{M}$	0	$1-\gamma$

Table 5.12 Conditional probabilities assigned to the node R. Factor M has two states, M, the suspect and the specimen have matching DNA profiles and $\bar{M}$, the suspect and the specimen do not have matching DNA profiles. Factor R has two states, R, the forensic scientist reports a match and $\bar{R}$ the forensic scientist reports a non-match between the suspect's profile and the profile of the sample.

	M:	M	$\bar{M}$
R:	R	1	fpp
	$\bar{R}$	0	$1 - fpp$

Figure 5.20 illustrates how the Bayesian networks earlier described in Sections 5.2 can be extended to handle the potential of error.

5.12 Sub-population and co-ancestry coefficient

5.12.1 Hardy-Weinberg equilibrium

The value of DNA evidence – at the source level – has been presented using a likelihood ratio which is just the reciprocal of the profile frequency, that is $1/2\gamma_i\gamma_j$ and $1/2\gamma_i^2$ for, respectively, a heterozygote and a homozygote suspect matching a bloodstain found on a crime scene where γ_i and γ_j are appropriate allele frequencies. This approach is based on the assumption that each of the alleles on a locus, one from the father and one from the mother, are independent of each other, which leads to an equilibrium distribution for the relative frequencies of alleles in a population. This is known as *Hardy-Weinberg* equilibrium or *random mating*. The relative frequencies for the results of mating for a locus with two alleles, A and B, with frequencies γ_A and γ_B such that $\gamma_A + \gamma_B = 1$, are

- $AA: \quad \gamma_A^2$,
- $AB: \quad 2\gamma_A\gamma_B$,
- $BB: \quad \gamma_B^2$,

and $\gamma_A^2 + 2\gamma_A\gamma_B + \gamma_B^2 = (\gamma_A + \gamma_B)^2 = 1$. In general, let γ_i and γ_j be the population proportions of alleles A_i and A_j for $i, j = 1, \ldots, k$ where k is the number of alleles at the locus in question. The expected genotypic frequencies P_{ij} are obtained from the equations assuming Hardy-Weinberg equilibrium (5.3).

Although this is expedient, it glosses over some issues and does not accommodate population structure (Balding 2005; Buckleton et al. 2004; Evett and Weir 1998). The assumption of the Hardy-Weinberg law is violated when populations do not mate at random.

5.12.2 Variation in sub-population allele frequencies

Simple calculation of profile frequencies is not sufficient when there are dependencies between different individuals involved in the case under examination. The most common source of dependency is a result of a membership in the same population and the consequential existence of similar evolutionary histories. The mere fact that populations are finite in size means that two people taken at random from a population have a non-zero chance of having relatively recent common ancestors.

A measure F_{ST} of inter-population variation in allele frequencies was introduced by Wright (1922). It can be considered as a measure of population structure.

> The relationship between alleles of different individuals in one sub-population when compared to pairs of alleles in different sub-populations, also known as the *co-ancestry coefficient*, is denoted F_{ST} (Wright 1922).

Extensive studies have been made of allele frequencies in many human populations to estimate values of F_{ST} for forensic use (Balding et al. 1996; Balding and Nichols 1997; Foreman et al. 1998, 1997; Lee et al. 2002).

In Table 5.13, F_{ST} is used to describe the deviation of the genotype frequencies from the Hardy-Weinberg frequencies. When $F_{ST} = 0$, the usual Hardy-Weinberg frequencies γ_i^2, γ_j^2 and $2\gamma_i\gamma_j$, for homozygous and heterozygous profiles, respectively, are obtained.

The formulae presented in Table 5.13 are derived as follows. Take an individual at random from the population. Consider a single locus, X say. The probability that one of the individual's two alleles at X is A is γ_A. The probability that the other allele is also A, given that the first is A, is $\gamma_A(1 - F_{ST}) + F_{ST}$. This probability is composed of two mutually exclusive components. Consider the first component; the second allele is A in the natural course of events and not because of special circumstances. The absence of special circumstances occurs with probability $(1 - F_{ST})$ and the probability that the second allele is A, absent the special circumstances, is γ_A. The probability that the allele is A *and* the special circumstances are absent is the product of these two probabilities and is $\gamma_A(1 - F_{ST})$. Consider the second component; the second allele is A because of the homozygosity due to special circumstances, which occur with probability F_{ST}. The probability of allele A, given special circumstances, is 1 and the probability of the special circumstances is F_{ST}. The probability that the allele is A *and* the special circumstances are present is the product of these two probabilities and is $1 \times F_{ST}$. The sum of the probabilities associated with these two components gives the result for the probability that the second allele is A, given that the first is A.

The joint probability of A for the first and second alleles in the chosen individual is now be found to be $\gamma_A\{(1 - F_{ST})\gamma_A + F_{ST}\} = \gamma_A^2(1 - F_{ST}) + \gamma_A F_{ST}$.

Table 5.13 Genotypic frequencies when Hardy-Weinberg equilibrium is relaxed where F_{ST} denotes the co-ancestry coefficient. Allele frequencies for alleles A_i and A_j are γ_i and γ_j, respectively.

Genotype	A_iA_i	A_iA_j	A_jA_j
Frequency	$\gamma_i^2(1 - F_{ST}) + \gamma_i F_{ST}$	$2\gamma_i\gamma_j(1 - F_{ST})$	$\gamma_j^2(1 - F_{ST}) + \gamma_j F_{ST}$

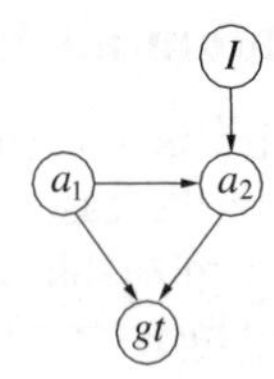

Figure 5.21 A Bayesian network for evaluating genotypic frequencies in situations where an abstraction needs to be made of the assumption of Hardy-Weinberg equilibrium. Node I represents *identical by descent, ibd* and has two states, *ibd* and *not-ibd*. Nodes a_1 and a_2 represent the paternal and maternal alleles, respectively, at the locus of interest and gt denotes the genotype.

5.12.3 A graphical structure for F_{ST}

The genotype of an individual, gt, is derived from the paternal and maternal alleles transmitted; call these alleles a_1 and a_2, respectively. A single locus with two alleles, A and B, is considered. The probability table for the three states of the genotype node gt, notably AA, AB and BB, may be completed logically. See Table 5.3 for an analogous example.

Figure 5.21 presents a graphical structure dealing with genotype frequencies when the Hardy-Weinberg equilibrium is relaxed. An additional node, named I, is introduced. I stands for *identical by descent*, that is, alleles that have descended from a single ancestral allele (Weir 1996). For a review of genetic models, the reader is directed to Graham et al. (2000) and Balding (2003, 2005). Two states are chosen for I, namely, *ibd* and *not – ibd*. Quantification is achieved with the co-ancestry coefficient F_{ST} and its complement, $(1 - F_{ST})$. A value for F_{ST} can be chosen according to the sub-population of interest.

Note that there is a link between the alleles a_1 and a_2. In fact, there are situations in which node a_2 should take into account knowledge about a_1. This is notably the case when allele a_2 is identical by descent with allele a_1. In the latter case, the probability table of the node a_2 contains the logical probabilities 0 and 1, otherwise allele frequencies of the general population are introduced.

Assume that $\gamma_A = 0.8$ and $\gamma_B = 0.2$. The genotype frequencies under Hardy-Weinberg equilibrium are: $P(AA) = 0.64$, $P(AB) = 0.32$ and $P(BB) = 0.04$. The same results can be obtained with the Bayesian network shown in Figure 5.21 if the node I is instantiated to $not - ibd$. This means that no co-ancestry is considered and the population is in Hardy-Weinberg equilibrium. Under this assumption, node a_2 is not influenced by a_1. This is a direct consequence of the way in which the node probabilities of a_2 have been completed, that is, $Pr(a_2 = A \mid a_1, I = not - ibd) = Pr(a_2 = A \mid I = not - ibd) = \gamma_A$. The genotype frequencies displayed in the node gt then are $P(AA \mid I = not - ibd) = 0.64$, $P(AB \mid I = not - ibd) = 0.32$ and $P(BB \mid I = not - ibd) = 0.04$.

When the node ibd is not instantiated, then the proposed Bayesian network will calculate genotypic frequencies in the node gt by taking account of the co-ancestry coefficient. For the purpose of illustration, consider a value of 0.1 for the probability of ibd. The genotypic frequencies in the node gt, with I not instantiated, can then be found to be:

- $P(AA) = 0.656$;

- $P(AB) = 0.288$;
- $P(BB) = 0.056$.

These results are in agreement with the formulae presented in Table 5.13 where F_{ST} is the probability of *ibd*.

Viewing alleles a_1 and a_2 as a sequence of alleles drawn one by one, a formula can be expressed in a recursive form. If, after n alleles have been drawn, n_i have been observed of allele A, then the probability that the next allele sampled is of type A is

$$Pr_{n+1}(A) = \frac{n_i F_{ST} + (1 - F_{ST})\gamma_A}{1 + (n-1)F_{ST}}.$$

By successively applying this equation to each allele in the sequence, and multiplying together the resulting expressions, a formula for the joint likelihood of the entire sample is obtained (Balding and Nichols 1995, 1997).

Assume a genotype of type AB. Each allele in turn is considered, as presented in Harbison and Buckleton (1998). Take allele a_1 to be A first. The recursive equation proposed by Balding and Nichols, with $n_A = 0$ and $n = 0$, that is, no alleles previously being sampled and no allele of type A previously being seen, becomes $(1 - F_{ST})\gamma_A/(1 - F_{ST}) = \gamma_A$.

Next, take allele B: one allele has been sampled but no alleles of type B have yet been seen. Thus $n_B = 0$ and $n = 1$, which gives $(1 - F_{ST})\gamma_B/1$. By instantiating node A to a_1, the probability of observing an allele of type B in the second allele sampled becomes 0.18. One can then find

$$Pr(gt = AB) = \frac{2(1 - F_{ST})\gamma_A(1 - F_{ST})\gamma_B}{(1 - F_{ST})} = 2\gamma_A\gamma_B(1 - F_{ST}) = 0.288,$$

where the factor 2 accounts for the ordering.

5.12.4 DNA likelihood ratio

Consider the evaluation of the likelihood ratio in a common situation involving DNA. Denote the genotype of the crime stain tgt and the genotype of a suspect's profile sgt, Denote the background information as I. Consider the propositions of interest, at the source level for example, to be

H_p : the suspect is the source of the stain;

H_d : another person, unrelated to the suspect, is the source of the stain; that is, the suspect is not the source of the stain.

Both profiles are found to be of type AB, for instance. The likelihood ratio can then be expressed as

$$\frac{Pr(tgt = AB \mid sgt = AB, H_p, I)}{Pr(tgt = AB \mid sgt = AB, H_d, I)}. \tag{5.32}$$

Assume that the DNA typing system is sufficiently reliable so that two samples from the same person will be found to match, for example, when the suspect is the donor of the stain (proposition H_p), and that there are no false-negatives. $Pr(tgt = AB \mid sgt = AB, H_p, I)$ will thus be set to 1.

The DNA profiles from two different people, for example, the suspect and the donor of the stain when proposition H_d is true, are dependent, so $Pr(tgt = AB \mid sgt = AB, H_d, I) \neq Pr(tgt = AB \mid I)$. The evidential value of a match between the profile of the recovered sample and that of the suspect needs to take into account the fact that there is a person, that is, the suspect, who has already been observed to have the profile of type AB.

Observing one allele in the sub-population increases the chance of observing another of the same type. Hence, within a sub-population, DNA profiles with matching allele types are more common than suggested by the independence assumption, even when two individuals are not directly related (Balding 1997). A conditional probability, also termed *random-match probability*, is thus used. It incorporates the effect of population structure or other dependencies between individuals, such as that imposed by family relationships (Weir 2001).

For the sake of simplicity, the likelihood ratio $1/Pr(tgt = AB \mid sgt = AB, H_d, I)$ is rewritten as

$$\frac{1}{Pr(tgt = AB \mid sgt = AB)} = \frac{Pr(sgt = AB)}{Pr(tgt = AB, sgt = AB)}.$$

Consider the numerator first. The probability of observing a genotype of type AB is $Pr(gt = AB) = 2(1 - F_{ST})\gamma_A(1 - F_{ST})\gamma_B/(1 - F_{ST})$, as developed in Section 5.12.3. This result applies then to sgt.

A similar argument – based on the recursive formula of Balding and Nichols (1995, 1997) – follows for the denominator, $Pr(tgt = AB, sgt = AB)$. Considering the first allele A ($n_A = 0, n = 0$), the first allele B ($n_B = 0, n = 1$), the second allele A ($n_A = 1, n = 2$) and the second allele B ($n_B = 1, n = 3$), one obtains

$$\frac{4(1 - F_{ST})\gamma_A(1 - F_{ST})\gamma_B[F_{ST} + (1 - F_{ST})\gamma_A][F_{ST} + (1 - F_{ST})\gamma_B]}{(1 - F_{ST})(1 + F_{ST})(1 + 2F_{ST})}.$$

Hence, the likelihood ratio for the scenario under investigation becomes

$$\frac{(1 + F_{ST})(1 + 2F_{ST})}{2[F_{ST} + (1 - F_{ST})\gamma_A][F_{ST} + (1 - F_{ST})\gamma_B]}. \tag{5.33}$$

The same line of reasoning leads to the random-match probability in situations involving homozygous profiles (AA or BB),

$$\frac{(1 + F_{ST})(1 + 2F_{ST})}{[2F_{ST} + (1 - F_{ST})\gamma_i][3F_{ST} + (1 - F_{ST})\gamma_i]}. \tag{5.34}$$

Details of the derivation of these formulae may be found in Balding and Nichols (1995, 1997).

A Bayesian network can be proposed that should allow scientists to reproduce the results implied *via* (5.33) and (5.34). Such a model is shown in Figure 5.22. The network structure is derived as follows: Start by considering the graphical model presented in Figure 5.21. There, the node gt represents the suspect's profile. This is denoted by sgt in Figure 5.22. A node I_2 in Figure 5.22 is used to model circumstances under which a_2 is affected by knowledge about a_1. The network fragment covering the nodes sgt, a_1, a_2 and I_2 is then duplicated with the aim of incorporating the genotype of the recovered (crime) stain. The two alleles,

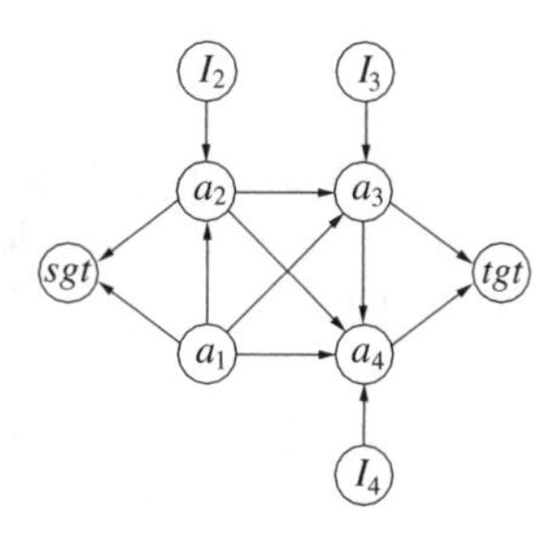

Figure 5.22 A Bayesian network for evaluating random-match probabilities. Nodes sgt and tgt represent the suspect and trace profiles. Nodes a_1 and a_2 represent the alleles for the suspect. Nodes a_3 and a_4 represent the alleles for the trace. Nodes I_2, I_3 and I_4 model circumstances in which alleles may be identical by descent to previously sampled alleles.

that is, a_3 and a_4, describing the trace genotype tgt of the crime stain, depend – if sub-population structure does exist – on knowledge of the previously sampled alleles a_1 and a_2. The following directed arcs are thus introduced: $a_1 \rightarrow a_3$, $a_2 \rightarrow a_3$, $a_1 \rightarrow a_4$, $a_2 \rightarrow a_4$ and $a_3 \rightarrow a_4$.

In analogy to the node I_2, nodes I_3 and I_4 represent circumstances under which the alleles a_3 and a_4 may be identical by descent to previously sampled alleles. The states of the nodes I_m $(m = 2, 3, 4)$ are defined as follows:

- I_2: the allele a_2 may be identical by descent or not with the first allele, a_1. Two states ibd_1 and $not - ibd$ are thus adopted for I_2.
- I_3: for the allele a_3, there is an additional possibility for being identical by descent, namely, with allele a_2. The states for I_3 thus are ibd_1, ibd_2 and $not - ibd$.
- I_4: in the same way, node I_4 covers the states ibd_1, ibd_2, ibd_3 and $not - ibd$.

In the current example, a ratio of 1:9 is considered between an allele identical by descent and an allele not identical by descent. Stated otherwise, a probability of $F_{ST} = 0.1$ is used to quantify the event by which an allele has descended from a single ancestral allele. Thus, the following unconditional probabilities are associated with the node tables of the nodes I_m:

- Node I_2: $Pr(idb_1) = 0.1$ and $Pr(not - idb) = 0.9$;
- Node I_3: $Pr(idb_1) = 0.0909$, $Pr(idb_2) = 0.0909$ and $Pr(not - idb) = 0.8182$;
- Node I_4: $Pr(idb_1) = 0.0833$, $Pr(idb_2) = 0.0833$, $Pr(idb_3) = 0.0833$ and $Pr(not - idb) = 0.7500$.

Note that subscripts refer to the allele potentially identical by descent. In the general case, the probabilities for the states ibd_x $(x = 1, 2, 3)$ are given by $F_{ST}/\{1 + (n - 1)F_{ST}\}$. Conversely, the probabilities for the states $not - ibd$ are given by $1 - [F_{ST}/\{1 + (n - 1)F_{ST}\}]$. As may be seen, as the number of alleles sampled increases, the probability to transmit an allele identical by descent increases.

Probability tables associated with the nodes a_x may be completed logically for as long as the corresponding node I_x is not in the state $not - ibd$. Tables 5.14 and 5.15 summarise

Table 5.14 Conditional probabilities applicable to the node a_2, which can be allele A or B. Factor a_1 can be allele A or allele B. Factor I_2 denotes a_2 ibd with a_1 (ibd_1) or not ($not-ibd$). Allele frequencies for A and B are γ_A and γ_B.

	I_2 :	ibd_1		$not-ibd$	
	a_1 :	A	B	A	B
a_2 :	A	1	0	γ_A	γ_A
	B	0	1	γ_B	γ_B

Table 5.15 Probabilities for the factor a_3, conditional on I_3, a_1 and a_2. Factors a_1, a_2 and a_3 can be allele A or allele B. Factor I_3 denotes a_3 ibd with a_1 (ibd_1), ibd with a_2 (ibd_2) or not ibd with either ($not-ibd$). Allele frequencies for A and B are γ_A and γ_B.

			a_3	
I_3	a_1	a_2	A	B
ibd_1	A	A	1	0
		B	1	0
	B	A	0	1
		B	0	1
ibd_2	A	A	1	0
		B	0	1
	B	A	1	0
		B	0	1
$not-ibd$	A	A	γ_A	γ_B
		B	γ_A	γ_B
	B	A	γ_A	γ_B
		B	γ_A	γ_B

the probabilities for the nodes a_2 and a_3 respectively. The probability table for the node a_4 is completed analogously.

In a Bayesian network set-up according to the above description the user can assess, for example, $Pr(a_3 = A \mid a_1 = A, a_2 = B)$ by instantiating $a_1 = A$ and $a_2 = B$, or, alternatively, by fixing sgt to AB. In either case $Pr(a_3 = A \mid a_1 = A, a_2 = B)$ can be found to be 0.7454, a value that is also obtained by the recursive formula. In a similar way, $Pr(a_4 = B \mid a_1 = A, a_2 = B, a_3 = A)$ can be found to be 0.2333.

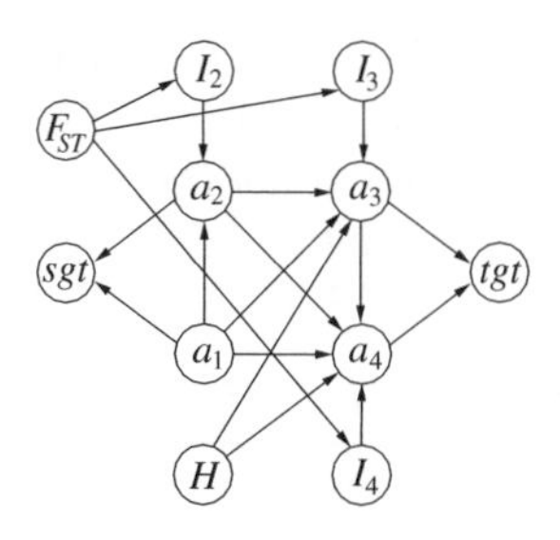

Figure 5.23 A Bayesian network for evaluating the DNA likelihood ratio with F_{ST} corrections. Nodes *sgt* and *tgt* represent the suspect and trace profiles. Nodes a_1 and a_2 represent the alleles for the suspect. Nodes a_3 and a_4 represent the alleles for the trace. Nodes I_2, I_3 and I_4 model circumstances in which alleles may be identical by descent to previously sampled alleles. Node H has two states: H_p, the suspect is the source of the crime stain, and H_d, some person other than the suspect is the source of the crime stain. The node F_{ST} allows one to specify a particular value for the factor F_{ST}; possible states are 0, 0.001, 0.01, 0.05 and 0.01.

Finally, $Pr(tgt = AB \mid sgt = AB)$ can be evaluated by instantiating the node *sgt* to AB. The value that is then indicated for the state AB of the node *tgt* is 0.3479. This result is just the reciprocal of (5.33).

Note that a more sophisticated mathematical development, taking into account the probabilities that any two, three or four alleles are identical by descent is presented in Weir (1996). (5.33) and (5.34) represent expressions where these probabilities are approximated by functions of F_{ST}.

The Bayesian network studied so far, Figure 5.22, assumes that the suspect and the source of the crime stain are different individuals. More generally, the network allows one to evaluate the probability of an individual's genotype if knowledge about the genotype of another individual from the same sub-population is available, and both individuals have been selected at random from that sub-population. This is relevant for assessing the denominator of the DNA likelihood ratio.

When the aim is to make an inference about H, a source level proposition as defined at the beginning of this Section, the current network can be extended as shown in Figure 5.23. A binary node H with states H_p and H_d is adopted as a parent for the allele nodes a_3 and a_4. As a consequence, the size of the probability tables for the latter nodes doubles. This is illustrated, for example, by comparing Tables 5.15 and 5.16. Notice that the right-hand side of Table 5.16 is identical to Table 5.15. This means that under H_d, the suspect and the true donor of the crime stain are taken to be different individuals, the network thus behaving in the same way as the network shown in Figure 5.22. On the other hand, if H_p is true, that is, the suspect being the source of the crime stain, then, necessarily, the suspect's actual genotype must equal that of the crime stain. In order for the network to reflect this property, the probability for node a_3 is specified in such a way that, if H_p is true, a_3 is solely determined by the states of the node a_2 (*see* left-hand side of Table 5.16). Analogously, the probability table of the node a_4 is specified, under H_p, so as to reflect the actual state of a_1.

Table 5.16 Probabilities for the factor a_3, conditional on H, I_3, a_1 and a_2. Factors a_1, a_2 and a_3 can be allele A or allele B. Factor I_3 denotes a_3 ibd with a_1 (ibd_1), ibd with a_2 (ibd_2) or not ibd with either ($not-ibd$). Allele frequencies for A and B are γ_A and γ_B.

H	I_3	a_1	a_2	a_3: A	a_3: B
H_p	ibd_1	A	A	1	0
			B	0	1
		B	A	1	0
			B	0	1
	ibd_2	A	A	1	0
			B	0	1
		B	A	1	0
			B	0	1
	$not-ibd$	A	A	1	0
			B	0	1
		B	A	1	0
			B	0	1

H	I_3	a_1	a_2	a_3: A	a_3: B
H_d	ibd_1	A	A	1	0
			B	1	0
		B	A	0	1
			B	0	1
	ibd_2	A	A	1	0
			B	0	1
		B	A	1	0
			B	0	1
	$not-ibd$	A	A	γ_A	γ_B
			B	γ_A	γ_B
		B	A	γ_A	γ_B
			B	γ_A	γ_B

Note a further extension to Figure 5.22 has been incorporated into Figure 5.23. A node F_{ST} allows the user to select a particular value for F_{ST}. Unlike in the model shown in Figure 5.22, one can avoid changing manually the unconditional probabilities of the nodes I_x ($x = 2, 3, 4$) in cases where another value for F_{ST} is needed. For the purpose of illustration, let the node F_{ST} cover states numbered 0, 0.001, 0.01, 0.05 and 0.1. It is assumed here that this node will be instantiated whilst using the model, so one need not care about the prior probabilities associated with this node. The node F_{ST} is a parent of the nodes I_2, I_3 and I_4, thus the conditional probabilities associated with the latter nodes are a function of the states of the node F_{ST}. In certain Bayesian network software, such as HUGIN, such a functional dependency can be exploited in order to complete automatically conditional probability tables. Table 5.17 contains expressions in HUGIN language applicable for the nodes I_x ($x = 2, 3, 4$).

5.13 Further reading

Most of the Bayesian networks described throughout this chapter, with the exception of those presented in the last section, do not take into account sub-population structure. In practice, the effect of shared ancestry may however often be important for casework. Disregarding the correlation of alleles in the calculation of the weight of the evidence results in an exaggeration of the strength of the evidence against the compared person, such as, for example, the suspect in a criminal case or the alleged father in civil paternity cases, even though the disregard for the correlation is not as important as the relatedness in the same population. Further relevant literature on this point can be found in Aitken and Taroni (2004) and Buckleton et al. (2004).

Table 5.17 Expressions in HUGIN language for specifying the conditional probability tables of the nodes I_x $(x = 2, 3, 4)$. The nodes I_x denote circumstances in which alleles may be identical by descent. Their states depend on the node F_{ST} (*see* also Figure 5.23) with states 0, 0.001, 0.01, 0.05 and 0.1. The expressions `Distribution( )` assign conditional probabilities to the states of the nodes I_x where `FST` denotes a particular outcome of the node F_{ST}.

Node	Expression
I_2	`Distribution(FST,1-FST)`
I_3	`Distribution(FST/(1+FST),FST/(1+FST),1-2*(FST/1+FST))`
I_4	`Distribution(FST/(1+2*FST),FST/(1+2*FST),FST/(1+2*FST), 1-3*FST/(1+2*FST))`

Notice also that in what is presented in this chapter, no uncertainty in estimates of allele frequencies is considered. In practice, γ_A and γ_B are not known exactly but are estimated from frequencies, say x_A and x_B, in a database of size n. In addition, previous developments do not take into account the uncertainty due to sampling error. For a discussion on this point, see Balding and Nichols (1994), Curran et al. (2002) and Curran (2005).

The analysis of DNA in forensic practice involves a wide range of challenging issues. For example, as noted by Evett et al. (2002a), the considerable increase in the sensitivity of DNA analysis over the last few years has initiated a tendency of courts to shift from questions of the kind 'whose DNA is this?' to 'how did this person's DNA get there?'. Evett et al. (2002a) studied such issues in the context of two casework examples which involved the results of the analysis of small quantities of DNA recovered from items such as a cigarette end or a watch. Bayesian networks are shown to allow probabilistic analyses to be made over a large number of variables involving differently interrelated issues, situations in which a full algebraic solution would otherwise appear to be extremely difficult.

It has been noted throughout this chapter that Bayesian networks for evaluating DNA evidence may involve repetitive submodels, such as those shown in Figure 5.14. Although networks can aid the construction of appropriate structures in more complex cases, it has also been found advantageous to use a hierarchical approach for specifying graphical models. This may be achieved, as shown by Dawid (2003) and Vicard and Dawid (2003) in a study on the problem of estimating mutation rates, through the application of object oriented Bayesian networks which may be implemented in version 6 of the HUGIN system. Object oriented Bayesian networks are an approach for describing complex fields of application in terms of interrelated objects (Koller and Pfeffer 1997). An object may either be a standard random variable as defined in Chapter 2, or consist of a set of attributes, each of which is an object. For example, the object 'suspect' can be characterised by several attributes, such as their race or the colour of their eyes, and so on. In turn, the attribute 'colour of the suspect's eyes' can be viewed as a simple object that can assume values in some finite range. On the other hand, it may also be possible that an attribute is itself a complex object with its own attributes.

Other applications of object oriented Bayesian networks cover identification and separation of DNA mixtures using peak area information (Cowell et al. 2004) and inference problems associated with searches in DNA databases (Cavallini et al. 2004).

Cowell (2003) has described an alternative way for constructing Bayesian networks for inference from genetic markers. He has developed a software tool, called FINEX (Forensic Identification by Network Expert systems), where inference problems based on DNA evidence can be expressed through the syntax of a graphical specification language. FINEX uses an algorithm for the automatic construction of an appropriate representation in terms of a Bayesian network. Compared to general purpose Bayesian network software, FINEX allows one to save time in setting up networks and reduces the potential of error while completing large probability tables. Evidence from several markers may also readily be combined in order to evaluate an overall likelihood ratio.

6

Transfer evidence

The aim of this chapter is to propose a methodology for modelling and developing Bayesian networks in the specific area of 'transfer evidence' such as glass, fibres or paint. The starting points are the general formulae proposed in scientific literature to deal with the assessment of such evidence.

The major features of Bayesian networks applicable to transfer evidence will be illustrated by examples involving propositions at the crime level and the activity level. Examples will be presented in which the scientist assesses the value of the evidence in terms of intrinsic and extrinsic features (Kind 1994). Further issues, such as cross-transfer and missing evidence (evidence that is expected but it is either not found or is not produced on request (Schum 1994)), are also developed and discussed.

Transfer evidence is an interesting area of research in forensic science, essentially because the phenomena of transfer, persistence and recovery of traces play an important role. However, care should be taken in order to account correctly for such factors. The aim here is to show that this accounting can be clarified substantially through the use of Bayesian networks, which provide an appropriate environment for the representation and evaluation of the various dependencies that may exist among parameters affecting a coherent evaluation.

Remember that when attempting to model large and complex domains, it may be preferable to limit the number of variables in the model, at least at the beginning, in order to keep the structure and probability calculus tractable. Throughout the remainder of this chapter, the number of variables will be limited as much as is possible. Some target nodes, for example the node representing the variable *transfer*, will be studied in more detail and possible extensions will be proposed.

Generally, the key point will be to determine the most important variables, described by nodes, in order to simplify the description of the situation and facilitate the elicitation of probabilities. Satisfaction of this point suggests that fewer nodes, fewer arcs and smaller state spaces are used, keeping in mind that the model should maximise the fidelity to the reality of the scenario investigated.

Bayesian Networks and Probabilistic Inference in Forensic Science F. Taroni, C. Aitken, P. Garbolino and A. Biedermann

6.1 Assessment of transfer evidence under crime level propositions

6.1.1 A single-offender scenario

Consider the following scenario proposed by Champod and Taroni (1999). An offender entered the rear of a house through a hole which he cut in a metal grille. Here, the offender attempted to force the entry but failed and a security alarm went off. He left the scene. About ten minutes after the offence, a suspect wearing a red pullover is apprehended in the vicinity of the house following information from an eyewitness who testified that they saw a man wearing a red pullover running away from the scene. At the scene, a tuft of red fibres was found on the jagged end of one of the cut edges of the grille.

Notice that in such a scenario, the eyewitness testified that they saw a man wearing a red pullover running away. It may thus be admitted that the offender wore a red pullover. This is one of the assumptions that will be introduced in the Bayesian network proposed for this scenario.

A suspect is then apprehended. Fibres coming from his pullover 'match', in some sense, the red fibres from the tuft recovered on the scene. Evaluation of this evidence means, in this context, the assessment, for example, of the correspondence in type and colour of the fibres given a pair of propositions H_p and H_d. When considering the case at the crime level, the propositions may be as follows:

- H_p: the suspect is the offender;
- H_d: the suspect is not the offender.

The scenario considered here closely resembles the one-stain scenario considered in Section 4.1. The Bayesian network for the evaluation of one-stain scenarios, shown in Figure 4.1 may thus serve as a starting point for approaching the fibre case outlined above.

Here, a new proposition is considered, as suggested by the eyewitness. The new proposition is denoted R and is that 'The suspect wore a red pullover at the time the crime was committed'. If it is admitted that this proposition R is independent of the crime level proposition H, 'The suspect is the offender', and G, 'The red fibres have been left by the offender', then R can be added as a new node from which only one arc originates. This arc is pointing to the source node F, 'The red fibres came from the suspect'. Event E is now 'The red fibres found at the grille match the fibres of a pullover belonging to the suspect'. A graphical representation of these assumptions is shown in Figure 6.1.

The probability of the new root node, $Pr(R)$, can be evaluated on the basis of the fact that the suspect was wearing a red pullover ten minutes after the attempted crime as well as on the basis of other information. In the present case, it is clear that this probability is very near to certainty. However, it represents an independent variable that might have a probability less than one, depending on particular circumstances; for example, if the suspect were apprehended several hours after the crime. This probability thus is subjective (epistemic) in nature.

From the probabilistic dependencies represented in Figure 6.1, it follows that conditional probabilities need to be assessed for the variable E. Consider the following:

1. If F is true, that is, the fibres come from the suspect, then the probability of a 'match' between suspect and control fibres is at its maximum, $Pr(E \mid F) = 1$.

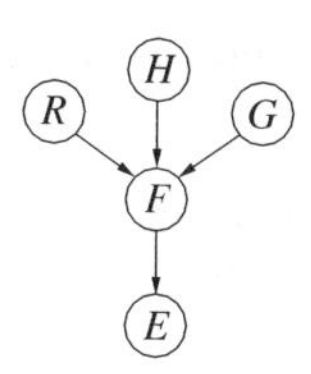

Figure 6.1 Bayesian network for evaluating a single-offender scenario involving fibres recovered on the scene of a crime. Each node has two states. For H, these are H_p, the suspect is the offender, and H_d, the suspect is not the offender. For R these are R, the suspect wore a red pullover at the time the crime was committed and $\bar{R}$, the suspect was not wearing a red pullover at the time the crime was committed. For G these are G, the red fibres have been left by the offender, and $\bar{G}$, the red fibres were not left by the offender. For F these are F, the red fibres came from the suspect, and $\bar{F}$, the red fibres did not come from the suspect. For E these are E, the red fibres found at the grille match the fibres of a pullover belonging to the suspect, and $\bar{E}$, the red fibres found at the grille do not match the fibres of a pullover belonging to the suspect. [Reproduced with permission from Elsevier.]

2. If F is false, that is, the fibres do not come from the suspect, then the probability of E is given by the chance occurrence of matching fibres. This may be estimated using the relative frequency γ of the matching characteristics among members of the relevant population: $Pr(E \mid \bar{F}) = \gamma$.

For the variable F, eight conditional probabilities need to be assigned, which correspond to the eight possible combinations of the outcomes of events H, G and R. The following are assumed.

3. If the suspect is the offender (H_p), he was wearing a red pullover (R), and the fibres came from the offender (G), then certainly the fibres came from the suspect: $Pr(F \mid H_p, R, G) = 1$.

4. If the suspect is not the offender (H_d), he was wearing a red pullover (R), and the fibres did not come from the offender ($\bar{G}$), then the probability that the fibres came from the suspect (F) is the same as the probability that he passed through the hole in the grille for innocent reasons. Indeed, this is a very highly improbable setting, but account is taken of it in order to consider the most general case. Denote $Pr(F \mid H_d, R, \bar{G}) = p$, in analogy to the notation used in Section 4.1.1.

5. All the other conditional probabilities are clearly equal to zero. In fact, if the suspect is the offender (H_p), matching fibres come from the offender (G) and he was not wearing a red pullover at the time the crime was committed ($\bar{R}$), then it is clearly impossible that the red fibres found at the grille match the fibres of a pullover belonging to the suspect. This is so, similarly, (a) if the suspect is the offender (H_p), he was wearing a red pullover at the time the crime was committed (R), but matching fibres did not come from the offender ($\bar{G}$), and (b) if the suspect is the offender (H_p), he was not wearing a red pullover at the time the crime was committed ($\bar{R}$), and the matching fibres did not come from the offender ($\bar{G}$). Under proposition H_d (the suspect is not the offender), the probability the red fibres came from the suspect are clearly an

impossible event. The sole exception is described in the situation presented under point 4.

Therefore, denoting $Pr(G) = r$, relevance, and $Pr(R) = q$, and using the laws of probability, the following likelihood ratio is obtained.

$$V = \frac{Pr(E \mid H_p, I)}{Pr(E \mid H_d, I)} = \frac{rq + (1 - rq)\gamma}{r\gamma + (1 - r)[pq + (1 - pq)\gamma]} . \tag{6.1}$$

The likelihood of H_p given E, $Pr(E \mid H_p, I)$, is now the sum of the probability of the scenario where the trace is relevant and the suspect was wearing a red pullover when the crime was committed (rq), and of the probability of the negation of that scenario $(1 - rq)$ times the profile frequency of that kind of fibres (γ).

The likelihood of H_d given E, $Pr(E \mid H_d, I)$, is the sum of the probability of the scenario where the trace is relevant ($r\gamma$) and of the probability of the scenario where the trace is not relevant $(1 - r)$ and either it has been left by the innocent suspect wearing the red pullover (pq) or by another innocent person $[(1 - pq)\gamma]$.

Comparing (6.1) with (4.11), one may see that if $q = 1$, then (6.1) equals the likelihood ratio described by (4.11). Notice also that if $r = 1$, that is, the fibres found at the crime scene came from the offender, then the likelihood ratio reduces to:

$$V = \frac{Pr(E \mid H_p, I)}{Pr(E \mid H_d, I)} = \frac{q + (1 - q)\gamma}{\gamma} . \tag{6.2}$$

For illustration, consider a numerical example with q and γ assuming the values 0.5 and 0.01, respectively. If the evidential material is considered to be relevant $[Pr(G) = 1]$, then the value of the evidence is measured by a likelihood ratio of 50:50. This can be deduced from the posterior probability $Pr(H_p \mid E, I) = 0.9806$ shown in Figure 6.2(i), but may also be verified using (6.2). Notice that equal prior probabilities have been specified for the states of the node H.

If both r and q are true, then the likelihood ratio reduces to $1/\gamma$ and the posterior probability of $Pr(H_p \mid E, I)$ is equal 0.9901. This situation is represented in Figure 6.2(ii).

6.1.2 A fibre scenario with multiple offenders

This scenario is the same as the previous one, but instead of one offender described by an eyewitness, a certain number of offenders, say k, were viewed entering through the hole which they cut in the metal grille, and one of the them was seen to wear a red pullover. A formal mathematical development of this scenario has previously been discussed in Section 3.3.3.

The two alternative propositions (states of node H) are now H_p, 'The suspect is one of the k offenders', and H_d, 'The suspect is not one of the k offenders'. Proposition G becomes, 'The red fibres came from one of the k offenders'.

Referring to a graph as shown in Figure 6.1, it may be accepted that the above-mentioned assumptions (3), (4) and (5) hold. Then (6.1) is still applicable in this case and, if one assumes that the fibres surely come from one of the offenders (that is, $r = 1$), (6.2) may again be obtained.

The value q for the probability the suspect was wearing the pullover (node R) depends again upon the available circumstantial information. If the suspect has been apprehended

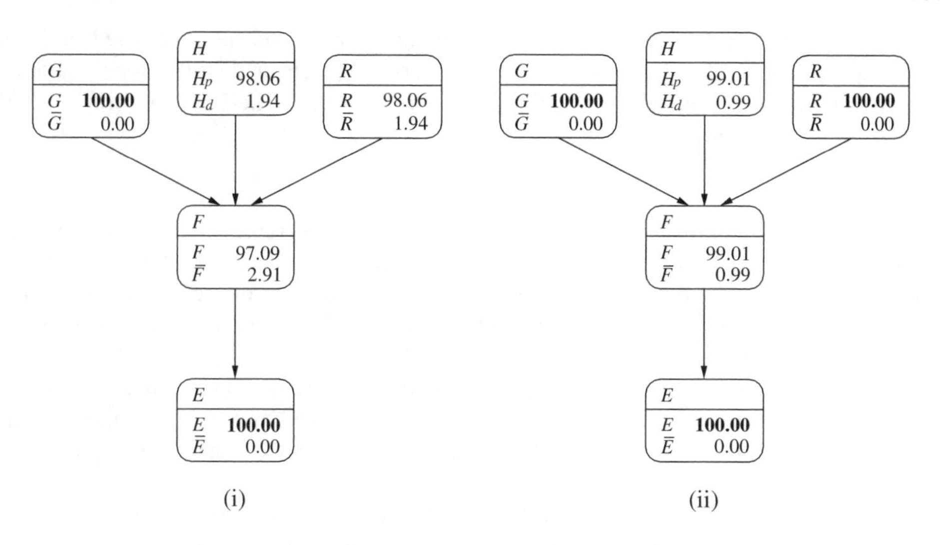

Figure 6.2 Bayesian network for the evaluation of a single-offender fibre scenario. Each node has two states. For H, these are H_p, the suspect is the offender, and H_d, the suspect is not the offender. For R these are R, the suspect wore a red pullover at the time the crime was committed and $\bar{R}$, the suspect was not wearing a red pullover at the time the crime was committed. For G these are G, the red fibres have been left by the offender, and $\bar{G}$, the red fibres were not left by the offender. For F these are F, the red fibres came from the suspect, and $\bar{F}$, the red fibres did not come from the suspect. For E these are E, the red fibres found at the grille match the fibres of a pullover belonging to the suspect, and $\bar{E}$, the red fibres found at the grille do not match the fibres of a pullover belonging to the suspect. Evaluation of the posterior probability of the proposition H when the red fibres found at the grille match the fibres of a pullover belonging to the suspect, that is, node E is instantiated to E. Assumptions are (i) relevance, $Pr(G) = r = 1$, the probability the suspect wore a red pullover at the time the crime was committed, $Pr(R) = q = 0.5$ and the frequency of the fibres, $\gamma = 0.01$, and (ii) $r = 1$, $q = 1$ and $\gamma = 0.01$.

shortly after the attempted robbery wearing a red pullover, a value close to 1 seems appropriate.

A different scenario would be one in which the suspect has been apprehended the following day or later. Here, a value for q of $= 1/k$ has been proposed (Evett 1993), in which case the following likelihood ratio is obtained,

$$V = \frac{1/k + (1 - 1/k)\gamma}{\gamma} . \tag{6.3}$$

As with any assumption of equal probabilities, the assumption can be used only after a careful scrutiny of any relevant additional information that scientists may have to assist in the identification of the person wearing the red pullover. When there is no eyewitness statement that one of the offenders was wearing a red pullover, the assumption that $q = 1/k$ is not warranted. Moreover, assumption (3) has to be changed as well, and an additional epistemic probability has to be assigned, namely, $P(F|H_p, R, G) = f$, for which a value < 1 appears appropriate.

Indeed, even if the suspect is one of the offenders (H_p), and he wore a red pullover (R), and the fibres came from one of the offenders (G), then it is still possible another offender could have worn similar clothing and left the trace. The likelihood ratio, V, if $r = 1$ is assumed, is then

$$V = \frac{fq + [(1-f)q]\gamma}{\gamma}. \tag{6.4}$$

The above developments are mathematical and graphical representations of evidence assessment under crime level propositions. Recall that H_p has been used to denote propositions such as 'the suspect is the offender' or 'the suspect is one of the k offenders'. Elicitation of probability plays an important role here; it has to be done carefully taking account of information available from the case under investigation.

The major advantage of the use of Bayesian networks here is to help forensic scientists to focus on the relationships among the propositions involved and the assessment of the respective target probabilities, while calculations can largely be automated with appropriate software tools.

An alternative approach in the evaluation of evidence is to consider findings under activity level propositions (Section 3.3.2). The application of such an approach to fibre evidence will be outlined in the next section.

6.2 Assessment of transfer evidence under activity level propositions

6.2.1 Preliminaries

When addressing a case at the activity level, the definitions of the propositions of interest include the description of an action (Section 3.3). Such propositions could be, for example, 'Mr X assaulted the victim', and 'Mr X did not assault the victim', meaning that some other man assaulted the victim and Mr X was not involved in the action. Another pair of propositions at the activity level could be 'Mr X sat on the car driver's seat', and 'Mr X never sat on the car driver's seat'. An important aspect or consequence of an activity like assaulting a victim or sitting on a driver's seat is the occurrence of a contact, for example, between two people involved in an assault, or between the driver and the seat of a car. A further consequence that needs to be considered in case of contact is the occurrence of the phenomenon of transfer of material, such as fibres, for example.

Note that the use of the term 'contact' is considered in the context of a logical consequence of the action. Although this rather vague definition do not tell one much about the *kind* of contact, it is clearly useful to incorporate this structurally in a Bayesian network, essentially because it allows the scientist to take into account the possibility of a legitimate contact between the persons involved in the scenario under investigation. Notice that this can be meaningful even though the suspect may claim not to be involved in the alleged offence.

6.2.2 Derivation of a basic structure for a Bayesian network

A standard Bayesian network that may be used to describe the transfer of material from the scene to the criminal or from the victim to the criminal is shown in Figure 6.3. Here, the

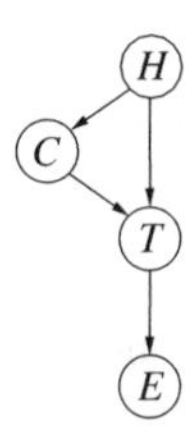

Figure 6.3 Bayesian network for transfer from the victim/scene to the criminal. Each node has two states. Node H is the probandum, H_p, the suspect assaulted the victim, H_d, some person other than the suspect assaulted the victim. Node C, C, the victim has been in contact with the suspect, $\bar{C}$, the victim has not been in contact with the suspect. Node T, T, there was a transfer from the victim, $\bar{T}$, there was not a transfer from the victim. Node E, E, the trace found on the suspect is similar to the control coming from the victim, $\bar{E}$, the trace found on the suspect is not similar to the control coming from the victim. [Reproduced with permission from Elsevier.]

material found on the suspect comes either from the victim/scene *or* it is there by chance alone because the suspect has no link with the action under investigation.

Let the node H be the probandum 'The suspect assaulted the victim', C the proposition 'The victim has been in contact with the suspect', T 'There was a transfer from the victim', and E 'The trace found on the suspect is similar to the control coming from the victim'.

Logically, if the suspect assaulted the victim, he had a physical contact with the victim. Such a contact may involve a possible transfer of evidential material, for example, fibres, from the victim to the suspect. Therefore, node H is directly linked to nodes C and T. Furthermore, a converging connection at the node T may be used so that $H \rightarrow T \leftarrow C$. In fact, different conditional probabilities can be imagined for T given different outcomes for the variables H and C. For further details on this refer to points 3 and 4 below.

Notice that the structural relations among H, C and T as shown in Figure 6.3 provide a modelling example for an indirect relationship. It is an expression of the belief that an individual assaulting the victim influences the probability of transfer both directly (through $H \rightarrow T$) and indirectly (through $H \rightarrow C \rightarrow T$). Such relationships need to be carefully reviewed in order to see whether they correctly represent one's perception of the problem. In the case at hand, the variables H, C and T are mutually dependent and any two of these will remain so upon observation of a third.

Descriptions of case scenarios may provide useful assistance in deciding whether a given set of dependence and independence statements is acceptable; for example, the probability of transfer given contact is assumed to be different given H_p and H_d. In a special case in which $Pr(T \mid C, H_p) = Pr(T \mid C, H_d)$ and $Pr(T \mid \bar{C}, H_p) = Pr(T \mid \bar{C}, H_d)$, the directed edge between the nodes H and T would become superfluous and may be removed from the network structure, leaving the serial connection $H \rightarrow C \rightarrow T$.

Generally, the respective conditional probabilities to be estimated are as follows:

1. The probability of contact given the appropriate proposition, for example the proposition that the suspect assaulted the victim: $Pr(C \mid H_p) = c$.

2. The probability of contact given the negation of H, for example the proposition that the suspect did not assault the victim. This probability can be different from zero,

depending upon specific circumstantial information. For example, the suspect could have been in contact with the victim for other reasons. The probability to be estimated here will be denoted as $Pr(C \mid H_d) = d$.

3. The probability the material trace has been transferred from the victim depends upon both events H and C. Information concerning the nature of the crime and the position of the trace may allow the assignment of different transfer probabilities: $Pr(T \mid C, H_p) = t$, $Pr(T \mid C, H_d) = s$.

4. It is obvious that there is no possibility for transfer when there is no contact. Therefore, $Pr(T \mid \bar{C}, H_p) = 0$, $Pr(T \mid \bar{C}, H_d) = 0$.

5. The transfer node 'screens off' the contact node C from E, the 'matching' node. It seems natural to postulate that $Pr(E \mid T, C) = Pr(E \mid T, \bar{C}) = 1$. If there is no transfer, $Pr(E \mid \bar{T}, C) = Pr(E \mid \bar{T}, \bar{C})$, and there is a match between fibres, the fibres should be there for another reason than transfer linked to a contact (innocent or not) with the victim. So, it can be assumed that $Pr(E \mid \bar{T}, C) = Pr(E \mid \bar{T}, \bar{C}) = b_{1,m}\gamma$ where the term $b_{1,m}\gamma$ represents the probability of transfer other than from the victim (background). This term is the product of two probabilities. Suppose one group of m fibres is on the suspect's clothes before the assault, for innocent reasons. The probability of this event is denoted $b_{1,m}$, where the first subscript, 1, denotes the number of groups and the second subscript, m, denotes the size of the one group. The probability γ denotes the probability that the characteristics of this group of fibres match those of the control fibres. For ease of notation, the probability $b_{1,m}\gamma$ of this joint event is denoted γ_b, that is, $\gamma_b = b_{1,m}\gamma$.

The likelihoods, calculated from these assumptions and the probabilistic dependence assumptions represented in Figure 6.3, are as follows,

$$Pr(E \mid H_p) = ct + [c(1-t) + (1-c)]\gamma_b \tag{6.5}$$

and

$$Pr(E \mid H_d) = ds + [d(1-s) + (1-d)]\gamma_b \, . \tag{6.6}$$

However, one of the conditional probabilities, γ_b, associated with node E is the product of two other probabilities, $b_{1,m}$ and γ (*see* point 5 above). This is not completely satisfactory since the determination of probabilities with several components (compound probabilities) is not only time-consuming, but is also prone to error. Later, it is shown that the Bayesian network can be modified in order to avoid the development of probability tables with compound probabilities, *see* Section 6.2.5.

6.2.3 Stain found on a suspect's clothing

A group of fibres has been found on Mr X's jacket who has been arrested by the police because it is suspected that he physically assaulted Miss Y. The characteristics of the fibres are different from those of fibres from the suspect's own jacket but indistinguishable from those on the clothing of the victim.

Reconsider the Bayesian network shown in Figure 6.3. Node H has two states, H_p denotes 'The suspect assaulted the victim', H_d denotes 'The suspect did not assault the victim'. Probabilities t and s denote the probabilities of transfer (node T) under H_p and H_d respectively, and are conditioned on background information concerning the circumstances of the case under examination. In fact, if the group of fibres has been found on the jacket, it is reasonable to assume that they are probably linked to the assault, so $t > s$, $Pr(T \mid C, H_p) > Pr(T \mid C, H_d)$. On the other hand, if fibres are found on the lower part of the trousers, then the assessment changes and the assumption $t = s$ could be accepted: the probability of transfer is independent from the main propositions, H_p and H_d. Notice also that, given the circumstances of the case, it could be reasonable to assume $t = 1$.

Another possibility may be considered. Let $c = 1$ and $t = 1$, and suppose the defence strategy is to assume that the suspect was at the scene of the crime for reasons unconnected to the crime and that he had contact with the victim, (*e.g.*, he assisted the victim after the assault whilst waiting for the police to arrive), that is, $d = 1$; then the likelihood ratio is

$$V = \frac{1}{s + (1 - s)\gamma_b} . \tag{6.7}$$

Assume $s = 0$, meaning that no transfer from the victim is possible under H_d (the suspect did not assault the victim). The likelihood ratio then becomes

$$V = \frac{1}{\gamma_b} . \tag{6.8}$$

The circumstances of the case thus turn out to be a fundamental element of the analysis; the epistemic probabilities c, d, t and s are crucial variables in the assessment of the evidence. Furthermore, the probability of a group of fibres being present on the receptor beforehand, denoted as $b_{1,m}$, also plays an important role.

6.2.4 Fibres found on a car seat

Consider a scenario involving a car that belongs to a man who is suspected of abducting a woman and attempting rape. There is a single group of foreign red woollen fibres that have been collected on the passenger seat of the car. The victim was wearing a red woollen pullover. According to the suspect, no one other than his wife ever sits on the passenger seat. In addition, the car seats have been vacuumed recently.

The suspect denies that the victim has ever been in contact with the car. Propositions may then be defined and associated probabilities determined as follows:

- H, the issue of concern, with two states:
  - H_p: The victim sat on the passenger seat of the suspect's car;
  - H_d: The victim has never sat on the passenger seat of the suspect's car.
- C: This is the contact node and in this context means, 'The victim has been in contact with the seat'. Note that, under the states of H, it is clear that $Pr(C \mid H_p) = c = 1$ and $Pr(C \mid H_d) = d = 0$. The likelihood (6.5) is then $t + (1 - t)\gamma_b$.
- Under H_d, the transfer probability $Pr(T \mid C, H_d) = s = 0$, so the likelihood (6.6) is then γ_b.

Therefore, the likelihood ratio is

$$\frac{t + (1 - t)\gamma_b}{\gamma_b}. \tag{6.9}$$

6.2.5 The *Background* node

Using the fibre scenario as described in Section 6.2.2 and assuming the variables of interest to be as before, that is, H, C, T and E, consider the addition of a new node, denoted B, to represent the event to be known as *Background*. This event is thought of as a *matching* group of fibres that may be present on the passenger seat by chance alone. Let this event assume two states: either it is true (B), in which case there is a matching group of background fibres, or it is false ($\bar{B}$), meaning that there is no background of matching fibres *or* there is no background at all, that is, there are no fibres at all on the receptor.

As a consequence of this, a new definition of the states of the 'matching' node E becomes necessary. Two different situations are now taken into account under $\bar{E}$:

1. the presence of fibres on the receptor, which do not match the control from the victim;
2. no fibres at all on the receptor.

The probabilities for the node E are presented in Table 6.1.

Notice that the proposed modification using a distinct background node B does not avoid the problem of compound probabilities described at the end of Section 6.2.2; it merely shifts the problem from node E to node B. Now, there is a compound probability in the unconditional probability table associated with node B. The probability of B is a combination of the probability $b_{1,m}$ (the presence of a single group of fibres of size m), and γ, which is an expression of the frequency of the fibres' characteristics.

Table 6.1 Conditional probability table associated with the node E, where E has two states, E, the presence of fibres on the receptor whose characteristics match those of the control fibres from the victim, $\bar{E}$, no matching fibres or no fibres at all on the receptor. Factor T has two states, T, there was a transfer from the victim to the car seat, $\bar{T}$, there was no transfer from the victim to the car seat. Factor B has two states, B, there is a matching group of background fibres, $\bar{B}$ there is no background of matching fibres or there is no background at all.

	T :	T		$\bar{T}$	
	B :	B	$\bar{B}$	B	$\bar{B}$
E :	E	1	1	1	0
	$\bar{E}$	0	0	0	1

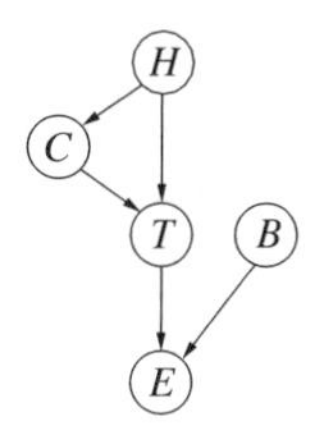

Figure 6.4 Extension of the Bayesian network for transfer from the victim/scene to the criminal (Figure 6.3): a node B has been added, representing the possibility of a matching group of background fibres being present on the receptor item. Each node has two states. Node H is the probandum, H_p, the suspect assaulted the victim, H_d, some person other than the suspect assaulted the victim. Node C, the victim has been in contact with the suspect (C), the victim has not been in contact with the suspect ($\bar{C}$). Node T, there was a transfer from the victim (T), there was no transfer from the victim ($\bar{T}$). Node E, the trace found on the suspect is similar to the control coming from the victim (E), the trace found on the suspect is not similar to the control coming from the victim, or, no fibres at all on the receptor ($\bar{E}$). Node B is the background, B, there is a matching group of background fibres, $\bar{B}$, there is no background of matching fibres or there is no background at all. [Adapted from Garbolino and Taroni (2002), Reproduced with permission from Elsevier.]

Figure 6.5 represents a comparison between the Bayesian network described in this section (Figure 6.4) and the one described in Section 6.2.2 (Figure 6.3). For both networks shown in Figure 6.5, probabilities have been assessed so that $c = s = t = 1$ and $d = 0$. The only differences are as follows:

- In the network represented in Figure 6.5(i), $Pr(E \mid \bar{T})$ is estimated by a parameter γ_b for which the value 0.05 has been assumed. Recall that γ_b is an abbreviation for the product of the probabilities $b_{1,m}$ and γ.
- In the network shown in Figure 6.5(ii), the probability table of the node E has been completed according to the indications given by Table 6.1. The prior probabilities specified for B and $\bar{B}$ are $b_{1,m}\gamma$, that is, γ_b, and $1 - b_{1,m}\gamma$, respectively.

As may be read from the labelled nodes, both evaluations provide the same result as far as the support for the proposition H is concerned. The example illustrates that structurally different Bayesian networks may offer the same results. However, it has become quite apparent that this requires considerable care in the way in which probabilities are interpreted, defined and specified.

The attentive reader may have noticed that there are still compound probabilities contained in the Bayesian network shown in Figure 6.5, notably those associated with the states of the node B. This inconvenience may be overcome by an extension of the current network, which is shown in Figure 6.6. Here, node *Background* is still present, but for convenience it is called P, for *Presence* (of matching fibres). This node has a somewhat different definition due to the fact that it is no longer a root node, but is conditioned on two separate nodes:

- Node N: This node represents the number of compatible fibre groups recovered on the receptor. In order to keep the arguments simple, two states will be assumed, that is, 0 groups (n_0) and 1 group (n_1).

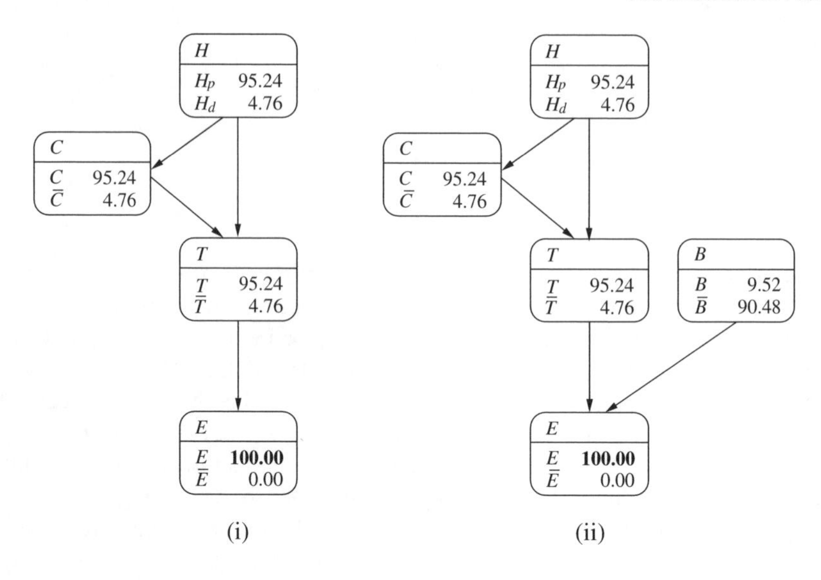

Figure 6.5 Comparison of the Bayesian networks shown in Figures 6.3 and 6.4 with the node E instantiated. Each node has two states. Node H is the probandum, H_p, the suspect assaulted the victim, H_d, some person other than the suspect assaulted the victim. Node C, the victim has been in contact with the suspect (C), the victim has not been in contact with the suspect ($\bar{C}$). Node T, there was a transfer from the victim (T), there was not a transfer from the victim ($\bar{T}$). Node E, the trace found on the suspect is similar to the control coming from the victim (E), the trace found on the suspect is not similar to the control coming from the victim, or, no fibres at all on the receptor ($\bar{E}$). Node B is the background, B, there is a matching group of background fibres, $\bar{B}$, there is no background of matching fibres or there is no background at all.

- Node Ty: This node is a representation of a background fibre group's characteristics, which may be assumed to be either 'matching' (Ty) or 'not-matching' ($\bar{T}y$).

It is now straightforward to see that the probabilities associated with the states n_1 and n_0 are $b_{1,m}$ and $1 - b_{1,m}$ respectively. For the states Ty and $\bar{T}y$ of the node Ty the probabilities are γ and $1 - \gamma$, respectively. In turn, the conditional probability table (Table 6.2) of node P may be completed logically.

A numerical example is given in Figure 6.6. Assume, as before, $c = s = t = 1$ and $d = 0$. Assume the parameters $b_{1,m}$ and γ take the values 0.5 and 0.01, respectively. Notice that the particular choice of the values for the latter parameters has been made in order to reproduce the initial values specified for B and $\bar{B}$ in the network shown in Figure 6.5(ii). The evaluation of the evidence using the network displayed in Figure 6.6 will thus yield the same result as each of the networks evaluated in Figure 6.4.

6.2.6 Background from different sources

In Section 6.2.2, a basic structure has been proposed for the evaluation of scenarios involving the one-way transfer of evidential material. The converging connection at the node T

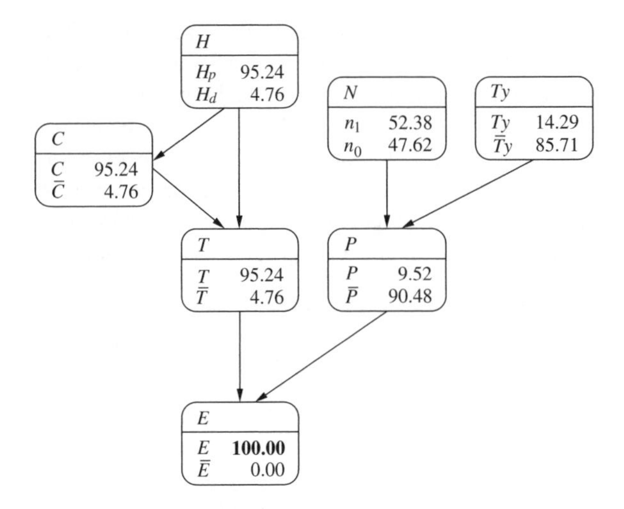

Figure 6.6 Extension of the Bayesian network shown in Figure 6.4 with the node E instantiated. Each node has two states. Node H is the probandum, H_p, the suspect assaulted the victim, H_d, some person other than the suspect assaulted the victim. Node C, the victim has been in contact with the suspect (C), the victim has not been in contact with the suspect ($\bar{C}$). Node T, there was a transfer from the victim (T), there was no transfer from the victim ($\bar{T}$). Node E, the trace found on the suspect is similar to the control coming from the victim (E), the trace found on the suspect is not similar to the control coming from the victim, or, no fibres at all on the receptor ($\bar{E}$). Node P, there is a presence of matching fibres (P), there is no presence of matching fibres ($\bar{P}$). Node N, n_0, no compatible groups recovered on the receptor, n_1, one compatible group recovered on the receptor. Node Ty, the characteristics of the background fibres match those of the control (Ty), the characteristics of the background fibres do not match those of the control ($\bar{T}y$).

($H \rightarrow T \leftarrow C$) is considered here in more detail. Some of the various conditional probabilities for C and T given different outcomes for the variable H will also be discussed.

Consider a case of suspected trafficking of illicit drugs. A suspect is apprehended and his clothing is examined for the presence of traces of drugs, such as cocaine, heroine, amphetamines and so on. The presence of such substances may indicate that the individual has somehow been in contact with such substances, for example, by a direct contact with a primary source (during preparation, packaging, and so on) or through exposure to a contaminated environment. This belief is based in part on the generally accepted assumption that microtraces of drugs are less likely to be found among persons unrelated to drug offences.

The suspect's clothing is vacuumed and the material collected is subjected to an analysis by some kind of detection apparatus. A portable ion mobility spectrometer (IMS) may be appropriate for such purposes. Imagine further that a positive result has been obtained for some target substance, for example, cocaine. In practice, such a result will need to be confirmed by a second, independent method such as gas chromatography/mass spectrometry (GC/MS). Notice, however, that the probabilistic analysis presented hereafter will solely be concerned with the results obtained by the IMS.

Table 6.2 Probability table associated to the node P. Factor P has two states, P, there is a presence of matching fibres, $\bar{P}$, there is no presence of matching fibres. Factor N has two states, n_0, no compatible groups are recovered on the receptor, n_1, one compatible group is recovered on the receptor. Factor Ty has two states, Ty, the characteristics of the background fibres match those of the control, $\bar{T}y$, the characteristics of the background fibres do not match those of the control.

	N :	n_1		n_0	
	Ty :	Ty	$\bar{T}y$	Ty	$\bar{T}y$
P :	P	1	0	0	0
	$\bar{P}$	0	1	1	1

The analytical results are assessed under activity level propositions:

- H_p: the suspect is involved in illegal drug trafficking, reselling and activities related thereto;
- H_d: the suspect is not involved in such activities (drug offences).

The definitions of the nodes C and T are taken to be slightly different from those in Section 6.2.2. Consider the variable C first. Here, C represents the proposition that the suspect has been in direct contact with illicit drugs. This may happen, for example, when packing or otherwise manipulating illicit drugs. These substances may be in different conditions, such as powder or compressed into blocs. The transfer that occurs during a direct contact with such items will be referred to as a 'primary transfer', whereas the item from which the transfer originates will be denoted as a 'primary source'.

Notice, however, that transfer may also be secondary or tertiary. These types of transfer may be illustrated as follows. Imagine a person whose clothing has been contaminated with cocaine through a primary transfer. That person takes a seat in a car, which may prompt a transfer of minute quantities of cocaine from the person's clothing to the car seat, for example. This sort of transfer would be termed *secondary transfer*. Some time later, another person sits on the same car seat. This is when a 'tertiary transfer' may occur, that is, from the car seat to the clothing of the second person.

For the node T, a rather general definition will be adopted. Factor T will denote the event 'there has been a transfer of a quantity Q of illicit drugs'. In the scenario above, conditional probabilities can be described as follows:

1. The probability of contact given the proposition that the suspect engages in drug trafficking: $Pr(C \mid H_p) = c$, where c can reasonably be assumed to be close to 1.

2. The probability of contact given the suspect has no relation at all with trafficking of illegal drug, $Pr(C \mid H_d) = d$, which can reasonably be assumed to be close to 0.

3. Assessment of transfer probabilities requires conditioning on both H and C. The assessments here depend crucially on the nature of the crime and the strategy chosen by the defence, as is illustrated by the following:

   - $Pr(T \mid C, H_p) = t$, where t represents the probability of transfer given that the suspect is involved in drug trafficking and given he has been in contact with a primary source. Generally, t may be set equal to 1.
   - $Pr(T \mid C, H_d) = s$, where s is the probability of transfer given that the suspect has been in contact with a primary source, but is not involved in drug trafficking. This parameter can also be set equal to 1. In fact, the probability of transfer given a contact with a primary source is independent of whether the individual is involved in illegal activities.
   - In Section 6.2.2, it has been assumed that there is no possibility for fibre transfer when there was no contact between the victim and the offender. Consequently, both $Pr(T \mid \bar{C}, H_p)$ and $Pr(T \mid \bar{C}, H_d)$ were set to 0. In the current scenario, this assumption may not necessarily hold as there may be a possibility of secondary or tertiary transfer. Let $Pr(T \mid \bar{C}, H_p)$ be denoted by t', the probability of transfer given no direct contact with a primary source despite involvement in drug trafficking. It is through consideration of t' as a non-zero probability that one may allow for the possibility of a secondary or tertiary transfer. Such transfer may occur, for example, through exposure to a contaminated environment (*e.g.*, a room or a vehicle). Denote $Pr(T \mid \bar{C}, H_d)$ by s'; this is the probability of a secondary or tertiary transfer given that the suspect is not involved in drug affairs. Probability s' also may assume a non-zero value. Notice, however, that the estimates t' and s' relate to two different populations. The parameter t' can be estimated through a study of contaminations among persons known to be involved in drugs affairs. The parameter s' may be estimated from a general population of innocent individuals. One also needs to be aware of alternative defence strategies. Consider, for example, the proposition put forward by the defence that 'the suspect is not involved in drug affairs but is on good terms with a third individual who is known to be involved in drug trafficking'. In such a situation, $Pr(T \mid \bar{C}, H_d)$ will be assessed in a population composed by individuals who have (non-criminal) relations with other individuals linked to drug affairs; the value of s' may thus be increased.

Notice that in scenarios in which $Pr(T \mid \bar{C}, H_p)$ and $Pr(T \mid \bar{C}, H_d)$ are not automatically set equal to 0, a node B, defined in Sections 6.2.2 and 6.2.5 as a possible matching background, is not needed. Similar situations involving second or tertiary transfer are frequently encountered in connection with other kinds of scientific evidence, such as fibres or gunshot residues.

It may appear difficult to obtain appropriate assessments for all of the various parameters mentioned above, rendering the evaluation of the evidence a complicated task. Indeed, scenarios involving microtraces of illicit drugs often relate to a highly specific set of circumstantial information for which perfectly suitable assessments may hardly ever be obtainable through surveys or simulations. Subjective expert judgments will therefore remain an integral part of evaluations. Methods exist that allow the treatment of such assessments in a probabilistically coherent way while the explicit specification of numerical probabilities is not a necessary requirement. Further details are given in Chapter 9. Section 9.1.5 provides a detailed discussion of an example involving microtraces of drugs.

6.2.7 A note on the *Match* node

In Section 6.2.2, a fibre scenario has been developed where the recovered fibres may either be associated with the offence or may have been present for reasons unconnected to the offence. In what follows, the situation in which the recovered evidence comes from the offender is developed through the construction of a Bayesian network which will show that (3.8) is a special case of a more general formula.

Imagine the following scenario: a stolen car is used in a robbery on the day of its theft. One hour later, the car is abandoned. During the night, the stolen vehicle is found by the police. On the polyester seats (lower and upper back), a group of n extraneous textile fibres is collected. The day following the robbery, a suspect is apprehended. His red woollen pullover is seized and submitted to the laboratory.

The *Match* node may be developed in more detail, as has been done previously for the *Background* node. In particular, instead of considering the presence (or not) of a match between the recovered material and the potential source, the scientist can consider characteristics of the recovered material and of the control material, separately. Consider this in terms of the example described in the next section.

6.2.8 A match considered in terms of components y and x

Using the notation introduced in Section 3.2, the evidential material M consists of two distinct parts: (a) the material recovered on the car seat, denoted as M_c, where c indicates the assumed relation to the *criminal* incident, and (b) the material extracted from the suspect's pullover, denoted as M_s, where s indicates the *suspect* as the source of the known material that is being used for comparative analysis. E consists of E_c and E_s, denoting the measurements taken on M_c and M_s respectively. The term 'match' is no longer used to describe the findings. The notation $Ev = (M, E)$ is used to summarise the overall available information.

For convenience, y will be used to denote the measurements E_c taken on the recovered material, whereas x will be used to denote the measurements E_s taken on the control, also termed *known material*. The practical example presented in Section 6.2.7 is still used. Thus, in this specific context, the forensic evidence is a consideration of the following attributes:

- y: represents a set of extrinsic (physical attributes such as quantity and position) and intrinsic features (chemical or physical descriptors such as analytical results) characterising the group of n recovered red woollen fibres;
- x: represents the extrinsic (physical attributes such as the sheddability) and intrinsic features of known red woollen fibres generated by the suspect's pullover.

The likelihood ratio is expressed as follows:

$$V = \frac{Pr(y, x \mid H_p, I)}{Pr(y, x \mid H_d, I)}$$

where

H_p: The suspect sat on the driver's seat of the stolen car;

H_d: The suspect has never sat on the driver's seat of the stolen car.

Note that H_d implies that *another person* sat on the driver's seat of the stolen car. This point is important in the assessment of the *transfer* probabilities as will be seen later. The previous equation can be expanded using the third law of probability.

$$V = \frac{Pr(y, x \mid H_p, I)}{Pr(y, x \mid H_d, I)} = \frac{Pr(y \mid x, H_p, I)}{Pr(y \mid x, H_d, I)} \times \frac{Pr(x \mid H_p, I)}{Pr(x \mid H_d, I)}. \quad (6.10)$$

It appears reasonable to assume that the probability of the characteristics of the suspect's pullover, x, does not depend on whether the suspect sat on the driver's seat of the stolen car. So, the second ratio of the right-hand side of (6.10) equals 1 and the likelihood ratio is reduced to:

$$V = \frac{Pr(y \mid x, H_p, I)}{Pr(y \mid x, H_d, I)} .$$

A further assumption is that, given H_d, x is not relevant when assessing the probability of y. The denominator of V may then be written as $Pr(y \mid H_d, I)$.

Next, a variable T with the following components (states) is defined:

- T_n: a group of n fibres has been transferred, has persisted, and has successfully been recovered on the driver's seat, implying that the group consisting of n fibres has not been there before;
- T_0: no group of n fibres has been transferred, persisted or recovered, meaning that the group of fibres found on the seat is unconnected with the offence.

Note that variable T accounts not only for the phenomenon of transfer but also for aspects such as persistence and recovery.

Considering these association propositions, the likelihood ratio V extends to the following equation, where the notation I for background information has been omitted for clarity:

$$\frac{Pr(y \mid x, H_p, T_n)Pr(T_n \mid x, H_p) + Pr(y \mid x, H_p, T_0)Pr(T_0 \mid x, H_p)}{Pr(y \mid H_d, T_n)Pr(T_n \mid H_d) + Pr(y \mid H_d, T_0)Pr(T_0 \mid H_d)} . \quad (6.11)$$

Consideration then needs to be given to eight conditional probabilities.

1. $Pr(y \mid x, H_p, T_n)$ represents the probability of observing a group of n red woollen fibres on the car seat given that the suspect wore a red woollen pullover, that he

sat on the driver's seat of the stolen car and that the group of fibres was transferred during the activity, had persisted and was recovered successfully. This probability can be estimated by $1 \times b_0$. This is an expression of the belief that y is the joint occurrence of a crime-related transfer (factor 1) and the probability of 0 groups being on the driver's seat beforehand (factor b_0).

2. $Pr(T_n \mid x, H_p)$ represents the probability that a group of n red woollen fibres was transferred, had persisted and was recovered successfully from the driver's seat, given that the suspect sat on the driver's seat of the stolen car. This represents the probability, say t_n, that the fibres had been transferred (from the suspect's pullover), remained and recovered. This probability depends on *physical* characteristics (*e.g.*, sheddability) of the suspect's pullover. It is assumed that the characteristics are from the control group because the scientist assesses the probability under H_p. Note that the phenomenon of *transfer* is not only characterised by the material that is transferred, but also by the phenomena of *persistence* and *recovery*.

3. $Pr(y \mid x, H_p, T_0)$ is the probability that a group of n red woollen fibres is observed on the driver's seat given that the suspect wore a red woollen pullover, that he sat on the driver's seat of the stolen car and that this group of fibres was not transferred during the activity, or did not persist or was not successfully recovered. If the group has not been transferred, this means that it was present on the seat before the activity. Let this be represented by a term $b_{1,n} \times \gamma$, where $b_{1,n}$ represents the probability of a single group of n, a comparable number, of fibres being present on the driver's seat by chance alone, and γ is the estimated frequency of the compared characteristics from y in extraneous groups (of similar sizes) of fibres found on car seats.

4. $Pr(T_0 \mid x, H_p)$ represents the probability that no group of fibres was transferred from the suspect's pullover to the driver's seat, had persisted and was recovered from there. This probability, t_0, is estimated given that the suspect sat on the driver's seat, H_p.

The numerator of the likelihood ratio is then $b_0 t_n + b_{1,n}\gamma t_0$.

Note that in Section 6.2.2, the denominator did not require a detailed development as it was assumed that the group of fibres, had it not been transferred by the suspect, was present on the receptor for reasons unconnected to the crime. A different approach will be followed here. Notably, the group of recovered fibres will be considered to be left potentially by the (true) offender, which, under H_p, is someone other than the suspect. The terms comprising the denominator of (6.11) are outlined below.

5. $Pr(y \mid H_d, T_n)$ represents the probability of observing a group of n red woollen fibres, given that the suspect never sat on the driver's seat of the stolen car and that the group of fibres was transferred during the activity, had persisted and was successfully recovered. If the suspect never sat on the driver's seat and the group has been transferred, this means that the driver's seat did not have this group of fibres before the commission of the crime and the event of the shared characteristics is one of chance. This probability is $b_0 \times \gamma$.

6. $Pr(T_n \mid H_d)$ represents the probability that a group of n red woollen fibres was transferred, had persisted and was successfully recovered from the driver's seat,

given that the suspect never sat on the driver's seat of the stolen car. This means that this probability, say t'_n, has to be estimated given that the fibres have been transferred from the (true) *offender's* garment, which is an item different from the suspect's garment. This probability thus depends on the *physical* characteristics of an *unknown garment.*

7. $Pr(y \mid H_d, T_0)$ is the probability that a group of n red woollen fibres is observed on the driver's seat, given that the suspect never sat on the driver's seat and that this group of fibres was not transferred during the activity, had not persisted or was not recovered successfully. If the group was not transferred, it was present on the seat before the commission of the crime. The probability of the joint occurrence of a group of foreign fibres on the driver's seat and a set of characteristics described by y then is $b_{1,n} \times \gamma$.

8. $Pr(T_0 \mid H_d)$ represents the probability that no group of fibres was transferred from the offender's garments to the driver's seat, had persisted and was recovered. This probability, denoted as t'_0, assumes that the suspect never sat on the driver's seat and, thus, another individual sat in the stolen car.

The denominator of the likelihood ratio is $b_0\gamma t'_n + b_{1,n}\gamma t'_0$. The likelihood ratio (6.11) is then

$$V = \frac{b_0 t_n + b_{1,n}\gamma t_0}{b_0\gamma t'_n + b_{1,n}\gamma t'_0}. \tag{6.12}$$

This expression illustrates that, $Pr(T_n \mid x, H_p) \neq Pr(T_n \mid H_d)$ and $Pr(T_0 \mid x, H_p) \neq Pr(T_0 \mid H_d)$. Therefore, as the probabilities are estimated throughout controlled experiments using the garments involved under propositions H_p and H_d, it is reasonable to assume that the estimates are different.

6.2.9 A structure for a Bayesian network

The proposed probabilistic approach involves four variables, covering various propositions and types of evidence, notably *H*, *T*, *X*, and *Y*. These variables are defined as follows:

H : The suspect sat on the driver's seat of the stolen car. This variable has two states: H_p, the suspect sat on the driver's seat of the stolen car, and H_d, the suspect has never sat on the driver's seat of the stolen car; another person sat on that seat.

T : This variable denotes the transfer, persistence and recovery of a group of fibres. The variable has two states, T_n and T_0, denoting the transfer of a group of n and 0 fibres, respectively.

X : This variable represents the intrinsic and extrinsic features of the known suspect's pullover. Let X have two states, x and $\bar{x}$. For example, x may be red woollen fibres having a specified set of extrinsic features. For the sake of simplicity, let $\bar{x}$ denote the union of all sets of possible features other than x.

Y : It is tempting to restrict the definition of Y to the intrinsic and extrinsic descriptors only. However, further considerations will be necessary in order to obtain a sensible definition of the states of Y. For example, the way in which the states of Y are defined is partially determined on what further variables Y depends on and how these further variables are defined. For the time being, let Y be simply denoted with the term *findings*, whereas y designates a specific outcome, described in terms of intrinsic and extrinsic features.

On the basis of the eight conditional probabilities previously developed (Section 6.2.8), the following structural dependencies may be defined

- Variables H, T and X are relevant for variable Y. This means that the four variables can be related structurally as shown in Figure 6.7(i).
- Variable T, denoting 'transfer', depends on both X and H. This dependency is illustrated by Figure 6.7(ii). Note the importance of the extrinsic features of X. The physical attributes, such as the sheddability of the control material, determine its capacity for transferring fibres. Given that H is relevant for both T and Y, and considering that the extrinsic features of X are relevant for assessing probabilities of transfer, Figure 6.7(ii) can be modified to the graph shown in Figure 6.7(iii). Here, Y is still dependent on H, but indirectly, through the serial connection $H \rightarrow T \rightarrow Y$. In fact, if node T is instantiated, the path between H and Y is blocked, meaning that Y is conditionally independent of H. Note that the transfer node of the Bayesian network takes into account three phenomena, transfer, persistence and recovery. A more detailed development of the 'transfer' node T is proposed in Section 6.4.
- It has been seen that node Y depends on H (through T), T and X. A new variable is necessary in order to account for the fact that the scientist may observe y even if no transfer occurred, that is, T_0 was true. One possibility for this is that a group of fibres described as y was present on the receptor before the event of interest (proposition H) occurred. Call this event, B, for *Background*. The variable B is introduced in Figure 6.7(iv), depicting the direct influence on Y. The states of the variable B are b_0 and $b_{1,n}$, representing, respectively, the probabilities of no foreign group being present, and of the joint occurrence of a single group of n foreign fibres on the driver's seat and a set of characteristics described by y.
- Given the definitions of the nodes X, T, and B, all relevant information is now available in order to define the states of node Y. In analogy to node X, node Y is characterised by y, representing the observed features, that is, red woollen fibres with specified extrinsic characteristics, and $\bar{y}$, representing all combinations of features other than those described by y. However, two further situations need to be considered and they are a direct consequence of the previously defined set of parental variables of Y. The first situation is concerned with the joint occurrence of T_n and $b_{1,n}$. Under these circumstances, the scientist is faced with two groups of fibres. The state of Y corresponding to this situation will thus be denoted as *Two groups*. The second situation is one in which both T_0 and b_0 are true, that is, no fibres at all are present. In summary, there are four different states for the node Y, notably y, $\bar{y}$, *Two groups* and *No group of fibres*. Notice that some of these states may be specified in more detail, as required. For example, the state *Two groups* may be divided into a state representing two groups of fibres being present, each of which has the features y, a state representing two groups of fibres, each of which has features different from y

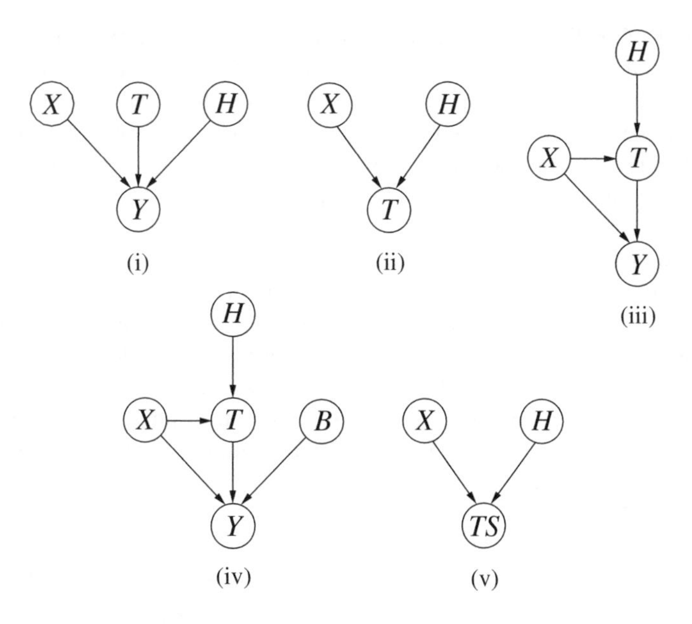

Figure 6.7 Local Bayesian networks for modelling selected aspects of a fibre transfer scenario. Node H has two states, H_p, the suspect sat on the driver's seat of the stolen car, H_d, the suspect never sat on the driver's seat of the stolen car, another person sat on that seat. Node T has two states, T_n and T_0, denoting the transfer of a group of n and 0 fibres, respectively. Node X represents the intrinsic and extrinsic features of the known suspect's pullover. It has two states, x and $\bar{x}$. For example, x may be red woollen fibres having a specified set of extrinsic features and $\bar{x}$ the union of all sets of possible features other than x. The node Y is denoted as *findings*, there are four states. State y designates a specific outcome, described in terms of intrinsic and extrinsic features, for a single group of fibres. State $\bar{y}$ represents all combinations of features other than those described by y. State *Two groups* is the situation in which there are two groups of fibres. State *No group* is the situation in which there are no groups of fibres. Node B has two states, b_0 when no foreign group of fibres is present and $b_{1,n}$ when there is the joint occurrence of a group of foreign fibres and a group with a set of characteristics described by y. Node TS has two states, x and $\bar{x}$, where $\bar{x}$ represents extrinsic characteristics other than those defined by the known source. [Adapted from Aitken and Taroni (2004), reproduced with permission from John Wiley & Sons Ltd.]

and so on. The consequence of this is that the size of the conditional probability table increases. However, as the Bayesian network described in this section is essentially intended to approach the scenario introduced in Section 6.2.7, that is, a scenario involving *one* group of n foreign fibres, no further states will be adopted here for the node Y. Conditional probabilities for node Y are presented in Table 6.3.

- Difficulties may arise in the assessment of the conditional probabilities $Pr(T_n \mid H_d)$ and $Pr(T_0 \mid H_d)$. These estimates assume the alternative proposition, H_d, to be true, that is, they are an expression of the transfer capacity of the true offender's garment,

Table 6.3 Conditional probabilities assigned to the factor Y. Factor Y is denoted as *findings*; there are four states. State y designates a specific outcome, described in terms of intrinsic and extrinsic features, for a single group of fibres. State $\bar{y}$ represents all combinations of features other than those described by y. State *Two groups* is the situation in which there are two groups of fibres. State *No group* is the situation in which there are no groups of fibres. Factor TS has two states, x and $\bar{x}$, where $\bar{x}$ represents extrinsic characteristics other than those defined by the known source. Factor T has two states, T_n and T_0, denoting the transfer of a group of n and 0 fibres, respectively. Factor B has two states, b_0, when no foreign group of fibres is present and $b_{1,n}$, when there is the joint occurrence of a group of foreign fibres and a group with a set of characteristics described by y. The probability that the characteristics of a group of fibres matches by chance alone the characteristics of the control group of fibres is γ. [Adapted from Aitken and Taroni (2004), reproduced with permission from John Wiley & Sons Ltd.]

	TS:	x				$\bar{x}$			
	T:	T_n		T_0		T_n		T_0	
	B:	b_0	$b_{1,n}$	b_0	$b_{1,n}$	b_0	$b_{1,n}$	b_0	$b_{1,n}$
Y :	y	1	0	0	γ	0	0	0	γ
	$\bar{y}$	0	0	0	$1-\gamma$	1	0	0	$1-\gamma$
	Two groups	0	1	0	0	0	1	0	0
	No group	0	0	1	0	0	0	1	0

different from the one of the suspect. Accordingly, it is no longer the extrinsic features (*i.e.*, sheddability) assumed under X that influence the potential of transfer, but the extrinsic features of an alternative source material. This issue may be handled by defining a new node, say TS for *True Source*. This node will allow the scientist to consider the influence of the extrinsic features defined under X and those defined by the alternative source. The states of the variable TS are x and $\bar{x}$, where the latter state represents extrinsic characteristics other than those defined by the known source, that is, the suspect's pullover. If H_p is true, then the characteristics of the source from which a transfer occurred are those of the suspect's garment, that is, the variables X and TS are in strictly the same state. If, on the other hand, H_d is true, then the characteristics of X may be different from those of TS. This may be translated graphically as a convergent connection (*see* Figure 6.7(v)). The probabilities associated to the node TS are presented in Table 6.4. As a consequence, the direct arrow from X to T disappears in favour of a link between TS and T. If H_p is true, then logically the extrinsic features of X are relevant for assessing t_n and t_0, the probabilities of transfer. Notice that it is assumed that $t_0 = 1 - t_n$. If H_d is true, then logically the extrinsic features of the alternative source are of importance for assessing the transfer probabilities, denoted as t'_n and t'_0, respectively. Table 6.5 summarises the probabilities for node T, and Figure 6.8 illustrates the complete Bayesian network for evaluating the fibres scenario under activity level propositions.

Table 6.4 Conditional probabilities assigned to the node TS. Factor TS has two states, x and $\bar{x}$, where $\bar{x}$ represents extrinsic characteristics other than those defined by the known source. Factor H has two states, H_p, the suspect sat on the driver's seat of the stolen car, H_d, the suspect never sat on the driver's seat of the stolen car; another person sat on that seat. Factor X represents the intrinsic and extrinsic features of the known suspect's pullover. It has two states, x and $\bar{x}$. For example, x may be red woollen fibres having a specified set of extrinsic features and $\bar{x}$ the union of all sets of possible features other than x. The probability that the characteristics of a group of fibres matches by chance alone the characteristics of the control group of fibres is γ.

	H:	H_p		H_d	
	X:	x	$\bar{x}$	x	$\bar{x}$
TS :	x	1	0	γ	γ
	$\bar{x}$	0	1	$1-\gamma$	$1-\gamma$

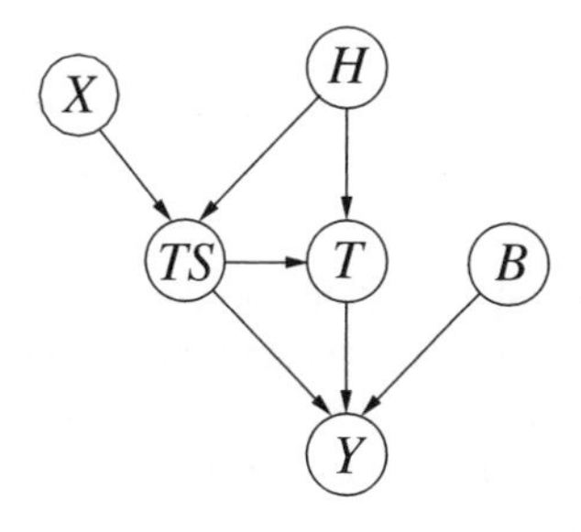

Figure 6.8 Bayesian network for evaluating fibre evidence under activity level propositions. Node H has two states, H_p, the suspect sat on the driver's seat of the stolen car, H_d, the suspect never sat on the driver's seat of the stolen car; another person sat on that seat. Node T has two states, T_n and T_0, denoting the transfer of a group of n and 0 fibres, respectively. Node X represents the intrinsic and extrinsic features of the known suspect's pullover. It has two states, x and $\bar{x}$. Node Y is denoted as *findings*; there are four states. State y designates a specific outcome, described in terms of intrinsic and extrinsic features, for a single group of fibres. State $\bar{y}$ represents all combinations of features other than those described by y. State *Two groups* is the situation in which there are two groups of fibres. State *No group* is the situation in which there are no groups of fibres. Node B has two states, b_0, when no foreign group of fibres is present and $b_{1,n}$, when there is the joint occurrence of a group of foreign fibres and a group with a set of characteristics described by y. Node TS has two states, x and $\bar{x}$, where $\bar{x}$ represents extrinsic characteristics other than those defined by the known source.

Table 6.5 Conditional probabilities assigned to the factor T. Factor T has two states, T_n and T_0, denoting the transfer of a group of n and 0 fibres, respectively. Factor H has two states, H_p, the suspect sat on the driver's seat of the stolen car, H_d, the suspect never sat on the driver's seat of the stolen car; another person sat on that seat. Factor TS has two states, x and $\bar{x}$, where $\bar{x}$ represents extrinsic characteristics other than those defined by the known source. Probability t_n is the probability that a group of n fibres had been transferred, persisted and recovered from the suspect's pullover. Probability t'_n is the probability that a group of n fibres had been transferred, persisted and recovered from an alternative source than the suspect's pullover.

	H:	H_p		H_d	
	TS:	x	$\bar{x}$	x	$\bar{x}$
T :	T_n	t_n	0	t'_n	t'_n
	T_0	$1 - t_n$	1	$1 - t'_n$	$1 - t'_n$

Referring to Table 6.5, note that:

1. If H_p is true and the state $\bar{x}$ characterises variable TS, then the probability of transfer equals 0, an impossible event, because if H_p is true then logically $X = x$ has to be true.
2. If H_d is true and the state $\bar{x}$ characterises variable TS, then the probability of transfer may be taken as t'_n. In this setting, the network reproduces the results that may be obtained from (6.12). This formula does not take into account situations in which $Pr(T \mid H_d, x) \neq Pr(T \mid H_d, \bar{x})$, as it addresses transfer phenomena on a more general level.

6.2.10 Evaluation of the proposed model

In situations in which the presence of the evidence given the alternative proposition arises from chance alone, because the alleged activity is assumed not to have happened (see also the scenario developed in Section 6.2.2), the numerator of the likelihood ratio is unchanged from (6.12), but the denominator changes. In fact, if it is assumed that the alleged activity has not occurred, there is no need to develop $Pr(y|H_d)$ using the association propositions T_n and T_0. For such a scenario, the likelihood ratio is

$$V = \frac{b_0 t_n + b_{1,n}\gamma t_0}{b_{1,n}\gamma}. \tag{6.13}$$

The aim is to try to incorporate into a single Bayesian network the scenario of the stolen car (6.12) and to be flexible enough to consider other evidence scenarios (6.13).

A Bayesian network as shown in Figure 6.8 allows the scientist to cope with both situations. A change from the situation involving an alternative donor of the evidence to the situation involving chance alone is operated through changes in the probability table associated to the node T. Notably, when the alternative proposition H_d is that the suspect never sat on the car seat, that is, the group of n foreign fibres is present by chance alone, then the value of t'_n will be set to zero. In fact, $Pr(T_n \mid H_d)$, the probability that a group of n red woollen fibres was transferred, has persisted and was recovered successfully from the driver's seat, given that the suspect *never* sat on the driver's seat of the stolen car, represents an impossible event, because the alleged activity is assumed not to have happened.

In order to use the model to evaluate the likelihood ratio, the following instantiations need to be made

- For evaluation of the numerator, the nodes H and X need to be set to the states H_p and x, respectively.

- For evaluation of the denominator, it is sufficient to instantiate the node H to H_d. Notice here that any instantiations made at the node X would not affect the probabilities associated with Y.

6.3 Cross- or two-way transfer of evidential material

The Bayesian network previously developed in Section 6.2.9 (Figure 6.8) can be used to approach another category of forensically relevant situations, notably those that occur when two persons or objects were in contact: the cross- or two-way transfer of (trace-)evidence. Consider an example discussed in Champod and Taroni (1999).

A stolen vehicle is used in a robbery on the day of its theft. An hour later it is abandoned. The vehicle is found by the police a few hours later. On the polyester seats, which were recently cleaned with a car vacuum cleaner, extraneous textile fibres are collected. The car owner lives alone and has never lent his vehicle to anyone. The owner wears nothing but cotton. The day following the robbery, a suspect is apprehended, his red woollen pullover and his denim jeans are confiscated.

On the driver's seat, one group of relevant foreign fibres is collected. It consists of a large number of, say, n red woollen fibres. This evidence, denoted as E_1, is a combination of the form $\{y_1, x_1\}$, where y_1 refers to the recovered fibres on the car seat and x_1 refers to known (control) material from the suspect's red woollen pullover. It is assumed here that the group of fibres on the driver's seat has been transferred from the offender's clothing.

One term that is introduced here is that of *foreign fibre groups* (FFG). An FFG consists of fibres that can be distinguished from fibres from a known source (either associated with the suspect or associated with an object such as a car seat).

On both the suspect's pullover and denim jeans there are many foreign fibre groups. One of them consists of twenty extraneous black fibres. These fibres correspond, in some sense, to the fibres of which the driver's seat is composed. This evidence, denoted as E_2, is a combination of the form $\{y_2, x_2\}$ where y_2 refers to the twenty recovered fibres on the suspect's clothes and x_2 refers to known material from the driver's seat.

Define the competing propositions at the activity level as follows:

H_p : the suspect sat on the driver's seat of the stolen car;

H_d : the suspect never sat on the driver's seat of the stolen car.

When two individuals or an individual and an object, such as a car seat, are in contact, a reciprocal transfer of material is usually involved. The two sets of recovered traces then have to be considered as dependent. In fact, if a transfer has occurred in one direction and the expert has recovered traces potentially characterising this transfer, then the expert would, in general, expect to find trace evidence characterising the transfer in the other direction. Stated otherwise, the presence of evidence transferred in one direction provides information about the presence of evidence transferred in the other direction. From Chapter 3, with I omitted for ease of notation, Bayes' formula for the combined set of evidence is

$$\frac{Pr(H_p \mid E_1, E_2)}{Pr(H_d \mid E_1, E_2)} = \frac{Pr(E_2 \mid H_p, E_1)}{Pr(E_2 \mid H_d, E_1)} \times \frac{Pr(H_p \mid E_1)}{Pr(H_d \mid E_1)}$$

$$= \frac{Pr(E_2 \mid H_p, E_1)}{Pr(E_2 \mid H_d, E_1)} \times \frac{Pr(E_1 \mid H_p)}{Pr(E_1 \mid H_d)} \times \frac{Pr(H_p)}{Pr(H_d)}. \quad (6.14)$$

The value of the evidence is then

$$V = \frac{Pr(E_2 \mid H_p, E_1)}{Pr(E_2 \mid H_d, E_1)} \times \frac{Pr(E_1 \mid H_p)}{Pr(E_1 \mid H_d)}. \quad (6.15)$$

The second ratio in (6.15) is equal to $1/\gamma_1$ where γ_1 is the estimated frequency of the characteristics from y_1 in extraneous groups of fibres of similar size found on seats of stolen cars (Champod and Taroni 1999). Notice that if the suspect has never sat on the driver's seat of the stolen car (H_d), another individual sat on it; so the transfer characteristics of the unknown garment of that individual, the true offender, are of importance. The parameters t and t' refer to the transfer probabilities from the suspect and from the true offender respectively. In cases in which these two parameters would be considered to be different, changes need to be adopted in Table 6.5.

The first ratio in (6.15) accounts for a group of twenty fibres (y_2) present on the suspect's clothing. In the numerator, it is assumed that the group of fibres has been transferred while the suspect has been in contact with the car's seat. A *conditional transfer probability*, denoted u_{20}, is estimated. This estimate is conditional on knowing E_1, that is, a group of foreign fibres recovered on the car seat. The conditional transfer probability is highly case-dependent as it is strongly influenced by the kind of textiles involved as well as their properties, such as sheddability. The conditional transfer probability thus requires a specific assessment in each case. The denominator is given by γ_2, the estimated frequency of the compared characteristics from y_2 in extraneous groups of fibres of similar size found on the clothing of potential offenders. The first ratio is then equal to u_{20}/γ_2. The value of the evidence is then (Champod and Taroni 1999)

$$V = \frac{u_{20}}{\gamma_1 \gamma_2}. \quad (6.16)$$

Figure 6.9 illustrates a possible Bayesian network for such a scenario. It consists of two separate networks previously presented in Figure 6.8. The assumed dependencies between the two sets of evidence are expressed in connections between the network fragments, which operate as follows:

- An arrow is used to link the transfer nodes, T_1 and T_2. In fact, if traces compatible with transfer from a potential offender's garment are found on a car's seat,

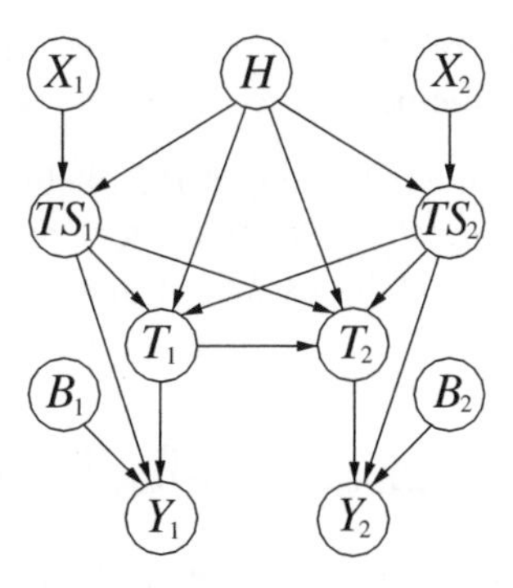

Figure 6.9 A Bayesian network for evaluating the cross-transfer of textile fibres. Node H has two states, H_p, the suspect sat on the driver's seat of the stolen car, H_d, the suspect never sat on the driver's seat of the stolen car; another person sat on that seat. Nodes T_1 and T_2 have two states, T_n and T_0, denoting the transfer of a group of n and 0 fibres, respectively. Node X_1 represents the intrinsic and extrinsic features of the known suspect's pullover. It has two states, x_1 and $\bar{x_1}$. Node X_2 represents the intrinsic and extrinsic features of the known fibres from the car seat. It has two states, x_2 and $\bar{x_2}$. Node Y_1 refers to the recovered fibres on the car seat. It has four states, y_1, $\bar{y}_1$, *Two groups* and *No groups*. Node Y_2 refers to the recovered fibres on the suspect's clothes. It also has four states, y_2, $\bar{y}_2$, *Two groups* and *No groups*. Node B_1 has two states, b_0, when no foreign group of fibres is present on the car seat and $b_{1,n}$, when there is the joint occurrence of a group of foreign fibres and a group with a set of characteristics described by y_1. Node B_2 has two states, b_0, when no foreign group of fibres is present on the suspect's pullover and $b_{1,n}$, when there is the joint occurrence of a group of foreign fibres and a group with a set of characteristics described by y_2. Node TS_1 has two states, x_1 and $\bar{x}_1$, where $\bar{x}_1$ represents extrinsic characteristics other than those defined by the known source of the suspect's pullover. Node TS_2 has two states, x_2 and $\bar{x}_2$, where $\bar{x}_2$ represents extrinsic characteristics other than those defined by the known source of the car seat.

should the scientist expect to find transfer evidence compatible with the car's seat on the suspect's garment? In other words, is reciprocal transfer predictive? What expectations should a scientist have? Generally, if a transfer has occurred in one direction and the scientist has recovered traces characterising this transfer, then the scientist may expect to find trace evidence characterising the transfer in the other direction. Therefore, in situations in which the two recovered traces (or sets of traces) should be considered as dependent, an arrow may be used to express this relationship.

- Arrows are also used to indicate that the physical attributes of the textiles involved are assumed to bear on the transfer, persistence and recovery of fibres in either direction: from the suspect or offender to the car seat and vice versa. The variables TS_1 and TS_2 thus influence both transfer nodes T_1 and T_2. Variables such as fibre length and sheddability of the suspect's garment influence the transfer of fibres to the car's seat. On the other hand, physical characteristics of the textile composition of the car seat influence the transfer of fibres to the suspect's (offender's) garments. Physical characteristics also affect the persistence of the transferred fibres.

It is often useful to examine the behaviour of the likelihood ratio for different values of its components (6.16). A cross-transfer of trace evidence does not necessarily increase the strength of the evidence; this point is discussed by Champod and Taroni (1999) and Aitken and Taroni (2004).

6.4 Increasing the level of detail of selected nodes

More elaborate Bayesian networks may be developed, for example, by expanding a particular node and taking into account additional information related to the node of interest. An example has been discussed in Section 6.2.5, where information summarised by node B has been split into the separate nodes P, N and Ty.

A second example can be considered at the node G (*see* Section 4.1), which has been taken as an expression of the *relevance* of evidential material. Stoney (1991, p. 126) defined the term 'relevance' using the following terms:

> During the collection of crime scene materials investigators seek traces that may have come from the offender. Often these are transferred materials in their particle form. Depending on the type of material, its location, condition and abundance, the chances that it has a true connection with the offender may range from very likely to practically nil. A true connection means simply that the trace material came from the offender and is, therefore, a valid predictor of some traits of the offender. Crime scene material that has this true connection is said to be relevant in that it is relevant to the evaluation of suspects as possible offenders.

Thus information concerning (a) the location, (b) the condition and (c) the abundance of the recovered stain – which represent information that the scientist needs to consider when assessing the relevance of the recovered trace – could be represented by three separate and independent nodes from which node G receives entering arcs. In Section 4.1, the assessment of node G was based on such detailed information but it has not been represented explicitly; the single node G contained all the information. By expanding the node G, more careful assessment of each parameter is possible. However, one would be required to think more deeply about how parameters such as location, condition and abundance affect one's assessment of the relevance term: The effect of adopting parental variables for G is that there is an increase in the size of G's probability table. The assessment of these probabilities is highly case-dependent and is not examined in further detail here.

Instead, the attention of the reader is drawn to an additional example. It is closely linked to the Bayesian networks developed in this chapter that themselves relate to T, a node describing transfer, persistence and recovery of material on some kind of receptor (a person or a scene). In situations in which more detailed information is available on these three distinct phenomena, the node T may be extended to a more elaborated Bayesian network. For illustration, some of the previously discussed fibre scenarios are used to approach a new scenario involving glass fragments, another frequently encountered kind of trace evidence.

Imagine the following scenario. A burglar has smashed the window of a house. A suspect is arrested and a quantity, say Q_r, of glass fragments is recovered from the surface of his clothing. These fragments match, in some sense, the type of glass of the house window. Instead of using a single transfer node T, as depicted, for example, in Figure 6.3, it is possible to extend the net by a level of detail as outlined below.

- The quantity of glass fragments recovered from the suspect's pullover, (Q_r), depends on both the quantity of glass fragments that have persisted (Q_p) and on the performance of the searching technique. The latter parameter may be assessed by the proportion of glass fragments lifted from the pullover, denoted by a variable P_l. For such a variable a variety of different states may be adopted as required. A more simple approach would consist in adopting, for example, a binary node with states *good* and *poor*.
- The quantity of glass fragments that have persisted on the suspect's clothing, Q_p, depends on the quantity Q_t of glass fragments transferred and on the proportion P_s of glass fragments that were shed between the time of transfer and the time of the examination of the pullover.
- The quantity of transferred glass fragments (Q_t) depends on the appropriate ultimate probandum (H).

The states of the variables can be *none, few, many*, or *none, small, large*.

Further distinctions may also be made at the node Q_t, which represents the quantity of transferred glass fragments. As noted above, Q_t is assumed to depend on the variable H, which denotes the action committed by the true offender. Estimates of Q_t seem logically different when assuming, for example, that the suspect broke the window, that he only stood nearby when someone else broke it, or, that the suspect had nothing to do with the event under investigation.

A variety of other variables can be theoretically proposed. Note that the term 'theoretically' is used to emphasise the idea that results from simulations can be used to support the argument that a variable affects or does not affect another variable. For the assessment of uncertainties in relation to the transfer of glass fragments parameters such as:

- the type of window under investigation,
- the size of the window,
- the distance between the window and the receptor (note that this information is closely related to the way the window has been broken),

may be considered (Curran et al. 2000).

Figure 6.10, inspired by Halliwell et al. (2003), illustrates a Bayesian network for some of the variables that have been discussed throughout this section.

Notice that, on the one hand, developments of the kind considered here tend to have fewer analogies with the traditional likelihood ratio formulae proposed in forensic literature. The reason for this is that more variables with varying definitions and interrelationships are employed. In part, this is a direct consequence of the graphical environment, which allows one to consider more variables (*i.e.*, sources of uncertainty) with less effort. On the

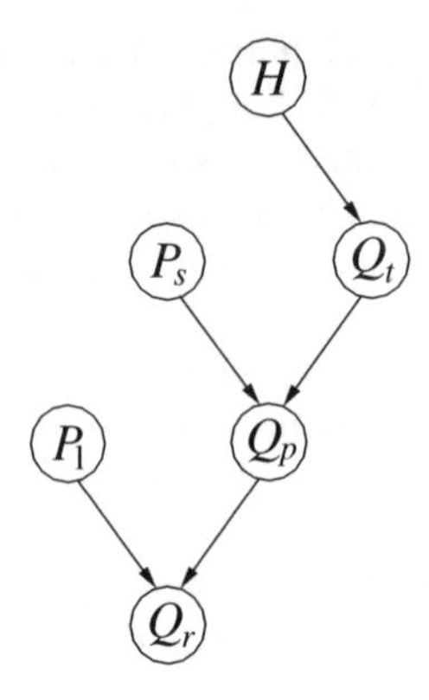

Figure 6.10 A Bayesian network for evaluating sources of uncertainty relating to a scenario involving a transfer of glass fragments. Node Q_r represents the quantity of glass fragments recovered from the suspect's pullover. Node Q_p represents the quantity of glass fragments that have persisted. Node Q_t represents the quantity of glass fragments transferred. Node P_l represents the proportion of glass fragments lifted from the pullover. It can be a binary node with states *good* and *poor*. Node P_s represents the proportion of glass fragments that were shed between the time of transfer and the time of the examination of the pullover. Node H represents the ultimate probandum, which is the action committed by the true offender. It has two states, H_p, the suspect broke the window, H_d, some other person broke the window.

other hand, the graphical structures become more tentative, as there may be fewer reference points, for example, in the form of likelihood ratio formulae, against which the validity of a model may be checked.

6.5 Missing evidence

Until now, discussion has mainly focused on evidence that is present. However, there are other forms of evidence that forensic scientists may encounter. One such form is missing evidence. Evidence has been defined as *missing* if it is expected, but is either not found or is not produced on request (Schum 1994).

A general Bayesian formula for the problem of missing evidence has been proposed by Lindley and Eggleston (1983). It will be used here to derive and discuss a structure for a Bayesian network. At the same time, an emphasis made in Chapter 4 is reiterated: A likelihood ratio formula that is already given can assist in the search for an appropriate network structure.

The example presented in Lindley and Eggleston (1983) relates to a collision between two motor cars. The scenario is as follows:

> The plaintiff sues the defendant, claiming that it was his car that collided with the plaintiff's. The evidence of identification is weak, and the defendant relies on the fact that, his car being red, the plaintiff has produced no evidence that

any paint, red or otherwise, was found on the plaintiff's car after the collision (Lindley and Eggleston 1983, p. 87).

6.5.1 Determination of a structure for a Bayesian network

There are three variables that can be derived from the above example on missing evidence.

1. The variable H represents the event that the defendant is guilty of the offence for which he has been charged. H may either be true or false, there are two states: H_p, the defendant is guilty and H_d, the defendant is not guilty.

2. The variable M represents the event that evidence is missing. This variable can take the values true or false and will be abbreviated with M, if the evidence is missing and $\bar{M}$, if the evidence is not missing.

3. The variable E represents the form of the evidence that is missing. For the example of the defendant's red car, three possible values are proposed for E:

   - e_1: there was red paint on the plaintiff's car;
   - e_2: there was paint, but it was not red;
   - e_3: there was no paint.

A likelihood ratio to assist the court in examining the effect that evidence is missing (M) has on the truth or otherwise of the proposition of interest (H) is given by (Lindley and Eggleston 1983):

$$\frac{Pr(M \mid H_p)}{Pr(M \mid H_d)} = \frac{Pr(M \mid e_1)Pr(e_1 \mid H_p) + Pr(M \mid e_2)Pr(e_2 \mid H_p) + Pr(M \mid e_3)Pr(e_3 \mid H_p)}{Pr(M \mid e_1)Pr(e_1 \mid H_d) + Pr(M \mid e_2)Pr(e_2 \mid H_d) + Pr(M \mid e_3)Pr(e_3 \mid H_d)} . \tag{6.17}$$

It is easily seen that the construction of a Bayesian network based upon an existing formula has the advantage that the number and definition of the nodes is already given. Here, there are three variables, H, M and E, for which nodes are required.

In order to find a graphical representation that correctly represents the conditional dependencies as specified by the likelihood ratio given above (6.17), it is helpful to follow a two-stage approach. According to Lindley and Eggleston (1983), (6.17) contains all the relevant considerations for the paint scenario, notably conditional probabilities for:

1. the various forms of the evidence given that the prosecution proposition (H_p) were true and given the defence proposition (H_d) were true,

2. the evidence being missing were it e_1, e_2 and then e_3, respectively.

Consider the first of the two points mentioned above. Acceptance that the probability of the evidence is conditioned on the truthstate of the variable H, would mean, graphically, that H should be chosen as a parental variable for E (*see* Figure 6.11(i)).

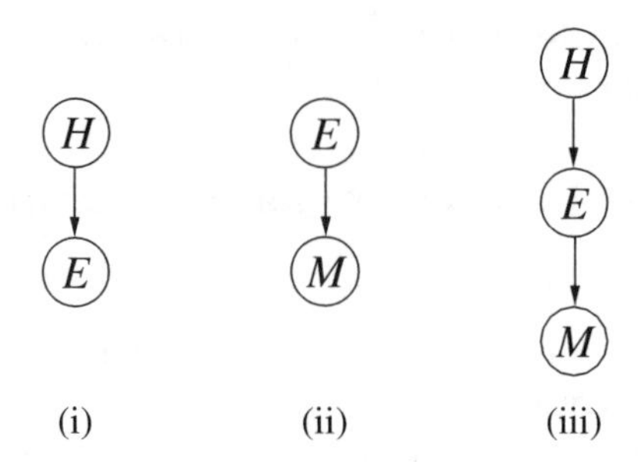

Figure 6.11 Bayesian network fragments representing the relation between (i) the variables E and H, and (ii) the variables M and E respectively; (iii) Bayesian network for missing evidence. Node H has two states, H_p, the defendant is guilty, H_d, the defendant is not guilty. Node E denotes the form of the evidence that is missing. Node M has two states, M, the evidence is missing, $\bar{M}$, the evidence is not missing. [Reproduced with permission from Elsevier.]

The situation is similar for the second point. If the event that evidence is missing (variable M) is conditioned on the form the missing evidence can take (variable E), then E can be chosen as a parental variable for M (*see* Figure 6.11(ii)).

Since the variable E shown in Figure 6.11(i) is the same as that in Figure 6.11(ii), these two network fragments combine to give the Bayesian network structure shown in Figure 6.11(iii).

While searching for an appropriate structure for a Bayesian network, on the basis of the three variables M, E and H, it would be legitimate to ask whether there could be an arc from H to M. As with any other graphical element employed in Bayesian networks structures, the absence of an arc must be justified as well. In the current example, the absence of a directed edge between H and M can be justified by the indications given by Lindley and Eggleston (1983, p. 90), who assume that 'were the form of the missing evidence known, then the view of the defendant's guilt would not be altered by knowing that this evidence had, or had not, been produced in court'. Stated otherwise 'the actual evidence suppresses any importance being attached to its omission' (Lindley and Eggleston 1983, p. 91).

Formally, this corresponds to the following: $P(H \mid E, M) = P(H \mid E)$, where E may take any of its three possible states e_1, e_2, or e_3, and M may either be M or $\bar{M}$. The proposed Bayesian network correctly encodes this property through its serial connection, where H and M are conditionally independent given E is known. It may also be said that the transmission of evidence between the nodes H and M is blocked whenever E is instantiated, or node E screens off M from H. This is a practical example of a d-separation property (Section 2.1.5).

7

Aspects of the combination of evidence

7.1 Introduction

In the previous chapters, Bayesian networks have been constructed and discussed with the aim of evaluating uncertainties associated with various forms of scientific evidence, such as trace evidence (fibres, glass, paint, *etc.*) and biological evidence (DNA). Such networks provide valuable assistance in addressing some of the wide range of issues that affect the coherent evaluation of evidence. Existing probabilistic solutions proposed in scientific literature may be used as a guide to elicit appropriate network structures. By providing an explicit representation of the relevant variables together with their assumed dependence and independence properties, Bayesian networks have the potential to clarify the rationale behind a probabilistic solution, in particular, formulae for likelihood ratios.

Many approaches proposed in the forensic literature focus on individual items of evidence and so do many of the Bayesian networks presented in the previous chapters. However, as noted by Lindley (2004, at p. xxiv), a '(...) problem that arises in courtroom, affecting both lawyers, witnesses and jurors, is that several pieces of evidence have to be put together before a reasoned judgement can be reached (...)'. Lindley (2004) also notes that 'probability is designed to effect such combinations but the accumulation of simple rules can produce complicated procedures. Methods of handling sets of evidence have been developed: for example, Bayesian networks (...)'. In the field of fact analysis, probabilistic studies have been provided on the various ways in which several items of evidence may interact. Detailed analyses of different forms of dissonant and harmonious evidence can be found, for example, in Schum (1994).

In forensic contexts, likelihood ratio formulae may attain critical levels of complexity, even for single items of evidence. One often needs to account for specific sources of uncertainty, such as phenomena of transfer, persistence, background presence, and so on, so it may become increasingly difficult to structure probabilistic analyses properly and to

Bayesian Networks and Probabilistic Inference in Forensic Science F. Taroni, C. Aitken, P. Garbolino and A. Biedermann

discern the relevant variables as well as their relationships. If, in addition, several items of evidence need to be combined, then further complications may be expected.

These, then, are the instances in which the use of Bayesian networks may assist forensic scientists in constructing coherent and acceptable arguments. The present chapter aims to illustrate this through a series of examples.

7.2 A difficulty in combining evidence

It has been emphasised throughout the previous chapters that forensic scientists should be careful in choosing the sort of questions they seek to address. For example, it is crucial to appreciate the differences that exist between the probability of the evidence and the probability of a proposition (Chapter 1). This distinction is particularly important when dealing with issues arising while combining evidence. As has been pointed out in forensic literature, attempts to evaluate the combined effect of separate items of evidence may be fraught with considerable difficulties, notably when evidence is presented in terms of posterior probabilities for the respective *probanda*.

Historically, the difficulty in combining evidence has been approached through a discussion following the problem called *difficulty of conjunction*: two pieces of evidence, when considered in combination, seem to produce a lower probability than when considered separately. This problem was the subject of a debate between Cohen (1977, 1988) and Dawid (1987). A summary of this debate and the solution proposed by Dawid (1987) can also be found in Aitken and Taroni (2004).

Let E_1 and E_2 denote two distinct items of evidence. These shall be used to draw an inference concerning some proposition of interest, say H for convenience. H has the two possible outcomes H_p and H_d, denoting the propositions proposed by the prosecution and the defence, respectively. Imagine further that some evaluator would retain a probability of 0.7 for H_p given the occurrence of either E_1 or E_2, that is, $Pr(H_p \mid E_1) = Pr(H_p \mid E_2) = 0.7$. The probability of interest is $Pr(H_p \mid E_1, E_2)$.

If E_1 and E_2 are considered to be independent, given H_p or H_d, their joint probability can be written as the product of the individual probabilities, for example, $Pr(E_1, E_2 \mid H_p) = Pr(E_1 \mid H_p) \times Pr(E_2 \mid H_p)$. It is tempting to believe that $Pr(H_p \mid E_1, E_2)$ is obtained analogously, that is, $Pr(H_p \mid E_1, E_2) = Pr(H_p \mid E_1) \times Pr(H_p \mid E_2)$. The apparent contradictory result of this (incorrect) procedure is $0.7 \times 0.7 = 0.49$, which is less than the probability of H_p given either E_1 or E_2.

At this stage, it is appropriate to consider Bayes' theorem. For two pieces of evidence E_1 and E_2 and propositions H_p and H_d,

$$\frac{Pr(H_p \mid E_1, E_2)}{Pr(H_d \mid E_1, E_2)} = \frac{Pr(E_1, E_2 \mid H_p)}{Pr(E_1, E_2 \mid H_d)} \times \frac{Pr(H_p)}{Pr(H_d)} \tag{7.1}$$

or

$$\text{posterior odds} = \text{likelihood ratio } (V) \times \text{prior odds}.$$

Assuming equal prior probabilities ($Pr(H_p) = Pr(H_d)$), the target probability $Pr(H_p \mid E_1, E_2)$ is given by $V/(1 + V)$. The likelihood ratio can be obtained as follows [assuming

$Pr(H_p) = Pr(H_d)$]:

$$\begin{aligned}
V &= \frac{Pr(E_1 \mid H_p)Pr(H_p)}{Pr(E_1 \mid H_d)Pr(H_d)} \times \frac{Pr(E_2 \mid H_p)Pr(H_p)}{Pr(E_2 \mid H_d)Pr(H_d)} \\
&= \frac{Pr(H_p \mid E_1)}{Pr(H_d \mid E_1)} \times \frac{Pr(H_p \mid E_2)}{Pr(H_d \mid E_2)} \\
&= \frac{0.7}{0.3} \times \frac{0.7}{0.3} \\
&= \frac{0.49}{0.09}.
\end{aligned}$$

From this, the probability of interest

$$Pr(H_p \mid E_1, E_2) = \frac{V}{1+V} = \frac{0.49/0.09}{1+0.49/0.09} = 0.84,$$

which is greater than 0.7. Thus, under the stated assumptions, the combination of the two pieces of evidence yields a higher probability for H_p than when considered separately.

This example illustrates that in cases in which two items of evidence are deemed to provide relevant information for the same pair of propositions, the value of the combined effect of the two pieces of evidence cannot be determined by the sole use of the posterior probabilities of the respective propositions. This is also one of the reasons why scales of conclusions based on posterior probabilities, as have been proposed for example in the field of shoe print (Katterwe 2003) or handwriting examination (Köller et al. 2004), are inadequate means for the assessment of scientific evidence (Taroni and Biedermann 2005).

7.3 The likelihood ratio and the combination of evidence

As noted in the previous section, inferential difficulties may be encountered when attempting to combine two or more items of evidence through posterior probabilities. Such difficulties may be avoided by following existing inferential procedures based on the likelihood ratio. Using a likelihood ratio (Section 1.2), one can successively add one piece of evidence at a time and examine the probability of a proposition of interest, H for example, given the available evidence. The posterior odds after considering one item of evidence, say E_1 for example, become the prior odds for the following item of evidence, E_2, say. In a more formal notation, one has for propositions H_p and H_d:

$$\frac{Pr(H_p)}{Pr(H_d)} \times \frac{Pr(E_1 \mid H_p)}{Pr(E_1 \mid H_d)} = \frac{Pr(H_p \mid E_1)}{Pr(H_d \mid E_1)} . \tag{7.2}$$

The term on the right-hand side of (7.2) represents the odds in favour of the proposition H_p given E_1. When E_2, a second item of evidence, becomes available, then one may proceed as follows:

$$\frac{Pr(H_p \mid E_1)}{Pr(H_d \mid E_1)} \times \frac{Pr(E_2 \mid H_p, E_1)}{Pr(E_2 \mid H_d, E_1)} = \frac{Pr(H_p \mid E_1, E_2)}{Pr(H_d \mid E_1, E_2)} . \tag{7.3}$$

Here, the posterior odds in favour of the proposition H_p incorporates knowledge about both the items of evidence, E_1 and E_2. The likelihood ratio for E_2, shown in the centre of (7.3), allows for a possible dependency of E_2 on E_1.

In what follows, two distinct situations are examined through examples, illustrating theoretical concepts earlier presented in Sections 2.1.4, 2.1.5, 2.1.6 and 2.1.7. In Section 7.3.1, items of evidence that are *independent given H* are considered. In Section 7.3.2, instances in which items of evidence may *not be independent given H* are discussed.

7.3.1 Conditionally independent items of evidence

Recall the likelihood ratio for two pieces of evidence E_1 and E_2 as given by the term in the centre of (7.1). If the two items of evidence are independent, then the likelihood ratio for the combination of E_1 and E_2, denoted V_{12}, is given by the product of the individual likelihood ratios for each item of evidence (V_1 and V_2, respectively):

$$\frac{Pr(E_1 \mid H_p)}{Pr(E_1 \mid H_d)} \times \frac{Pr(E_2 \mid H_p)}{Pr(E_2 \mid H_d)} = \frac{Pr(E_1, E_2 \mid H_p)}{Pr(E_1, E_2 \mid H_d)} , \tag{7.4}$$

or,

$$V_1 \times V_2 = V_{12} . \tag{7.5}$$

An appropriate representation in terms of a Bayesian network would be a diverging connection, where the three variables H, E_1 and E_2 combine to give $E_1 \leftarrow H \rightarrow E_2$. This would be an expression of the belief that E_1 and E_2 are conditionally independent given H. One is thus allowed to write, for example:

$$Pr(E_1 \mid H, E_2) = Pr(E_1 \mid H, \bar{E}_2) = Pr(E_1 \mid H) . \tag{7.6}$$

Analogously, it may be written for E_2:

$$Pr(E_2 \mid H, E_1) = Pr(E_2 \mid H, \bar{E}_1) = Pr(E_2 \mid H) . \tag{7.7}$$

Verbally stated, conditional independence between the two variables E_1 and E_2 is assumed to mean here that one's belief in, for example, E_1 would not change upon learning the truth of E_2 if, in addition, H were known.

As an example, consider again the conjunction problem discussed in Section 7.2. Its representation in terms of a Bayesian network requires the specification of node probabilities, which may be chosen as follows:

- One of the assumptions made was that the possible states of the proposition H are equally likely prior to consideration of either E_1 or E_2, so $Pr(H_p) = Pr(H_d) = 0.5$.

- Another assumption was that the occurrence of either E_1 or E_2 would incline some kind of fact finder to retain a probability of 0.7 for H_p. From the assumption $Pr(H_p) = Pr(H_d)$, one is then allowed to write, for example:

$$\frac{Pr(H_p \mid E_1)}{Pr(H_d \mid E_1)} = \frac{Pr(E_1 \mid H_p)}{Pr(E_1 \mid H_d)} = \frac{0.7}{0.3} . \tag{7.8}$$

  Equation (7.8) defines the conditional probabilities assigned to the table of the node E_1. The probability assignments for the node E_2 can be derived analogously.

When implemented in an appropriate software tool, the proposed Bayesian network exhibits the following properties:

- If either E_1 or E_2 is true, then the probability of H_p would increase from 0.5 to 0.7.
- If E_1 and E_2 are both true, then the probability of the node H being in state H_p is 0.84. This is just the value that was determined earlier for $Pr(H_p \mid E_1, E_2)$ (Section 7.2).
- If H_p is true, then the probability of E_1 remains unaffected if in addition knowledge about E_2 becomes available. Analogously, E_2 remains unaffected by knowledge about E_1 if H_p is known.

This last property is an explicit representation of the independence of E_1 and E_2 given H. This is also an instance when H would be said to 'screen off' E_1 from E_2.

7.3.2 Conditionally non-independent items of evidence

In the previous section, single items of evidence (E_1 and E_2) have been combined in a way that assumes their independence conditional on knowing H. However, such an assumption may not be appropriate in the general case, as pointed out by Lempert (1977, at p. 1043). Consider the following:

> [...] the defendant's thumb print was found on the gun the killer used. [...] assume that the factfinder believes that the presence of this evidence is 500 times more likely if the defendant is guilty than if he is not guilty. [...] Now suppose that the prosecution wished to introduce evidence proving that a print matching the defendant's index finger was found on the murder weapon. If this were the only fingerprint evidence in the case, it would lead the factfinder to increase his estimated odds on the defendant's guilt to the same degree that the proof of the thumb print did. Yet, it is intuitively obvious that another five hundredfold increase is not justified when evidence of the thumb print has already been admitted.

When represented in terms of a Bayesian network, this scenario provides an illustrative example of the notion of conditional non-independence. Consider three binary variables defined as follows:

- H: the suspect is the killer.
- F_1: Friction ridge marks of the suspect's thumb are present on the murder weapon.
- F_2: Friction ridge marks of the suspect's index finger are present on the murder weapon.

Lempert argues that, given that the marks of the suspect's thumb are found on the murder weapon, the probability of finding marks of his index finger if he is guilty is not very different from finding the same evidence if he were not guilty. A direct dependency of F_2 on F_1 is thus assumed. Graphically, a structure as shown in Figure 7.1(i) could be proposed. Note that such a network structure has been discussed earlier in the context of the Baskervilles' case (Figure 2.7, Section 2.1.9).

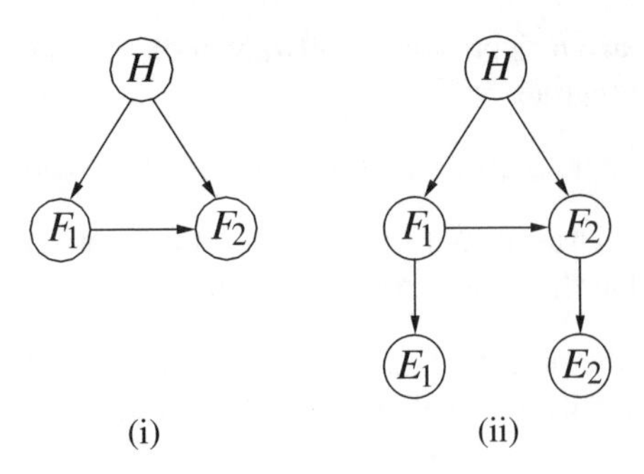

Figure 7.1 Bayesian network for combining two items of evidence F_1 and F_2 that cannot, given H, be assumed to be independent. For the example in the text, the three nodes are binary. Node H: H_p, the suspect is the killer, and H_d, the suspect is not the killer. Node F_1: friction ridge marks of the suspect's thumb are, or are not, present on the murder weapon. Node F_2: friction ridge marks of the suspect's index finger are, or are not, present on the murder weapon. In the network (ii), a distinction is made between a (source level) proposition (F) and evidence for that proposition (E). [Adapted from Taroni and Biedermann (2005), reproduced with permission from Oxford University Press.]

Table 7.1 Conditional probabilities associated to the node F_1. Factor H has two states: H_p, the suspect is the killer, and H_d, the suspect is not the killer. Factor F_1, has two states: F_1 and $\bar{F}_1$, the friction ridge marks of the suspect's thumb are, or are not, present on the murder weapon. [Adapted from Taroni and Biedermann (2005), reproduced with permission from Oxford University Press.]

	H:	H_p	H_d
F_1:	F_1	0.2	0.0004
	$\bar{F}_1$	0.8	0.9996

The evidence F_1 is taken to provide a likelihood ratio of 500 for the proposition of the defendant being guilty, as Lempert (1977, at p. 1043) considers that a '(...) mathematically inclined juror might, for example, believe that there is a 0.2 probability that the print would be found if the defendant were guilty (the probability is considerably less than one because guilty people often have taken the trouble to wipe their prints from weapons and, even if they had not, not all prints are identifiable) and a 0.0004 probability that the evidence would be found if the defendant were not guilty'. Consequently, the node probability table for F_1 may be completed as shown in Table 7.1.

For completion of the probability table associated with the node F_2, the following may be considered:

- If the thumb mark evidence were absent ($\bar{F}_1$), then evidence of marks of the suspect's index finger may be associated to a weight that has the same order of magnitude as that

Table 7.2 Conditional probabilities associated to the node F_2. The three factors are each binary. Node H: H_p, the suspect is the killer, and H_d, the suspect is not the killer. Node F_1: friction ridge marks of the suspect's thumb are, F_1, or are not, $\bar{F}_1$, present on the murder weapon. Node F_2: friction ridge marks of the suspect's index finger are, F_2, or are not, $\bar{F}_2$, present on the murder weapon. The probability of finding marks of the suspect's index finger, F_2, given he is guilty, H_p, and F_1 is also true, that is, the thumb mark evidence is present, is denoted as α. The probability of finding marks of the suspect's index finger, F_2, given he is innocent, H_d, and F_1 is also true, that is, the thumb mark evidence is present, is denoted as β. Other probabilities are taken from Table 7.1. [Adapted from Taroni and Biedermann (2005), reproduced with permission from Oxford University Press.]

	G :	H_p		$\bar{H}_d$	
	F_1 :	F_1	$\bar{F}_1$	F_1	$\bar{F}_1$
F_2 :	F_2	α	0.2	β	0.0004
	$\bar{F}_2$	$1-\alpha$	0.8	$1-\beta$	0.9996

associated to the thumb print evidence alone. One may thus set $Pr(F_2 \mid \bar{F}_1, H_p) = Pr(F_1 \mid H_p)$ and $Pr(F_2 \mid \bar{F}_1, H_d) = Pr(F_1 \mid H_d)$. In such a case, the values contained in columns 2 and 4 in the body of Table 7.2 may be set equal to the values in the columns 1 and 2 in the body of Table 7.1.

- Knowing that marks of the suspect's thumb are present on the weapon (F_1), the capacity of the second item of evidence to discriminate between H_p and H_d is crucially dependent on the ratio of the probabilities $Pr(F_2 \mid F_1, H_p)$ and $Pr(F_2 \mid F_1, H_d)$. Denote these two probabilities as α and β, respectively. Then, if one believes, as mentioned above, that the probability of finding marks of the suspect's index finger, F_2, given he is guilty, H_p, (and F_1 is also true, as the thumb mark evidence is present), does not greatly differ from the probability of F_2, given the suspect's innocence, H_d, and F_1, that is, $\alpha \approx \beta$, then a likelihood ratio of approximately 1 is indicated for the evidence of F_2 given F_1.

The assessment of the parameters α and β is a matter of subjective judgment and is largely dependent on one's views and beliefs held in a particular case. The following quotation from Lempert (1977, at p. 1043) provides a clear illustration of this:

> The presence of the second print depends largely on the way the defendant held the gun when he left the thumb print. Unless murderers hold guns differently from non-murderers or are more likely to wipe off some, but not all, of their fingerprints, the finding of the second print is no more consistent with the

> hypothesis that the defendant is guilty than with its opposite. Indeed, because a murderer is more likely to attempt to wipe off fingerprints from a gun than one with no apprehension of being linked to a murder and since an attempt to wipe off fingerprints might be only partially successful, there is a plausible argument that the presence of the second print should lead jurors to be somewhat less confident that the defendant is the murderer than they would be if only one of the defendant's fingerprints were found.

Note that, strictly speaking, the variables F_1 and F_2 do not denote items of evidence in the sense of Section 7.3.1. Here, F_1 and F_2 merely represent source level propositions, that is, propositions of the kind 'the friction ridge mark comes from the suspect'. For this reason, the propositions are denoted here by a variable F_i rather than by E_i.

No distinction has thus been made between evidence for a proposition, for example, 'a certain number of corresponding minutiae are observed in the friction ridge mark found on the weapon and in the prints obtained from the suspect under controlled conditions', and a proposition itself, for example, 'the friction ridge mark comes from the suspect'. In order to consider both aspects, a Bayesian network as shown in Figure 7.1(ii) could be adopted. For more detailed discussions on the use of distinct propositional levels, see Schum (1994).

7.4 Combination of distinct items of evidence

General aspects of combining evidence that have been considered in the previous sections are now studied further in a series of examples that involve selected items of scientific evidence. In order to illustrate the rationale behind the proposed Bayesian networks, a sequential procedure is followed. First, network structures are elicited for reasoning separately about each item of scientific evidence. Then, in a second step, ways of combining network fragments logically are examined.

7.4.1 Example 1: Handwriting and fingermark evidence

The first example is concerned with a disputed signature. Consider the following scenario:

> Two people, Mr Adams and Miss Brown, have been arrested in a shop trying to pass a stolen cheque. Mr Adams admits to writing the signature on the cheque, but claims he has just acquired the chequebook and certainly has not written out any more cheques from the book. Miss Brown claims to know nothing of the whole affair. On investigation, it is found that sixteen further transactions have taken place using the chequebook and accompanying bank card, which was reported stolen a few days earlier. The loser, Mr Constantine, claims the card must have been stolen in the post and the loser's signature does not look anything like the signature on the card.
>
> Handwriting samples, including the signatures in the name of Mr Constantine, are obtained from both Mr Adams and Miss Brown.

This is a variation of an example in Stockton and Day (2001, at p. 3) with 'credit card' in the original changed to 'chequebook and accompanying bank card'.

Different analyses have been proposed by Stockton and Day (2001), such as evaluation of single signatures or groups of signatures. Here, an abstraction will be made of the scenario in the sense that consideration will only be given to the evaluation of a single signature present on one of the sixteen cheques. This cheque may have been selected because of the particularly high amount of money involved, for example.

Comparative examinations may be performed between the disputed signature and various samples obtained from the suspect, Mr Adams. The question of authorship may then be addressed by defining a pair of source level propositions:

- S_p: Mr Adams wrote the signature.
- S_d: Mr Adams did not write the signature.

As in previously discussed scenarios, the subscripts p and d are used to denote the propositions of the prosecution and the defence, respectively.

Different ways may be used to describe the findings, according to the desired level of detail. Following Stockton and Day (2001), the findings may be summarised in terms of *few*, *some* and *many* similarities. Thus, let E_1 denote the outcomes of the comparative analyses and e_1, e_2 and e_3 stand for, respectively, *few*, *some* and *many* similarities. An assessment is then necessary for expressing one's expectations of observing each of these outcomes, given that the suspect is or is not the author of the disputed signature. These are assessments that are largely dependent on the circumstances of the case at hand. As it may be difficult to find appropriate data from surveys or literature, assessments would consist mostly of an examiner's judgment, based upon past experience. The following provides an idea as to the range of considerations that may be involved when assessing the various outcomes, assuming that the suspect is the author of the signature in question.

> If Mr Adams is particularly naive, he might write the samples in exactly the same way as the signature and then we may find many similarities in detail and a few differences. However, this is very unlikely. He is more likely to vary his writing in some way (either in his samples or in the signature in question), so we may find some similarities but some differences will also be present. It is most likely that he will disguise his writing so much that while we may find a few similarities there will be many differences (Stockton and Day 2001, at p. 4).

If the suspect were not the author of the signature in question, then few similarities may be more likely to be observed than many similarities. Table 7.3 displays values that may be assigned to each possible outcome of E_1 given S_p and S_d, respectively. The values are approximations derived from areas in pie charts proposed by Stockton and Day (2001). Notice that the values retained here mainly serve the purpose of illustration. Methods exist that focus on the relative order of magnitude of probabilistic assessments rather than specific numerical values. For a discussion of such methods, see Chapter 9.

So far, the evaluation has been based on the variable E_1, denoting the number of observed similarities, and S, denoting the proposition of authorship of the disputed signature. In terms of a Bayesian network, a simple two-node representation with $S \rightarrow E_1$ could be adopted for this scenario (Figure 7.2(i)).

A number of scenarios may be imagined in which comparative handwriting analysis may only provide poor evidence for discriminating between the propositions put forward by the prosecution and the defence. This may be in part due to inappropriate comparative

Table 7.3 Probabilities for the outcome E_1 of the comparative analysis of handwriting with three states, the finding of few (e_1), some (e_2) and many (e_3) similarities given different propositions of authorship. Factor S has two states: S_p, Mr Adams wrote the signature, and S_d, Mr Adams did not write the signature. [adapted from Stockton and Day (2001)].

	S:	S_p	S_d
E_1:	e_1	0.60	0.78
	e_2	0.27	0.20
	e_3	0.13	0.02

material, for example. On other occasions, the disputed handwriting may consist only of a few lines. Difficulties may also arise if the incriminated writing exhibits only a few comparative features, such as with an abbreviated signature.

Further evidence may thus be needed in order to associate the suspect with the incriminated cheque. Fingerprints are one such type of evidence. An inferred contact between the suspect and the cheque in question may, in some situations, be considered a relevant piece of information for constructing an argument to some ultimate probandum, for example, that the suspect is the author of the signature.

Such a scenario is now studied in further detail. Assume, for example, that various kinds of detection techniques have been applied to the document in question (cheque, in this example), a process during which two fragmentary friction ridge marks became visible. A total of, say, 10 minutiae (*e.g.*, 4 in one mark and 6 in the other mark) are found to correspond to features present in prints obtained, under controlled conditions, of the thumb and the index finger of Mr Adams' left hand. No unexplainable differences are observed. Imagine further that the respective position and sequence of the marks present on the document are such that they form what may be called an *anatomical sequence*. Moreover, the marks are present in a position in which one would naturally expect the author of the signature in question to leave such marks. Let the evidence of the friction ridge marks be denoted as E_2.

Evidence E_2 will first be considered separate from the handwriting evidence. The current discussion does not aim at addressing aspects of the wide range of issues pertaining to the friction ridge mark identification process, for which extensive literature is available to the interested reader [*e.g.*, Champod et al. (2004)]. The assumption made here is that the value of fingermark evidence can be evaluated and assessed in probabilistic terms (Champod and Evett 2001).

The evidence of friction ridge marks is dependent on a proposition that will be defined here by a variable F:

- F: The fingermarks come from the suspect, Mr Adams.
- $\bar{F}$: The fingermarks come from an individual other than the suspect.

Note that a distinction is drawn between fingerprints and fingermarks. A fingermark is usually a latent and may be found, for example, at a crime scene or on an object thought to be relevant to the case. A fingerprint is an inked impression of a finger of a suspect, obtained under controlled conditions.

It may be expected that the observed correspondences (without any unexplainable differences), E_2, would be more likely to occur when the fingerprints have been left by the suspect (proposition F), rather than some other individual. Specification of numerical values will be avoided here as parameters such as the clarity of the marks, the kind of corresponding minutiae, their relative position, and so on, and are not discussed in further detail here. For the current level of discussion, it will suffice to note that one's beliefs may be represented by $Pr(E_2 \mid F) \gg Pr(E_2 \mid \bar{F})$, implying that the correspondences in the fingermarks have been taken to be good evidence for the proposition that the suspect is the source of the friction ridge marks.

An argument then needs to be constructed from the source level proposition F about fingermarks to the propositions S_p and S_d, relating to the authorship of the signature in question. To this end, a Bayesian network described in Section 4.1 (Figure 4.1) may be used. Recall that this Bayesian network allows for the evaluation of the evidence of one-stain scenarios where uncertainties may exist in relation to the relevance of the evidential material.

In the context of a scenario involving a disputed signature, the relevance of fingermarks could be taken as an expression of one's belief that the marks had been left by the author of the signature in question. Accordingly, a variable G is defined as follows:

- G: The fingermarks come from the author of the signature in question.
- $\bar{G}$: The fingermarks do not come from the author of the signature in question.

Usually, the probabilities assigned to the outcomes of G would be different from certainty and impossibility. From the position in which the marks are found on the cheque, one may, for example, retain a fairly high initial probability for G.

The node probabilities needed for the variable F can be specified in the same way as explained in Section 4.1. A summary of the values is given in Table 7.4. The variable p expresses the probability that the suspect would have left the friction ridge marks for innocent reasons (*e.g.*, on another occasion).

In the previous paragraphs, distinct network fragments have been proposed for separate evaluations of the handwriting and fingermark evidence [*see* Figures 7.2(i) and (ii)]. For a joint evaluation of these two items of evidence, a logical combination needs to be found between the two network fragments. A natural way to proceed could be as follows. Consider that both the network fragments shown in Figures 7.2(i) and (ii) contain a variable S that has the same definition. The two networks could thus be combined to give a single network by retaining only a single node S, as shown in Figure 7.2(iii).

Notice that this form of combination assumes the items of evidence E_1 and E_2 to be independent upon knowledge of S. The likelihood ratio for the combined items of evidence, V_{12}, may thus be written as the product of the individual likelihood ratios:

$$V_{12} = V_1 \times V_2 \ , \tag{7.9}$$

where V_1 denotes the likelihood ratio for the handwriting evidence and V_2 denotes the likelihood ratio for the fingermark evidence. Notice that the latter is an abbreviated form

Table 7.4 Conditional probabilities associated with factor F. There are two states for F: F, the suspect is the source of the fingermarks, and $\bar{F}$, the suspect is not the source of the fingermarks. Factor S has two states: S, the suspect is the author of the signature in question, and $\bar{S}$, the suspect is not the author of the signature in question. Node G has two states: G, the fingermarks have been left by the author of the signature in question, and $\bar{G}$, the fingermarks have not been left by the author of the signature in question. The probability that the suspect would have left the friction ridge marks for innocent reasons is denoted as p.

	S:	S_p		S_d	
	G:	G	$\bar{G}$	G	$\bar{G}$
F :	F	1	0	0	p
	$\bar{F}$	0	1	1	$1-p$

of the following:

$$V_2 = \frac{r + \{(1-r)\gamma\}}{(r\gamma) + (1-r)\{p + (1-p)\gamma\}} . \qquad (7.10)$$

Note that in deriving 7.10

- $Pr(E_2 \mid F) = 1$, an assumption essentially made with the aim of simplifying the form of the result (as there may be various numbers of corresponding minutiae given F, the term $Pr(E_2 \mid F)$ can take values lower than 1, according to a probability distribution over possible numbers of observed corresponding minutiae);
- $Pr(E_2 \mid \bar{F}) = \gamma$, the relative frequency of the corresponding features.

The variable r denotes the relevance term, that is, $Pr(G)$. It is a special case of (6.1) with the probability q omitted, which relates to an activity of the suspect at the time of the crime. The analogy can also be seen with a comparison of Figures 6.2 and 7.2.

7.4.2 Example 2: Issues in DNA analysis

An important aspect of the procedure described in the previous section is that a joint evaluation of several items of evidence can be obtained by combining local networks that separately represent existing likelihood ratio formulae. The user can thus be guided towards a coherent evaluation that would otherwise become increasingly difficult to achieve. This section will address a further example to illustrate this point.

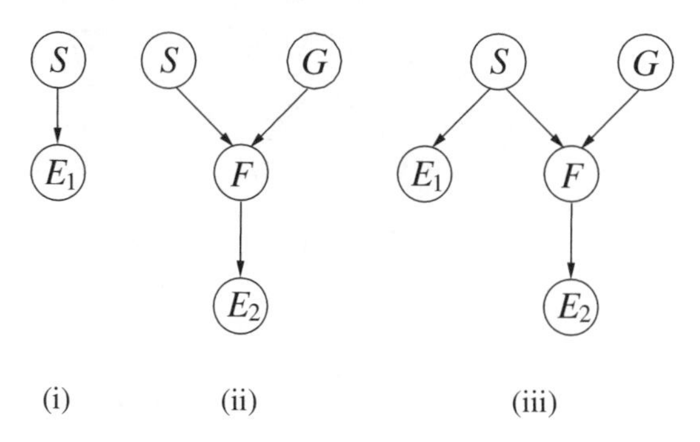

Figure 7.2 Bayesian networks for evaluating (i) handwriting evidence, (ii) fingermark evidence, and (iii) both, handwriting and fingermark evidence. There are two states for node F: F, the suspect is the source of the fingermarks, and $\bar{F}$, the suspect is not the source of the fingermarks. Node S has two states: S, the suspect is the author of the signature in question, and $\bar{S}$, the suspect is not the author of the signature in question. Node G has two states: G, the fingermarks have been left by the author of the signature in question, and $\bar{G}$, the fingermarks have not been left by the author of the signature in question. Node E_1 has three states: the finding of few (e_1), some (e_2) and many (e_3) similarities given different propositions of authorship for the comparative analysis of handwriting. Node E_2 has two states: E_2, there are correspondences between the fingermarks on the document in question and the fingerprints of the suspect, and $\bar{E}_2$, there are no correspondences between the fingermarks on the document in question and the fingerprints of the suspect.

Imagine a scenario similar to the one described in Section 4.1. A crime has been committed by one offender and a blood stain has been found at the scene. There is a report of a forensic scientist stating that the blood of a suspect shares the same DNA profile as the stain from the scene. Consideration also needs to be given to the following:

- The suspect has been selected through a search in a database containing DNA profiles.
- There is a potential for error in the analytical result.
- There is some doubt as to whether the recovered blood stain truly came from the offender.

What inference can be drawn to the proposition that the suspect is the offender?

This scenario can be broken down into several distinct issues, each of which may be represented by a 'network fragment'. The term network fragment is borrowed here from Neil et al. (2000), and refers to a set of related random variables that could be constructed and reasoned about separate from other fragments.

The following network fragments may be distinguished:

- There is a crime stain that matches a sample provided by the suspect. This is a proposition that depends on whether the suspect is the source of the crime stain. In turn, the latter is a proposition that may be used to infer something about whether the

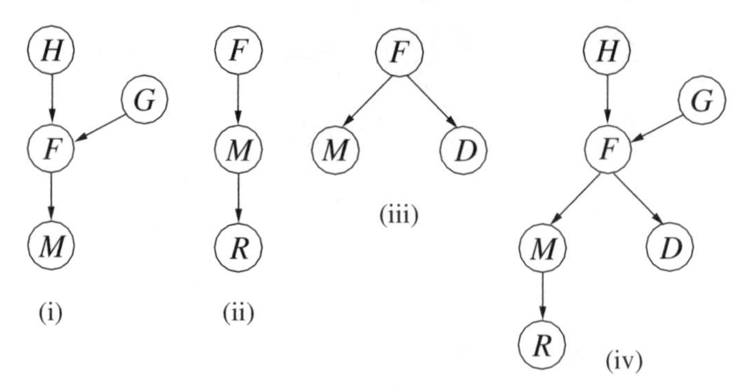

Figure 7.3 Bayesian networks for (i) a one-trace scenario of transfer from the suspect (including the concept of relevance), (ii) the problem of error rates, (iii) a database search scenario, and (iv) a combination of the issues represented in (i) to (iii). Definitions of the nodes are given in Table 7.5. [Adapted from Taroni et al. (2004), reproduced with permission from Elsevier.]

suspect is the criminal if consideration is given to the relevance of the crime stain. Figure 7.3(i), earlier described in Section 4.1, provides an appropriate representation for these issues.

- A forensic scientist's report of a match is a relevant piece of information for inferring whether there could be a true match and whether the suspect could be the source of the crime stain. A network that is usable here is shown in Figure 7.3(ii). It allows one to account for the possibility that the scientist erroneously reports a match. See Section 5.11 for further details on this submodel.

- Figure 7.3(iii) depicts a model for evaluating the results of a database search together with the information that a sample provided by the suspect matches the crime stain. This model has been discussed in Section 5.10.

A summary of all propositions used is given in Table 7.5. From the number of variables involved as well as the various possible dependencies among them, it may be seen that a formal development of a likelihood ratio would become an increasingly complex endeavour. However, if it can be accepted that the Bayesian network fragments shown in Figures 7.3(i) to (iii) are appropriate representations of their respective issues, then a Bayesian network for representing all of these issues can be proposed by combining the network fragments such that one obtains a structure as shown in Figure 7.3(iv).

This method of combination suggests that separate models may be examined for the presence of nodes with the same definition that appear in more than one model. A combination between two or more network fragments may then be operated via the node(s) that have the same definition(s). In the resulting network, only one node with the same definition will remain. This stems from the definition of Bayesian networks, according to which each variable can appear only once within the same model.

Notice, however, that when combining network fragments, one also needs to examine whether there could be additional links among nodes from the separate networks. In the example considered in this section, this does not appear to be the case.

Table 7.5 Definitions of the binary nodes used in Figure 7.3. [Adapted from Taroni et al. (2004), reproduced by permission of Elsevier.]

Node	Definition
H	The suspect is the offender
F	The crime stain came from the suspect
G	The crime stain came from the offender
M	The suspect's blood sample and the crime stain have matching profiles
R	Reported match between the suspect's profile and the profile of the crime stain
D	The other $(N-1)$ profiles of the database do not match

7.4.3 Example 3: Scenario with one offender and two corresponding items of evidence

Some scenarios involving more than one stain have already been considered in Section 4.3. For example, in Section 4.3.1, a case with one offender was examined where two stains were found at the scene of a crime. The two items of evidence were of the same kind (*e.g.*, blood) but of different types and had various degrees of relevance. A question of interest was how an argument can be constructed for crime level propositions, that is, the suspect is the offender. In Section 4.3.2, an approach was described for situations in which two items of evidence were found on two distinct crime scenes. The items of evidence were of the same kind and of the same type. The aim was to analyse if the two pieces of evidence could have come from the same source.

In this section, yet another two-stain scenario will be considered. Imagine the following scenario: during night time, an offender entered a house by forcing a back door. Visibly, several rooms were searched for the presence of objects of value; drawers were emptied and diverse objects such as clothing, paper, and so on, were scattered all over the floor. In one of the rooms, a safe had been opened using brute force. All objects of value in the safe had been stolen.

Upon examination of the scene, attention is drawn to two items of evidence. One of them is a blood stain found at the point of entry. From the position in which the stain was found, its apparent freshness and abundance, it is believed that it may have been left by the offender while gaining access to the premises. A second item of evidence is a partial fingermark found on the outer surface of the safe's front door. A few weeks later, a suspect comes to the attention of the police for reasons completely unrelated to the scientific evidence. It is found that the suspect's blood and the blood found on the crime scene share the same characteristics. It is also observed that the fingermark lifted on the safe corresponds to one of the prints obtained from the suspect, under controlled conditions.

The scenario thus involves two different kinds of evidence found on the same scene and correspondences were observed with a single potential source. One of the issues that needs to be addressed is how consideration may be given to possible dependencies that may exist among arguments affecting the separate evaluation of each item of evidence. A step-wise construction of a Bayesian network can help clarify this point.

Table 7.6 Conditional probabilities assigned to F_1, the source-level propositions referring to the first item of evidence. Factor F_1 has two states: F_1, the suspect is the source of the crime stain, and $\bar{F}_1$, the suspect is not the source of the crime stain. Factor H has two states: H_p, the suspect is the offender, and H_d, the suspect is not the offender. Factor G_1, relevance, has two states: G_1, the blood stain came from the offender, and $\bar{G}_1$, the blood stain did not come from the offender. The probability the suspect left the stain while being innocent is p_1.

	H :	H_p		H_d	
	G_1 :	G_1	$\bar{G}_1$	G_1	$\bar{G}_1$
F_1 :	F_1	1	0	0	p_1
	$\bar{F}_1$	0	1	1	$1 - p_1$

Start by considering the bloodstain. Let E_1 stand for the match between the characteristics of the crime stain and those of a sample provided by the suspect. As was pointed out in Section 4.1, such a correspondence is a relevant piece of information for inferring something about whether the suspect is or is not the source of the crime stain (proposition F_1). In turn, the latter proposition together with information on the relevance of the crime stain (G_1) allows the construction of a line of reasoning to a crime level proposition, that is, the suspect is or is not the perpetrator of the alleged offence (H). It has been seen that a Bayesian network as shown in Figure 4.1 can be used for this purpose.

Table 7.6 provides a summary of the various conditional probabilities associated with the node F_1. Recall from Section 4.1 that the first three columns may be completed logically:

- If the suspect is the offender and the crime stain comes from the offender, then necessarily the crime stain comes from the suspect: $Pr(F_1 \mid H_p, G_1) = 1$.

- If H_p is true and G_1 is false, or vice versa, then the suspect certainly is not the source of the crime stain: $Pr(\bar{F}_1 \mid H_p, \bar{G}_1) = Pr(\bar{F}_1 \mid H_d, G_1) = 1$.

Probabilities different from 0 and 1 may be applicable in column 4 of Table 7.6. If it is believed that there could be a possibility of the suspect leaving a stain even though he is innocent, then a non-zero probability is required for the state F_1. Denote such an assessment as p_1. Consequently, the state $\bar{F}_1$ is assigned a probability equal to $1 - p_1$.

Next, consider the second item of evidence, E_2, a correspondence between the friction ridge mark found on the safe and one of the prints obtained from the suspect under controlled conditions. The existing Bayesian network can be extended using a duplication of the structure already developed. Besides E_2, the nodes thus added are as follows:

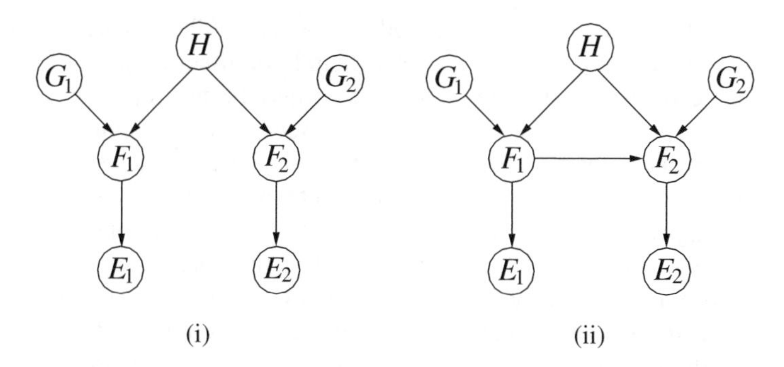

Figure 7.4 Bayesian networks for the joint evaluation of two items of evidence found on the same crime scene. Node H has two states: H_p, the suspect is the offender, and H_d, the suspect is not the offender. Node F_1 has two states: F_1, the suspect is the source of the crime stain (blood), and $\bar{F}_1$, the suspect is not the source of the crime stain. Node F_2 has two states: F_2, the fingermark has been left by the suspect, and $\bar{F}_2$, the fingermark has not been left by the suspect. Node G_1, relevance of the crime stain, has two states: G_1, the stain came from the offender, and $\bar{G}_1$, the stain did not come from the offender. Node G_2, relevance of the fingermark, has two states: G_2, the mark came from the offender, and $\bar{G}_2$, the mark did not come from the offender. Node E_1 has two states: E_1, there is a match between the characteristics of the crime stain and those of a sample provided by the suspect, and $\bar{E}_1$, there is not a match between the characteristics of the crime stain and those of a sample provided by the suspect. Node E_2 has two states: E_2, there is a correspondence between the friction ridge mark at the crime scene and a print provided by the suspect, and $\bar{E}_2$, there is not a match between the friction ridge mark at the crime scene and a print provided by the suspect.

- G_2: The fingermark has or has not been left by the offender, denoted as G_2 or $\bar{G}_2$.
- F_2: The fingermark has or has not been left by the suspect, denoted as F_2 or $\bar{F}_2$.

The Bayesian network constructed so far is shown in Figure 7.4(i). Next, one of the questions that the scientist should consider is whether this is a satisfactory structure or whether there are any properties for which this network does not yet account.

Notice that the structure shown in Figure 7.4(i) assumes the two items of evidence to be independent given knowledge about H. A more close examination of the probability table for node F_2 will show whether this represents an appropriate assumption:

- If the suspect is the offender and the second item of evidence (fingermark) has been left by the offender, then certainly the second item of evidence comes from the suspect: $Pr(F_2 \mid H_p, G_2) = 1$.
- If the suspect is the offender but the second item of evidence has not been left by the offender, then certainly the suspect is not the source of the mark: $Pr(F_2 \mid H_p, \bar{G}_2) = 0$.
- If the suspect is not the offender and the second item of evidence comes from the offender, then certainly the suspect is not the source of the mark: $Pr(F_2 \mid H_d, G_2) = 0$.

These are reasonable assumptions, and knowledge about any of the nodes used for modelling the first item of evidence (*i.e.*, G_1, F_1 and E_1) should not affect these assessments.

However, there is one situation in which a need may be felt for some kind of dependency: when the suspect is not the offender and the second item of evidence (fingermark) did not come from the offender, that is, a situation in which the suspect could have left the stain for innocent reasons. Here, knowledge about whether the suspect has left the first item of evidence appears to be potentially informative. For example, one might be inclined to be more confident in the belief that the suspect is the source of the second item of evidence for innocent reasons if he were also known to be the source of the first item of evidence. Notice that it is still assumed here that the suspect is not the offender. Thus, were he the source of the first item of evidence, then he must be so for innocent reasons. This stems from the probability assigned to $Pr(F_1 \mid H_d, G_1)$, which has been said to be 0 (*see* Table 7.6).

So, if it is believed that for some assessments of F_2 knowledge about the truth state of F_1 is relevant, then a directed edge may be drawn between these two nodes. This situation is shown in Figure 7.4(ii) and the node probability table associated with F_2 is shown in Table 7.7. As may be seen from this table, knowledge about F_1 is allowed to affect one's expectations of the suspect being the source of the second item of evidence for innocent reasons, that is, when both H_p and G_2 are false (*asymmetric independence*, Section 2.1.9):

- A probability denoted as p_2 is used to describe the expectation that the suspect is the source of the second item of evidence given that he is the source of the first item of evidence: $Pr(F_2 \mid H_d, \bar{G}_2, F_1) = p_2$.

- A probability denoted as p_2' is used to describe the expectation that the suspect is the source of the second item of evidence given that he is not the source of the first item of evidence: $Pr(F_2 \mid H_d, \bar{G}_2, \bar{F}_1) = p_2'$.

The former of these may be thought of as a conditional probability of innocent acquisition, meaning that the object on the scene has acquired a stain from the suspect in a situation unrelated to the crime under investigation. The particular nature of this assessment is that it is conditioned on the knowledge about the presence of another stain left by the suspect for innocent reasons. Notice, however, that for the link $F_1 \rightarrow F_2$ to have some meaning,

- the probability of innocent acquisition of the first item of evidence, denoted as p_1 (*see* Table 7.6), must be different from 0, and

- the probabilities of innocent acquisition of the second item of evidence, p_2 and p_2' (*see* Table 7.7), must have different values.

Whenever one of these conditions is not satisfied, then the link is superfluous in the sense that the network would yield the same result as the one shown in Figure 7.4(i).

An interesting aspect of the Bayesian network for combining conditionally independent items of evidence (Figure 7.4(i)) is that it may be shown to be a logical extension of the basic network $E_1 \leftarrow H \rightarrow E_2$ discussed in Section 7.3.1. Compare, for example, the properties of these two networks when the specifications are as follows:

- For the network shown in Figure 7.4(i): $Pr(E_1 \mid F_1) = 1$, $Pr(E_1 \mid \bar{F}_1) = 0.01$, $Pr(F_1 \mid H_p, G_1) = 1$, $Pr(F_1 \mid H_p, \bar{G}_1) = Pr(F_1 \mid H_d, G_1) = Pr(F_1 \mid H_d, \bar{G}_1) = 0$ (analogously for E_2 and F_2).

Table 7.7 Conditional probabilities assigned to F_2, the source-level propositions referring to the second item of evidence. Factor F_2 has two states: F_2, the suspect is the source of the second item of evidence, and $\bar{F}_2$, the suspect is not the source of the second item of evidence. Factor H has two states: H_p, the suspect is the offender, and H_d, the suspect is not the offender. Factor F_1 has two states: F_1, the suspect is the source of the first item of evidence, and $\bar{F}_1$, the suspect is not the source of the first item of evidence. Node G_2, relevance of the second item of evidence, has two states: G_2, the second item came from the offender, and $\bar{G}_2$, the second item did not come from the offender. The probability that the suspect is the source of the second item of evidence, given he is the source of the first item, he did not commit the crime and the second item of evidence did not come from the offender is p_2. The probability that the suspect is the source of the second item of evidence, given that he is not the source of the first item, he did not commit the crime and the second item of evidence did not come from the offender is p_2'.

	H :	H_p				H_d			
	G_1 :	G_2		$\bar{G}_2$		G_2		$\bar{G}_2$	
	F_1 :	F_1	$\bar{F}_1$	F_1	$\bar{F}_1$	F_1	$\bar{F}_1$	F_1	$\bar{F}_1$
F_2 :	F_2	1	1	0	0	0	0	p_2	p_2'
	$\bar{F}_2$	0	0	1	1	1	1	$1 - p_2$	$1 - p_2'$

- For the network $E_1 \leftarrow H \rightarrow E_2$: $Pr(E_1 \mid H_p) = 1$, $Pr(E_1 \mid H_d) = 0.01$ (analogously for E_2).

If, in addition, the nodes G_1 and G_2 in Figure 7.4(i) are set to 1, then the joint occurrence of E_1 and E_2 supports the proposition H_p by a likelihood ratio of 10^4, the product of the likelihood ratios provided by the individual items of evidence. What does this mean? Verbally, it may be stated that, for example, knowledge about E_1 (match between the characteristics of the crime stain and a sample provided by a suspect) makes F_1 (the suspect being the source of the crime stain) $Pr(E_1 \mid F_1)/Pr(E_1 \mid \bar{F}_1) = 1/0.01 = 100$ times more probable. As it is assumed that the crime stain came from the offender ($Pr(G_1) = 1$) and the suspect could not have left the stain for innocent reasons ($Pr(F_1 \mid H_d, \bar{G}_1) = 0$), evidence that increases the odds in favour of the suspect being the source of the crime stain by a factor of 100 is also evidence that increases the odds in favour of the suspect being the offender by a factor of 100. The same holds for E_2.

An analogous result may be obtained with the network $E_1 \leftarrow H \rightarrow E_2$. Here, the likelihood ratio for the joint occurrence of the two items of evidence is just given by $[Pr(E_1 \mid H_p)/Pr(E_1 \mid H_d)] \times [Pr(E_2 \mid H_p)/Pr(E_2 \mid H_d)] = 100 \times 100 = 10^4$.

What then is the aim of using the network shown in Figure 7.4(i)? One of the advantages of this structure is that the relevance of each stain can be specified in a range of values between 0 and 1. As mentioned above, in the special case in which $Pr(G_1) = Pr(G_2) = 1$, the network 7.4(i) provides the same result as the network $E_1 \leftarrow H \rightarrow E_2$. Whenever the relevance of an item of evidence is smaller than 1, then the degree to which knowledge of E_1 will affect or induce changes in the probability of H via F_1, for example, is reduced. In the special case in which a relevance node assumes the value 0 (*i.e.*, stain did not come from the offender), the likelihood ratio falls down to 1. Notice, however, that the degree of support that some item of evidence, for example, E_1, provides for its respective source node, for example, F_1, would remain unaffected by the value assumed by the relevance node. Imagine, for example, a situation in which knowledge is available on the occurrence of the event E_1. This makes the occurrence of F_1, the suspect being the source of the crime stain, 100 times more probable than before considering that information ($Pr(E_1 \mid F_1)/Pr(E_1 \mid \bar{F}_1) = 1/0.01 = 100$). However, when assuming that the crime stain did not come from the offender, then this would not affect the probability of the proposition H_p, the suspect is the offender. All of these are reasonable results.

7.4.4 Example 4: Scenarios involving firearms

7.4.4.1 Marks present on fired bullets

In Section 4.2.6, a graphical model has been discussed for evaluating mark evidence as it may be encountered, for example, when examining fired bullets. The basic idea has been adopted from an approach initially described by Evett et al. (1998b) in the context of footwear mark evidence.

Notice, however, that the Bayesian network described in Section 4.2.6 (*e.g.*, Figure 4.6) did not go as far as to propose an explicit incorporation of any particular identification criteria. On the contrary, the model has intentionally been kept on a general level of detail. This should allow for an evaluation of subjective degrees of beliefs, leaving room for the scientist to retain those criteria considered relevant in a case at hand. A criterion by which a subjective opinion may be informed is, for example, the counting of consecutive matching striations (CMS), a statistic which has been suggested by Biasotti (1959) and Biasotti and Murdock (1997). For a review of this method, see Bunch (2000). General reviews of criteria for firearm and tool mark identification are given by Nichols (1997, 2003).

In many of the discussions on firearms and tool mark identifications, focus is mainly on the problem of evaluating matching striations. Correspondences between features of manufacture are regarded only as a preliminary requirement for the extension of comparative analyses to acquired features. However, features of manufacture bear their own evidential value as they allow for a reduction in the size of the population of firearms that could have fired an incriminated bullet. With the proposed Bayesian network, a logical combination of information pertaining to acquired characteristics and information pertaining to features of manufacture can be achieved.

The combination of these levels of information may become important, for example, when only limited information is available on acquired characteristics. For instance, there may be cases in which bullets are badly damaged so that only limited information is available on the number and quality of corresponding striations. This then is an instance in which information on features of manufacture, such as calibre, the numbers of lands and grooves, their twist, and so on, may be useful for enabling some reduction in the size of the population of firearms that could have fired the bullet in question. It is assumed, however, that suitable statistics are available to aid the scientist's judgement.

7.4.4.2 Gunshot residues

So far, consideration has been given here to different levels of detail that may be encountered within the same kind of evidence, for example, marks present on fired bullets. Observations made during comparative analyses may be used to make inferences on whether a suspect's gun has been used to fire an incriminated bullet.

Besides characterising the link that may exist between a suspect's gun and an incriminated bullet, firearms examiners may also be called on to conduct a range of fire evaluations. To this end, the firearms specialists may examine the distribution pattern of certain gunshot residues (GSR) in the area of the entrance hole. It is thought that, for a given weapon and ammunition, there is a dependence between the distribution of GSR and the distance from which the firearm has been discharged (Lichtenberg 1990). When GSR patterns are not clearly visible, for example, when present on a dark target medium, various techniques may be applied for visualisation of the residue patterns. Also, the examiner would test fire the suspect weapon into appropriate targets arranged at different distances. Parameters such as size and density of the patterns thus obtained are visually compared with the pattern in question. Results of such experiments are thought to be amenable for making inferences on the distance from which the pattern in question was shot.

Firearms examiners would use some terminology to characterise the range of fire (Rowe 2000). A distant shot, for example, refers to a distance that is such that no residue would reach the target. Close-range shots refer to distances close enough for GSR to reach the target. When a shot is fired at a range of 1 inch (2.5 cm) or less, it may be referred to as a near-contact shot and when the shot is fired with the muzzle in contact with the target surface, it may be referred to as a contact shot (Rowe 2000). Certain kinds of observations are typically encountered with near-contact and contact shots, such as the singeing of hairs or textile fibres. The examination of gunshot wounds during autopsy may also reveal distinctive defects, such as the splitting of tissue.

As an example, consider the Modified Griess Test (MGT), a technique used for the visualisation of nitrite compounds, formed when smokeless powder burns, or partially burns (Dillon 1990). Using a suspect's gun and ammunition in combination, certain residue patterns can be reproduced at known distances. A certain amount of subjectivity is involved, however, when approximating a pattern in question by patterns obtained from firings at known distances. Generally, parameters such as size and density may be used to characterise a pattern. For the purpose of the current discussion, the characterisation of a pattern will be focused solely on a simple measure, such as the number of visualised nitrite residues. These will appear in an orange colour subsequent to the application of the MGT.

Imagine a scientist test firing a suspect weapon at distances of, say, 30, 50, 70 and 90 cm. Usually, several firings would be made for each distance in order to get an idea of the pattern's variability. Conditions for test firing would be chosen so as to reproduce as closely as possible the circumstances assumed to have existed at the moment the pattern in question was shot. A hypothetical set of data are shown in Table 7.8.

From personal experience, it is reasonable to assume a Normal probability density function with parameters μ and σ^2 for the number of particles, denoted as $N(\mu, \sigma^2)$. The parameters are estimated by the sample mean and variance. For example, let D_{30} denote the number of GSR particles found for firings made from a distance of 30 cm. The distribution may be written as $(D_{30} \mid 500, 80^2) \sim N(500, 80^2)$, or $D_{30} \sim N(500, 80^2)$ for short, with similar notations for other distances. Y represents a pattern in question, with y denoting a

Table 7.8 Mean and standard deviation of the number of visualised nitrite residues for firings at known distances. The number of firings at each distance is 10. [Reproduced by permission of Elsevier.]

	Distance (cm):	30	50	70	90
Number of residues:	mean	500	350	120	55
	standard deviation	80	60	35	20

particular outcome, that is, a certain number of visualised particles. For the purpose of the current example, let y be 400. It is accepted here that the approximation of the distribution of a discrete random variable (*i.e.*, the number of particles) by a continuous random variable is reasonable because of the large number of particles.

Next, imagine that a scientist seeks to evaluate the pattern in question given specific distributional assumptions. For example, the scientist may consider to what degree the finding of 400 GSR particles in a pattern in question is what may be expected if the distance of firing was 30 cm. However, it is also necessary to consider the evidence given some alternative proposition, for example, the distance of firing being 50 cm.

The evidence thus is evaluated given the following pair of propositions:

- H_p: The pattern in question was shot from a distance of 30 cm.
- H_d: The pattern in question was shot from a distance of 50 cm.

Notice that the approach described here does not lead scientists to propose any specific value for the distance from which the pattern in question was shot. Rather, evidence is used to discriminate between a set of specified, discrete propositions. The choice of these propositions may essentially depend on the positions held by the prosecution and the defence, respectively. An advantage of this approach is that the scientist may conduct carefully designed experiments – firings under controlled conditions, for example, at specific distances – so as to obtain suitable data for the evaluation of the evidence.

A likelihood ratio V can be determined in order to express the degree to which the evidence favours one proposition over another. For the scenario considered here, V may be written as a ratio of two Normal probability densities, denoted as $f_i(y \mid \mu, \sigma^2)$, where μ and σ^2 are replaced by estimates appropriate to the propositions $i = p(H_p)$ and $i = d(H_d)$.

$$V = \frac{f_p(y \mid \bar{x}_{D30}, \sigma^2_{D30})}{f_d(y \mid \bar{x}_{D50}, \sigma^2_{D50})} = \frac{f_p(400 \mid 500, 80^2)}{f_d(400 \mid 350, 60^2)} = \frac{0.0023}{0.0047} \approx 0.49. \quad (7.11)$$

The numerator represents the probability density at the point $y = 400$ when the distribution is $D_{30} \sim N(500, 80^2)$. Analogously, the denominator is given by the probability density at the point $y = 400$ when the distribution is $D_{50} \sim N(350, 60^2)$. In the current example, the evidence supports, weakly, H_d, the proposition according to which the pattern in question was shot from a distance of 50 cm. The likelihood ratio supporting H_d is V^{-1}, or, approximately 2.

Generally, the likelihood ratio for some evidence $Y = y$ (assuming Y is normally distributed) and propositions $H_1 : N(\mu_1, \sigma_1^2)$ and $H_2 : N(\mu_2, \sigma_2^2)$ may be written as

$$\frac{\sigma_2}{\sigma_1}\exp\frac{1}{2}\left[\left(\frac{y-\mu_2}{\sigma_2}\right)^2 - \left(\frac{y-\mu_1}{\sigma_1}\right)^2\right]. \tag{7.12}$$

An evaluation of alternative scenarios shows that the value of V depends crucially on the specified propositions, the parameters of the associated probability density functions as well as on the characteristics of the evidence (*i.e.*, the number of visualised GSR particles). It is, for example, little surprising that the likelihood ratio found using (7.11) is small. The distributions of D_{30} and D_{50} partially overlap, and the number of GSR particles of the pattern in question lies in the interval of overlap. For illustration, contrast this finding with the value of the likelihood ratio in the following settings:

- The number of GSR particles found in the pattern in question is 250. The two propositions compared are as before. The likelihood ratio becomes

$$V = \frac{f_p(250 \mid \bar{x}_{D_{30}}, \sigma^2_{D_{30}})}{f_d(250 \mid \bar{x}_{D_{50}}, \sigma^2_{D_{50}})} = \frac{f_p(250 \mid 500, 80^2)}{f_d(250 \mid 350, 60^2)} = \frac{3.78 \cdot 10^{-5}}{1.66 \cdot 10^{-3}} = 0.0228 . \tag{7.13}$$

  Here, the evidence still supports H_d, but the value has increased to $1/0.0228 \approx 44$.

- When the propositions H_p and H_d considered are D_{50} and D_{70}, respectively, for example, V becomes

$$V = \frac{f_p(250 \mid \bar{x}_{D_{50}}, \sigma_{D_{50}})}{f_d(250 \mid \bar{x}_{D_{70}}, \sigma_{D_{70}})} = \frac{f_p(250 \mid 350, 60^2)}{f_d(250 \mid 120, 35^2)} = \frac{1.66 \cdot 10^{-3}}{1.15 \cdot 10^{-5}} = 144 . \tag{7.14}$$

Note that it is not always necessary to have two discrete propositions for comparison. The following pairs of propositions may be considered.

(*i*) H_p: $\mu = 30$;

(*ii*) H_d: $\mu \neq 30$.

(*i*) H_p: $\mu = 30$;

(*ii*) H_d: $\mu < 30$.

(*i*) H_p: $\mu = 30$;

(*ii*) H_d: $\mu > 30$.

(*i*) H_p: $\mu \leq 30$;

(*ii*) H_d: $\mu > 30$.

For each of these propositions in which μ does not have a specific value, it is possible to determine the value of the appropriate density function by integrating over the relevant range of μ, with the variability of μ represented by some appropriate probability density function, for example, a Normal probability density function, parameterised by a mean θ and variance η. It will also be necessary to represent the variability of the variance, σ^2, and this may be done with a chi-squared distribution, parameterised by a measure of location τ^2 and degrees of freedom ν. For example, consider the proposition that $\mu \geq 30$ (and, implicitly, $\sigma^2 > 0$) and denote this region of (μ, σ^2) by A; that is, $A = \{(\mu, \sigma^2); \mu \geq 30, \sigma^2 > 0\}$. Then, the component of the likelihood ratio corresponding to this proposition for Y will be given by

$$\int_A f(y \mid \mu, \sigma^2) f(\mu \mid \theta, \eta) f(\sigma^2 \mid \tau^2, \nu) d\mu d\sigma^2.$$

The mathematics for the derivation of this result is beyond the scope of this book; see, for example, Lee (2004) for further details.

7.4.4.3 *Bayesian network for evaluating GSR*

The previously described procedure for evaluating GSR may be implemented in a Bayesian network (Biedermann and Taroni 2005). Consider two variables defined as follows:

- Node D: This variable is discrete and represents the various propositions relating to the distance from which a pattern in question may have been shot. Four states are proposed: D_{30}, D_{50}, D_{70} and D_{90}, denoting distances of firing of 30, 50 70 and 90 cm, respectively.

- Node Y: This variable is continuous and is used to represent the quantity of particles observed in a pattern in question.

The probabilities assigned to the node D are unconditional and represent prior beliefs held about the distance from which the pattern in question may have been shot. Assume, *a priori*, that there is no preference for any of the four distances so that $Pr(D_{30}) = Pr(D_{50}) = Pr(D_{70}) = Pr(D_{90}) = 0.25$. Note that this is a discrete probability distribution.

For the time being, the continuous variable Y will be used here without providing any further details. More details on the technicalities of the use of continuous variables can be found in Chapter 10. For the node Y, means and variances need to be specified given each of the possible states of the respective parent node D. The parameter values in Table 7.8 are used for illustration. A summary is provided in Table 7.9. As the quantity of visualised GSR particles is thought to depend on the distance of firing, the node D is chosen as a parent node for the node Y. A graphical representation is provided by Figure 7.5(i).

Figure 7.5(ii) displays the Bayesian network in its initial state, that is, when no evidence is entered. Each of the four possible states of the variable D are equally likely. The variable Y indicates the mean number of GSR particles that may be expected to be found given the specified prior beliefs. This value is given by

$$E(Y) = \mu_{30} Pr(D_{30}) + \mu_{50} Pr(D_{50}) + \mu_{70} Pr(D_{70}) + \mu_{90} Pr(D_{90}) \,. \quad (7.15)$$

This is a particular example of a general result that, given two dependent random variables Y and X, $E(Y) = E(E(Y \mid X))$. Further use of this result is made in Chapter 10.

Table 7.9 Means, μ_i, and variances, σ_i^2, of the probability density functions assigned to the variable Y, and the number of particles of gunshot residue for various values of D, represented by D_i, where i is the distance in centimetres from which the gun was fired. [Reproduced by permission of Elsevier.]

	i	30	50	70	90
	D:	D_{30}	D_{50}	D_{70}	D_{90}
Y:	mean (μ_i)	500	350	120	55
	variance (σ_i^2)	6400	3600	1225	400

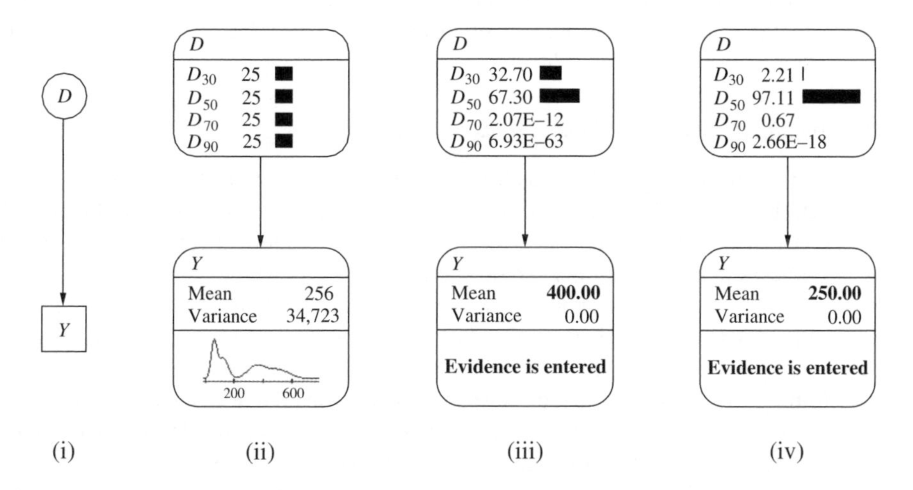

Figure 7.5 Bayesian network for evaluating GSR: (i) abstract representation of the graphical structure (a rectangle is used to represent a continuous variable, a circle is used to represent a discrete variable); (ii) initial state of the numerically specified network; note that the marginal distribution of Y is multimodal. (iii) and (iv) are states of the network after entering the evidence at the node Y, that is, observation of 400 GSR and 250 GSR, respectively. [Reproduced by permission of Elsevier.]

Note that while the distribution of Y, conditional on a particular value of D, is Normal, the marginal distribution of Y is multimodal.

Consider some of the properties of the proposed Bayesian network when entering evidence. When instantiating the variable D to any of its four possible states, the node Y displays the parameters of the appropriate Normal distribution specified in Table 7.9. For example, if $Pr(D = D_{90}) = 1$, then the node Y indicates a Normal distribution with a mean of 55 and a variance of 400. The value 55 is also obtained via (7.15), which reduces to $E(Y) = \mu_{90} = 55$ (owing to $Pr(D = D_{90}) = 1$).

Next, consider an inference in the opposite direction. Figure 7.5(iii) shows a situation in which 400 GSR particles are observed. This evidence is entered at the node Y. The node D

displays the posterior probability of each hypothesis given the evidence. For example, the probability associated with the state D_{30} is the posterior probability of the distance of firing being 30 cm given that 400 particles have been observed, which is formally written as $Pr(D = D_{30} \mid y = 400)$. The latter probability can be obtained as follows:

$$Pr(D = D_{30} \mid y = 400) = \frac{f(y \mid D_{30})Pr(D_{30})}{\sum_i f(y \mid D_i)Pr(D_i)}, \text{ with } i = (30, 50, 70, 90)$$

and $f(y \mid D_i)$ being the probability density function for y at distance D_i. The value for $Pr(D = D_{30} \mid y = 400)$ thus obtained is 0.3270. The posterior probabilities for D_{50}, D_{70} and D_{90} may be obtained analogously.

Consider, for example, the ratio of the posterior probabilities of D_{30} and D_{50}, respectively, given y, which yields $0.327/0.673 = 0.49$. This value corresponds to the likelihood ratio earlier obtained in Section 7.4.4.2. Notice, however, that this agreement requires equal prior probabilities for D_{30} and D_{50}. In Section 7.4.4.2, likelihood ratios have also been proposed for situations in which the number of visualised GSR is 250. An evaluation of this scenario is represented by Figure 7.5(iv).

7.4.4.4 Combining gunshot residue and marks evidence

In the previous sections, a procedure has been described for discriminating among a specific set of discrete hypotheses. Roughly speaking, the procedure combines observed data – the quantity of visualised GSR particles – with previous knowledge and can be seen as an aid to rank hypotheses according to their credibility.

Consider now a scenario in which a forensic firearms specialist is needed to conduct certain examinations. Imagine that a bullet has been extracted from a dead body. Assume that only one bullet has been fired and that the GSR around the entrance hole on the victim's clothing are from that firing. A suspect gun is available and comparisons have been made between both, the bullets fired from that weapon and the incriminated bullet. Suppose that the scientist judges the quality of the observations to be sufficient for forming the (subjective) opinion that the incriminated bullet has been fired from the suspect gun.

At a subsequent stage, the scientist may be asked to conduct an estimation of the range of fire. Ammunition that is thought to come from the same lot as that used to fire the incriminated pattern is available. The suspect's weapon, reasonably believed to have fired the pattern in question, is then used for the firings at known distances and under controlled conditions. The GSR patterns thus obtained are compared with the pattern surrounding the entrance hole on the victim's clothing. A procedure, as described in Sections 7.4.4.2 and 7.4.4.3, may be used to assist the inferential task associated with these experiments.

So far, the evaluation of the marks evidence has been considered quite independently of the evaluation of the GSR evidence, and vice versa. Usually, the scientist would not proceed to a range of fire estimation until he has attained a certain subjective degree of belief that some specified weapon is the one used to fire an incriminated pattern. However, this opinion is probabilistic in nature, meaning that some uncertainty may usually exist about whether that weapon is in fact, the one used during the shooting incident. Hence, it may be felt that this uncertainty should be accounted for in a range of fire estimation. This is an interesting issue in the joint evaluation of marks evidence and GSR evidence.

Consider the Bayesian network described in Section 7.4.4.3 (Figure 7.5(i)). For each hypothesis relating to a specified distance of firing (node D), a probability distribution for modelling the number of GSR particles visualised on the target is defined (node Y). An implicit assumption of this approach is that the evidence is assessed only under varying propositions for distances of firing. For example, the evidence y is assessed under the proposition that the distance of firing was 30 cm ($D = D_{30}$) and under the proposition that the distance of firing was 50 cm ($D = D_{50}$). However, in both the settings, the suspect's weapon is taken to be the one that has fired the pattern in question, an assumption that may not necessarily reflect the defence position. Also, the proposition that the suspect's weapon has fired the incriminated pattern may be questionable, for example, if the mark evidence is weak.

This inconvenience may be overcome by the addition of a further node, F say. F is a binary node that takes two values F_s and $\bar{F}_s$. These are defined as 'the bullet was or was not fired by the suspect's weapon'. Notice that, according to the assumption made earlier in this section, this is equivalent to saying that the GSR are the result of firing the suspect's weapon. The node F is chosen as a parent for the node Y, so the Bayesian network shown in Figure 7.5(i) now becomes $D \rightarrow Y \leftarrow F$ (Figure 7.6(i)). Consequently, the probability table of the node Y needs to be extended as shown in Table 7.10. When the first four columns of that table are specified in the same way as Table 7.9, then, with F instantiated to F_s, the Bayesian network $D \rightarrow Y \leftarrow F$ yields strictly the same results as the network $D \rightarrow Y$ (Figure 7.5(i)).

The parameters specified for Y given $\bar{F}_s$ characterise the expected Normal distributions for Y when a firearm other than that of the suspect was used to fire the incriminated bullet. Note that these parameters are labelled with a ' $'$ ' as they may be different from the parameters assumed for Y given F_s. The parameters for Y given $\bar{F}_s$ tend to affect the evaluation of some observation y if, for a given distance D_i of firing, the parameters of the Normal distributions specified for Y given F_s and $\bar{F}_s$ differ, and F_s is not true with certainty. When F_s were known to be true, then the evaluation is not affected whatever be the value that is specified for Y given $\bar{F}_s$.

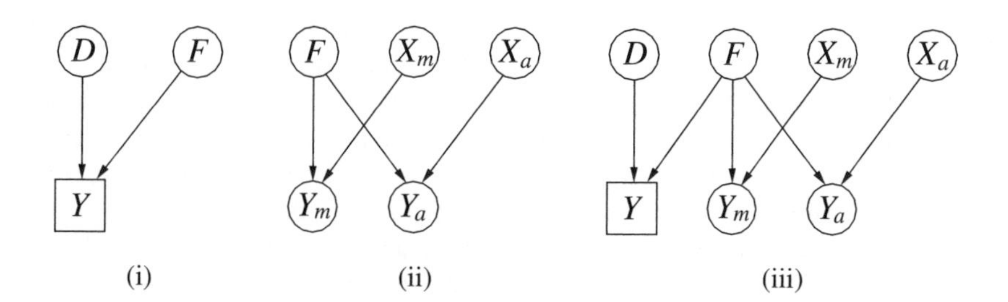

Figure 7.6 Bayesian networks for (i) evaluating GSR evidence, (ii) evaluating marks evidence at the source level, and (iii) the joint evaluation of GSR evidence and marks evidence. Node Y is a continuous node representing the number of gunshot residue particles found from a firing. Node D is a discrete node for the firing distance, with four states, representing distances 30, 50, 70 and 90 cm. Node F has two states, F_s and $\bar{F}_s$, that is, the bullet was or was not fired from the suspect weapon, respectively. Factors Y_m, Y_a, X_m and X_a denote the observations made on the incriminated bullet (Y_m, Y_a) and the bullets (X_m, X_a), respectively, from a known source. The subscripts m and a indicate that the marks originate because of the features of manufacture of the firearm and because of the acquired characteristics, respectively. [Reproduced by permission of Elsevier.]

Table 7.10 Parameters of the Normal distributions associated with the variable Y. Means, μ_i, μ'_i, and variances, $\sigma_i^2, \sigma_i'^2$, of the probability density functions assigned to the variable Y, the number of particles of gunshot residue for various values of D, represented by D_i, where i is the distance in centimetres from which the gun was fired. Factor F has two states, F_s and $\bar{F}_s$, that is, the bullet was or was not fired from the suspect's weapon. [Reproduced by permission of Elsevier.]

	F:	F_s				$\bar{F}_s$			
	i	30	50	70	90	30	50	70	90
	D:	D_{30}	D_{50}	D_{70}	D_{90}	D_{30}	D_{50}	D_{70}	D_{90}
Y:	mean (μ_i)	μ_{30}	μ_{50}	μ_{70}	μ_{90}	μ'_{30}	μ'_{50}	μ'_{70}	μ'_{90}
	variance (σ_i)	σ_{30}^2	σ_{50}^2	σ_{70}^2	σ_{90}^2	$\sigma_{30}'^2$	$\sigma_{50}'^2$	$\sigma_{70}'^2$	$\sigma_{90}'^2$

Let the prior probabilities for F_s and $\bar{F}_s$ be denoted as f_0 and f_1 respectively. Then, the mean number of GSR particles expected to be found, given prior beliefs specified for the variables D and Y, is given by

$$\begin{aligned} E(Y) &= E(E(Y \mid D))f_0 + E(E(Y \mid D))f_1 \\ &= \sum_i \mu_i Pr(D_i) f_0 + \sum_i \mu'_i Pr(D_i) f_1 \\ &= \sum_i Pr(D_i)(\mu_i f_0 + \mu'_i f_1), \text{ for } i = (30, 50, 70, 90)\ . \end{aligned} \tag{7.16}$$

If it can be accepted that uncertainty may exist with respect to the variable F, the proposition that the suspect's weapon has fired the incriminated bullet, then it may be desirable to base one's belief in the truth or otherwise of that variable, based on other evidence. The marks evidence appears to be suitable for this task. Thus, imagine that comparative examinations have been performed between the incriminated bullet and the bullets fired (under controlled conditions) through the barrel of the suspect's weapon. During the examination process, a specific set of similarities and differences is observed. Consider an evaluation of such evidence at the source level, for which part of a Bayesian network constructed and discussed earlier in Section 4.2.6 may be appropriate [Figure 7.6(ii)].

The factors Y_m, Y_a, X_m and X_a in the model shown in Figure 7.6(ii) refer to the observations made on the incriminated bullet (Y_m, Y_a) and the bullets (X_m, X_a), respectively, from a known source, for example, the suspect's weapon. The subscripts m and a refer to marks that originate from the features of manufacture of the firearm and from acquired characteristics, respectively.

Figure 7.6(iii) depicts a combination of the networks for evaluating GSR evidence and marks evidence. The two network fragments are linked *via* the node F, defined as 'the incriminated bullet has been fired by the suspect's weapon'. As mentioned earlier, the importance of incorporating F when evaluating GSR evidence is that one may account for the uncertainty about whether the weapon used for the test firings at known distances, that is, the suspect's weapon, is in fact the one used to fire the incriminated pattern. An advantage of the Bayesian network shown in Figure 7.6(iii) is that one's beliefs about the truthstate of F can be more than just a guess: they may coherently be informed by the knowledge available from marks evidence.

The Bayesian network shown in Figure 7.6(iii) can be used to analyse the quite different positions that may be held by the prosecution and the defence. For example, the prosecution's case may be that the suspect's weapon has fired the incriminated bullet (F_s) and that the distance of firing was about 30 cm. The defence may argue that a weapon other than that of the suspect was used ($\bar{F}_s$) and that the distance of firing was about 70 cm. In such a setting, the likelihood ratio for some observation $Y = y$ is given by

$$\frac{\sigma'_{70}}{\sigma_{30}} \exp \frac{1}{2} \left[\left(\frac{y - \mu'_{70}}{\sigma'_{70}} \right)^2 - \left(\frac{y - \mu_{30}}{\sigma_{30}} \right)^2 \right].$$

The proposed Bayesian network clarifies the potential scenarios as to how the pattern in question may have been produced, and the assumptions made by the expert during the evaluative process. A difficulty, however, is that experts may not have knowledge of the nature of the outcomes in all scenarios. The nature of the outcomes, for example, may include the values of the parameters assigned to Y given $\bar{F}$ in Table 7.9, which are the kind of GSR patterns produced by firearms other than that of the suspect. The scientist may be able to make reasonable assumptions such as about values for μ' and σ' that relate to the nature of the GSR deposition of an alternative weapon because such a weapon is available for test firing. Alternatively, he may make the explicit assumption that it was the suspect's weapon that has been used to fire the incriminated pattern.

7.4.4.5 Comments

A particular aspect of the scenario involving GSR evidence and marks evidence is that the two items of evidence are used to draw inferences about different propositions. In the examples discussed through Sections 7.4.1 to 7.4.3, two items of evidence have each time been used to make inferences about one major proposition. Here, however, the marks evidence is used to make inferences about whether the suspect's weapon has been used to fire an incriminated bullet. The GSR evidence is used to make inferences about the distance of firing. The evaluation is interrelated owing to propositions relevant for both types of evidence.

It may be felt that the counting of single GSR is too laborious a task to be used in practice. Also, scientists may consider that the phenomenon of GSR transfer on a target is not directly amenable to a specific numerical description. As far as chromophoric tests are concerned, examiners may prefer a visual inspection based on characteristics such as the spread and density of an incriminated pattern. Observations are then judged regarding whether they are more 'compatible' with one distance of firing rather than with another. Here, experts mostly rely on qualitative expression of their findings. Notice that Bayesian networks can address inferences on such a level of detail as well. Further details on this topic can be found in Section 9.1.1.

The approach to distance evaluation described here, which is based on a Bayesian network, avoids categorical statements of specific distances of firing. In the proposed model (Figure 7.6), the value of the evidence is dependent on specified distributional assumptions. The value can provide indications as to which propositions are favoured by the evidence, a common way to look at evidence in Bayesian data analysis (D'Agostini 2003).

8

Pre-assessment

8.1 Introduction

An evaluation process starts when the scientist first meets the case. It is at this stage that the scientist thinks about the questions that are to be addressed and the outcomes that may be expected. The scientist should attempt to frame propositions of interest and think about the value of evidence that is expected (Evett et al. 2000). There is a wide tendency to consider evaluation of evidence as the final step of a casework examination, notably at the time of preparing the formal report. This is so even if an earlier interest in the process would enable the scientist to make better decisions about the allocation of resources. One of the first approaches to decision making in an operational forensic science unit was proposed by Cook et al. (1998b). It is based on a model embodying the likelihood ratio as a measure of the weight of evidence. The proposed model is intended to enhance the cost-effectiveness of casework activity from initial contact with the customer. The aim is to enable the customer to make better decisions.

In routine work, forensic scientists often require an estimate of the expected likelihood ratio before performing any tests. Such an estimate can help scientists to support a better decision for the customer.

Imagine, for the sake of illustration, a situation in paternity testing where the alleged father is unavailable but a cousin of the alleged father could potentially be considered and tested. In such a case, the two propositions of interest may be of the form of:

H_p : The tested person is a cousin of the true father;

H_d : The tested person is unrelated to the child.

Two questions may be of interest in this scenario: (1) can a value of the evidence be obtained such that the hypothesis H_p is favoured over H_d, and (2) how can the laboratory or the customer take a rational decision on the necessity of performing DNA testing after an estimate of possible values of the likelihood ratio?

Bayesian Networks and Probabilistic Inference in Forensic Science F. Taroni, C. Aitken, P. Garbolino and A. Biedermann

The first question refers to a process known as *case pre-assessment*. The second question is one that refers to *decision making*. An approach to the first question has been proposed by Cook et al. (1998b), and Taroni et al. (2005a,c) proposed answers to the second question.

Bayesian networks are intended to support human reasoning and decision-making through the formalisation of expert knowledge. They can be used to deal with both case pre-assessment and problems of decision making. The present chapter will essentially focus on the former. Aspects of decision making will be addressed briefly in Chapter 11.

8.2 Pre-assessment

In order to follow a logical procedure, the scientist requires an adequate appreciation of the circumstances of the case. This should allow a framework to be set up for considering the kind of examinations that may be carried out as well as the expected outcomes (Cook et al. 1998b).

The choice of a propositional level according to Cook et al. (1998a) has earlier been discussed in Chapter 3. The evaluation of scientific evidence is carried out within a framework of circumstances, and these circumstances have to be known before any examination of the evidence is made so as to propose relevant propositions. Such a process provides a basis for consistency for scientists who are thereby encouraged to consider carefully factors such as circumstantial information and data that are to be used for the evaluation of evidence.

The scientist should proceed by considering an estimation of the probability of whatever evidence will be found given each proposition of interest. Consider, for example, a case potentially involving glass fragments due to a smashed window. Assume that the prosecution and defence propose propositions at the activity level such as 'the suspect is the man who smashed the window' (H_p), and, 'the suspect did not smash the window' (H_d). The examination of the suspect's pullover leads to the recovery of a quantity Q of glass fragments. For the purpose of illustration, let the possible states that may be assigned to Q be *none, few* and *many*. Then,

- for assessment of the numerator of the likelihood ratio, the question to be answered is 'what is the probability of finding a quantity Q of 'matching' glass fragments if the suspect is the man who smashed the window'?;

- for assessment of the denominator of the likelihood ratio, one needs to address a question of the kind 'what is the probability of finding a quantity Q of 'matching' glass fragments if the suspect is not the man who smashed the window'?.

Initially, the scientist is asked to assess a total of six different probabilities, that is, each of the three states of the variable Q given H_p and H_d. These assessments may be informed by various sources of information, for example, surveys, relevant publications on the matter or subjective assessments (Taroni et al. 2001).

Note that these probabilities may not be easy to derive because of the possible lack of information. It may be difficult, for example, to assess transfer probabilities when the scientist has no indications about parameters pertaining to the *modus operandi*: how was the window smashed? How close was the perpetrator with respect to that window at the moment it was smashed?

Table 8.1 Probabilities of finding quantities Q of matching glass. The propositions are H_p and H_d, the suspect is the man who smashed, or, who did not smash, respectively, the window. The corresponding values of the likelihood ratio V are indicated in the right-hand column. [Adapted from Cook et al. (1998b).]

Q	$Pr(Q \mid H_p)$	$Pr(Q \mid H_d)$	V
none	0.05	0.95	1/19
few	0.30	0.04	7.5
many	0.65	0.01	65

For the sake of illustration, consider a scenario described by Cook et al. (1998b). Probability distributions for finding quantities Q of matching glass under the two competing propositions H_p and H_d are proposed. A summary is given in Table 8.1. These estimates can serve as a basis for calculating likelihood ratios as a measure of the value of finding a quantity Q of glass fragments. Possible conclusions may be that:

> [...] on the basis of this assessment, if the suspect is indeed the person who committed the burglary, there is a 65% chance that the result of the examination will provide moderate support for that proposition; and a 30% chance that it will provide weak support. If, on the other hand, the suspect is truly *not* the offender, then there is 95% chance of moderate evidence to support his innocence although there is a 5% chance of evidence that will falsely tend to incriminate him (Cook et al. 1998b, p. 155).

From these results, it is suggested that the scientist is in a position to help the customer take an informed decision. While this stage is a fundamental one in the decision-making process, it does not offer clear criteria, however, as to whether to perform a test (*i.e.*, an analytical test in the laboratory). Section 11.2 offers a discussion of this critical point, which is essential because scientists routinely take decisions and they do so generally under uncertainty. A more extended logical framework should be employed for this task.

8.3 Pre-assessment for a fibres scenario

8.3.1 Preliminaries

In Chapter 6, scenarios have been approached that involve the evaluation of a group of fibres associated with the investigation of an offence. Two main categories of situations have been studied in which evidence was assessed under activity level propositions, notably:

- situations in which the recovered fibres were assumed to come from the offender;
- situations in which the recovered fibres were taken to be present by chance alone.

The formal development has shown that parameters of interest are transfer, persistence and recovery, the presence by chance (background probabilities) and the relative frequency γ of the incriminated fibre's features in a relevant population.

This section provides a more detailed study of a scenario involving textile fibres. The following objectives are considered (Champod and Jackson 2000): defining the information the scientist may need together with relevant propositions used to assess the findings, performing a case pre-assessment, determining the examination strategy, assessing the likelihood ratio and its sensitivity, and evaluating the effect of a change in the propositions.

The scenario is as follows. Two armed and masked men burst into a post office and threaten the staff. The takings for the day were handed over and the men left. Witnesses reported that one of the men was wearing a dark green balaclava mask and the other man was wearing a knotted stocking mask. Witnesses further declared that the two men left the scene in a car driven by a third man. Some way along the presumed getaway route, a dark green balaclava was found. The following day Mr U was arrested. He denied all knowledge of the incident. Reference samples of his head hairs and blood were taken as well as combings from his head hair. Mr U has not yet been charged with the robbery because there is very little evidence against him.

8.3.2 Propositions and relevant events

At some stage in the investigative proceedings, it may be of interest to know whether Mr U has worn the incriminated mask. Initially, the scientist may define propositions at the *source* level:

- H_{p1}: Hairs in the mask came from Mr U;
- H_{d1}: Hairs in the mask came from someone else;
- H_{p2}: Saliva in the mask came from Mr U;
- H_{d2}: Saliva in the mask came from someone else;
- H_{p3}: Fibres in U's hair combings came from the mask;
- H_{d3}: Fibres in U's hair combings came from some other garment or fabric.

The scientist may also be asked to assess the findings under *activity* level propositions. These may be, for example, 'Mr U wore the mask at the time of the robbery' (H_p), and, 'Mr U has never worn the mask' (H_d). Such propositions can be more relevant for the court because a closer link with the alleged offence is offered. However, scientists will need more profound knowledge of

(a) background information on the incident, such as the time the alleged offence took place, circumstances of the suspect's arrest, the time when samples were taken, and so on. These data enable the scientist to characterise more precisely the factors that relate to transfer and persistence of the recovered items of evidence.

(b) data on transfer, persistence and recovery of hairs, fibres and saliva of individuals wearing masks, and survey data on masks.

Published data on hairs and saliva are very limited; data on the transfer of fibres to head hair including persistence are more readily available. So, if requirements (a) and (b) are both satisfied, the scientist may consider fibre evidence first. If, however, the background information is not sufficient for considering the evidence (fibres, saliva, hairs) under *activity* level propositions, then scientists may be obliged to stay at the *source* level. In the scenario described above, a feasible strategy would be to offer an assessment at the activity level for fibres but to consider the other evidence at the *source* level.

The second step in a case pre-assessment is to think about the possible *findings*. Considering fibres on hair combings, scientists may encounter results of the following kind:

1. no fibres are recovered,
2. a small number of fibres are recovered (*i.e.*, 1-3), or
3. a large number of fibres are recovered (*i.e.*, greater than 3).

The definitions of these categories can be considered flexible and may depend on the available data. It is also possible that more than one group of fibres could eventually be found.

An assessment of the activity level requires a definition of *events*. In order to help the scientist to determine which events are relevant for a pre-assessment in such a scenario, it appears useful to consider a question of the kind 'What could happen if Mr U wore the mask at the time of the robbery'?. If Mr U wore the mask, then three main events may be framed

- T_0: No fibres have been transferred. Arguably, there is no persistence and no recovery of fibres.
- T_s: A small number of fibres have been transferred. The fibres have persisted and have successfully been recovered.
- T_l: A large number of fibres have been transferred. The fibres have persisted and have successfully been recovered.

These events will represent the states of a variable in a Bayesian network developed later in this section. Denote this variable as 'transfer', or T for short.

Two main propositions may be considered for explaining the presence of fibres: a crime-related transfer, or the presence by chance prior to the crime. A possible background presence is defined as:

- P_0: No group of fibres is present by chance;
- P_1: One group of fibres is present by chance.

When a group of fibres is present by chance, it may either be a small or a large group. Let these events be denoted as follows:

- S_s: The group of fibres present by chance is small;
- S_l: The group of fibres present by chance is large.

Note that in Chapter 6, outcomes for P_i and $S_{i,j}$ have been aggregated within a single variable $b_{g,m}$. Background probabilities $b_{g,m}$ account for the chance occurrence of g group(s) of m foreign fibres on the receptor. This probability can be considered as a composition of probabilities p_i and $s_{i,j}$ as is currently practiced in the evaluation of glass evidence (*see*, for example, Curran et al. 2000). Values for p_i denote the probabilities by which i (≥ 0) groups of fibres are present. The parameter $s_{i,j}$ denotes the probability that group i of recovered material is of size j, where j may take a positive integer value. Notice that j may also be replaced by l or s in order to refer to a large or small group, respectively. Literature on the evaluation of glass evidence considers that (a) there is no association between the number of groups found on surfaces of interest and the sizes of those groups, and (b) there is no association between the frequency of a given type of glass fragment with either the number of groups or the size of the group (Curran et al. 2000). In the context of fibre evidence, such assumptions may be more difficult to maintain. Properties such as transfer and persistence may be affected by physical characteristics of the type of fibres involved. For clarity, the present discussion will, however, consider similar assumptions as acceptable.

When a recovered group of fibres of unknown origin is compared with fibres from a known source, that is, a control, two outcomes are possible

- M: The recovered fibres match the control with respect to some features of interest;
- $\bar{M}$: The recovered fibres do not match the control with respect to these features of interest.

Note that the assumption will be made here, for discrete variables, that the probability of a match will be taken to be 1 when a sample, known to originate from a particular source, is compared with that source.

8.3.3 Expected likelihood ratios

When searching hair combings for the presence of fibres, and, when analysing fibres thus recovered, scientists could reasonably meet any of the four situations summarised in Table 8.2. Note that this listing of possible outcomes does not take into account other scenarios like one group of fibres being present because of a transfer, and a second group of fibres being present as background.

For clarity, the current analysis will consider a possible presence of at most one group of fibres. Table 8.3 provides a summary of the various events given, respectively, H_p and H_d, and the associated probabilities. Note that $Pr(E|H_p)$ and $Pr(E|H_d)$ in outcomes C and D correspond to (6.12), Section 6.2.8.

The next step in the pre-assessment process is concerned with the estimation of these probabilities using data from published literature, case-specific experiments or subjective assessments based on the scientist's experience.

- *Transfer probabilities*: In order to find appropriate values for the parameters t_0, t_s, t_l, it may be useful to answer questions of the kind 'If the suspect wore a mask, what is the probability of, respectively, no fibres, a small or a large number of fibres being transferred, having persisted and successfully being recovered'? An appropriate assessment should also consider information pertaining to the suspect (*e.g.*, type and

Table 8.2 Possible outcomes of a procedure for searching and analysing fibres present in hair combings.

Outcome	Number of groups	Number of non-matching groups	Number of matching groups	Size of matching groups
A	0	0	0	–
B	1	1	0	–
C	1	0	1	Small
D	1	0	1	Large

Table 8.3 Events and probabilities relating to findings under H_p and H_d, propositions that the fibres in Mr U's hair combings came from the mask. Factors T_0, T_s and T_l relate to no (0), a small number (s) of, and a large number (l) of, fibres having been transferred, with corresponding probabilities, t_0, t_s and t_l. Factors P_0 and P_1 correspond to no (0) or one (1) group of, fibres being present by chance, with corresponding probabilities, p_0 and p_1. Factors S_s and S_l correspond to the size of the group of fibres being present by chance being small (s) or large (l), with corresponding probabilities, s_s and s_l. Factor M has two states, the recovered fibres match, M, or do not match, $\bar{M}$, the control fibres with respect to some features of interest, with corresponding probabilities, m and $1 - m$.

Outcome from Table 8.2	Events to occur if H_p is true	$Pr(E\|H_p)$
A	T_0, P_0	$t_0 p_0$
B	T_0, P_1, $\bar{M}$	$t_0 p_1 (1 - m)$
C	T_0, P_1, S_s, M or T_s, P_0	$t_0 p_1 s_s m + t_s p_0$
D	T_0, P_1, S_l, M or T_l, P_0	$t_0 p_1 s_l m + t_l p_0$
Outcome from Table 8.2	**Events to occur if H_d is true**	**$Pr(E\|H_d)$**
A	P_0	p_0
B	P_1, $\bar{M}$	$p_1 (1 - m)$
C	P_1, S_s, M	$p_1 s_s m$
D	P_1, S_l, M	$p_1 s_l m$

length of his hair), the material involved (*e.g.*, the sheddability of the mask), the methods used to search and collect fibres as well as the circumstances of the case (*e.g.*, alleged activities, time delays).

- *Background probabilities*: The parameters p_0, p_1, s_s and s_l are used to assess situations involving the presence, by chance, of no fibres or one group of fibres (which could be small or large, as previously defined), assuming that no fibres have been transferred, or, alternatively, that the suspect never wore the mask. Notice that if the alternative proposition changes, for example, to one that the suspect wore a similar mask two days before the alleged facts, then background probabilities change and new assessments may be required.

- *Match probabilities*: The probabilities abbreviated by m represent, in some way, an assessment of the rarity of extraneous fibres found in the head hair of a person innocently accused of wearing a mask. These are fibres that match by chance the control fibres originating from the mask. This rarity may be assessed in various ways. For example, the scientist may refer to studies on textile fibres recovered from head hair and consider the relative proportions of fibres presenting the features of interest. Eventually, they may also consult so-called 'target fibre studies'. Note, however, that such studies provide different estimates of probabilities, such as the probabilities of observing by chance one target group of fibres that match the control: $Pr(p_1, s_s, m|H_d)$ and $Pr(p_1, s_l, m|H_d)$, the latter previously referred to as γ_b (*see* Section 6.2.2). To some extent, databases may also be used, assuming that the potential sources of fibres are hats, neckwear, bedding or jumpers, so that the scientist is able to assess the relative frequency of the matching fibres in these populations.

Likelihood ratios using probabilities proposed by Champod and Jackson (2000) are given in Table 8.4.

'Is it useful to conduct analyses of textile fibres'?. Likelihood ratios obtained as a result of the pre-assessment of fibres evidence can assist in the provision of an answer to such a question. It has been shown, for example, that all situations offer a likelihood ratio different from the 'inconclusive' value of 1. If no fibres at all are recovered, or if a group of fibres are recovered and this group does not match the control object, likelihood ratios supporting the defence's proposition are obtained. On the other hand, if one group (small or large) of fibres is recovered and the group matches the control group, then a likelihood ratio greater than 1 is obtained, and the prosecution's case is supported.

Table 8.4 Likelihood ratios, V, for the outcomes from Table 8.2 with $t_0 = 0.01$, $t_s = 0.04$, $t_l = 0.95$; $p_0 = 0.78$, $p_1 = 0.22$; $s_s = 0.92$, $s_l = 0.08$; $m = 0.05$, as proposed by Champod and Jackson (2000).

Outcome	V
A	0.01
B	0.01
C	3.09
D	842.05

Table 8.5 Conditional probabilities assigned to the outcome node O of Figure 8.1. Node O, the outcome, has five states, the discovery of no group of fibres, o_1, a group of non-matching fibres, o_2, a small group of matching fibres, o_3, a large group of matching fibres, o_4, and two groups of fibres, o_5. Factor T has three states, no transfer, t_0, transfer of a small amount of fibres, t_s, and transfer of a large amount of fibres, t_l. Factor P has two states, P_0 and P_1 corresponding to no (0) or one (1) group of, fibres being present by chance, with corresponding probabilities, p_0 and p_1. Factor S has two states, S_s and S_l, corresponding to the group of fibres being present by chance being small (s) or large (l), with corresponding probabilities, s_s and s_l.

	T:	t_0				t_s				t_l			
	P:	p_0		p_1		p_0		p_1		p_0		p_1	
	S:	s_s	s_l	s_s	s_l	s_s	s_l	s_s	s_l	s_s	s_l	s_s	s_l
O:	o_1	1	1	0	0	0	0	0	0	0	0	0	0
	o_2	0	0	0.95	0.95	0	0	0	0	0	0	0	0
	o_3	0	0	0.05	0	1	1	0	0	0	0	0	0
	o_4	0	0	0	0.05	0	0	0	0	1	1	0	0
	o_5	0	0	0	0	0	0	1	1	0	0	1	1

8.3.4 Construction of a Bayesian network

Recall the Bayesian network presented in Figure 6.4. A slight modification of this network allows the scientist to approach the case pre-assessment discussed so far in this chapter. Two new nodes P, denoting the presence of a group of fibres, and S, the size of a group, are used as a substitute for node B. A node O, for 'outcome', is also introduced. This node covers five states: no group of fibres (o_1), a group of non-matching fibres (o_2), a small group of matching fibres (o_3), a large group of matching fibres (o_4) and two groups of fibres (o_5). As mentioned earlier, this last state is not considered by Champod and Jackson (2000). The state 'two groups of fibres' enables consideration of situations involving the joint presence of transferred and background fibres.

All other nodes are binary except the transfer node T. The states of this variable are defined as: no transfer (t_0), transfer of a small amount of fibres (t_s) and transfer of a large amount of fibres (t_l). Suggested values for these parameters given H_p are given in the legend of Table 8.4. Given H_d, no transfer can occur, implying that $t_0 = 1$ and $t_s = t_l = 0$. The corresponding Bayesian network is shown in Figure 8.1.

Using probability values proposed in Section 8.3.3 and conditional probabilities linked to node O given in Table 8.5, the same likelihood ratios as given by Table 8.4 can be obtained.

8.4 Pre-assessment in a cross-transfer scenario

As a basic approach to decision making, case pre-assessment can also be applied to more sophisticated cases. For instance, imagine a case involving a cross transfer, also known as 'two-way transfer' (Aitken and Taroni 2004; Champod and Taroni 1999). The scenario

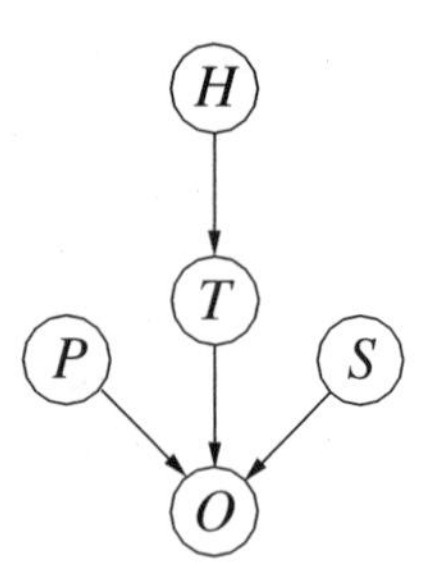

Figure 8.1 Bayesian network usable for pre-assessment in cases involving transfer evidence. Node O, the outcome, has five states, the discovery of no group of fibres, o_1, a group of non-matching fibres, o_2, a small group of matching fibres, o_3, a large group of matching fibres, o_4 and two groups of fibres, o_5. Node T has three states, no transfer, t_0, transfer of a small amount of fibres, t_s, and transfer of a large amount of fibres, t_l. Node P has two states, P_0 and P_1 corresponding to no (0) or one (1) group of, fibres being present by chance, with corresponding probabilities, p_0 and p_1. Node S has two states, S_s and S_l, corresponding to the group of fibres being present by chance being small (s) or large (l), with corresponding probabilities, s_s and s_l. Node H has two states, H_p, Mr U wore the mask at the time of the robbery, and H_d, Mr U has never worn the mask. [Adapted from Aitken and Taroni (2004), reproduced with permission of John Wiley & Sons, Ltd.]

is as follows. An assault has been committed. There is one victim and one criminal. The criminal and the victim have, in some way, been in contact with each other. The evidence under consideration is such as to yield DNA profiles of the victim and the criminal. In a rape case, the evidence could be semen and vaginal fluids. The evidence could also be blood, for example, in the case of an assault. Transfer can occur from the victim to the criminal, for example, in the form of vaginal fluids, or, from the criminal to the victim, for example, in the form of semen. It is possible that there may be transfer in one direction only. Suppose a victim has been killed with a knife and there is no evidence of a fight. The probability of a transfer of blood from the criminal to the victim is low. The probability of a transfer of blood from the victim to the criminal is high. Generally, the two sets of recovered traces have to be considered as dependent. In fact, if a transfer has occurred in one direction and the expert has recovered traces characterising this transfer, then the expert would generally expect to find trace evidence characterising the transfer in the other direction. The presence of one set of transfer evidence gives information about the presence of the other set of transfer evidence. Obviously, the absence of the other set of transfer evidence would in itself be significant as presented in Sections 6.3 and 8.5.

The dependence between the two sets of evidence plays an important role in the assessment of a case and allows scientists to *predict* future events. In particular, an assessment can be updated when a staged approach is followed. In a scenario involving fibres, for example, the victim's pullover may be examined first, and, in a second step, the suspect's pullover examined (Cook et al. 1999). The results of the examination of one of the garments are used to inform the decision about whether the second garment should be examined. Analogous reasoning could be applied to other kinds of cases. For example, if the crime involves the smashing of a sheet of glass and if clothing is submitted from a suspect, then the phased approach could be applied to the order in which the garments are examined. If examination of the jacket reveals no glass, then this information may be useful to inform the decision about whether to examine the trousers or shoes. Such a scenario will be approached in Section 8.4.3. Extension to a full decision analysis is presented in Chapter 11.

8.4.1 Preliminaries

Consider a case described by Cook et al. (1999). Mr A disturbed a man who had broken into his home. He attempted to restrain the intruder and struggled with him without being able to prevent him from making his escape. A man, Mr B, is arrested some time later.

Given the extent of available case-related information, police investigators are interested in the following two activity level propositions:

H_p : Mr B is the man who struggled with Mr A;

H_d : There has been no physical contact between Mr B and Mr A.

The possible findings consist of a quantity, say Q, of transferred fibres. The quantity Q concerns, independently, both Mr A and Mr B. For the sake of clarity, let Q_A and Q_B denote the quantities found on Mr A and Mr B, respectively. Then, when a quantity Q_B of fibres is found on the surface of the clothing of Mr B, indistinguishable from the fibres composing Mr A's garment, then the scientist should answer questions like 'what is the probability of finding a quantity Q_B of fibres on the clothing of Mr B if he is the man who struggled with Mr A'?, and, 'what is the probability of finding a quantity Q_B of fibres on the clothing of Mr B if there has been no physical contact between him and Mr A'?. The states of the variable Q may be defined to cover *none*, *few* and *many* fibres. The potential findings, or outcomes, are generally defined as follows:

Q_A : A quantity Q_A of fibres has been transferred, has persisted and was successfully recovered from the surface of Mr A's garment. The fibres were found to be of the same type as a control sample taken, for example, from the jumper of the suspect, Mr B.

Q_B : A quantity Q_B of fibres has been transferred, has persisted and was successfully recovered from the surface of Mr B's garment. Observation indicates that the fibres are of the same type as a comparison sample taken from the clothing of the victim Mr A.

As mentioned by Cook et al. (1999), one may allow for different intensities of contact between the victim and the aggressor. In fact, the extent of contact is a major factor affecting fibre transfer. It may be incorporated in the analysis as a distinct source of uncertainty. A variable 'contact' may be used for this purpose assuming states referring to, for example, *light*, *medium*, and *heavy* contact.

8.4.2 A Bayesian network for a pre-assessment of a cross-transfer scenario

A basic network covering four nodes as shown in Figure 8.2 may suffice to model a cross-transfer scenario. The four nodes are defined as follows:

H : This node represents the two main propositions, namely, 'Mr B is the man who struggled with Mr A' (H_p), and 'there has been no physical contact between Mr B and Mr A' (H_d).

C : A node C is required to represent the event 'contact'. It depends directly on the main propositions. Conditional probabilities need to be assessed for each state of C given, respectively, H_p and H_d. Unconditional probabilities of 0.1, 0.3 and 0.6 for a light, medium and heavy contact, respectively, have been suggested (Cook et al. 1999). Note that if H_d is true, that is, there has been no physical contact between Mr B and Mr A, then the variable C cannot be light, medium or heavy, but should assume a new state, denoted as 'no contact'. The variable C is thus described by four states whose conditional probabilities are summarised in Table 8.6.

Q_A : This node represents the quantity of fibres recovered from Mr A's garment. These fibres match those composing Mr B's garment. For the sake of simplicity, the node Q_A (and Q_B, see the following text) is taken as a compact representation including phenomena of transfer, persistence and recovery. The direct dependency of node Q_A on C, the intensity of contact, is represented by $C \rightarrow Q_A$. Node H influences Q_A in two distinct ways: directly, through $H \rightarrow Q_A$, and indirectly, through $H \rightarrow C \rightarrow Q_A$. Thus, the probability distribution of Q_A is not dictated by C alone. An assumption made by Cook et al. (1999) is that both garments have similar shedding and retentive properties. So, given that, for both garments, the extent of contact was the same and the lapse of time between incident and fibre recovery is approximately similar, it may be acceptable to assign the same probabilities to both Q_A and Q_B (*see* Table 8.7). This assumption appears reasonable as long as H_p is assumed to be true. However, when H_d is true, then the intensity of contact is not relevant and the scientist is concerned with the presence by chance of a quantity Q of matching fibres. For the purpose of illustration, conditional probabilities of 0.97, 0.02 and 0.01 may be proposed for none, few and many fibres, respectively. Note that these values are not match probabilities but represent, in some way, an assessment of the rarity of the extraneous fibres found on the receptor, that is, fibres that match by chance

Table 8.6 Conditional probabilities for the variable 'contact' (C) given, respectively, the prosecution proposition, H_p, that Mr B is the man who struggled with Mr A and the defence proposition, H_d, that there has been no physical contact between Mr B and Mr A.

	H :	H_p	H_d
C:	*light*	0.1	0
	medium	0.3	0
	heavy	0.6	0
	no contact	0	1

Table 8.7 Probability distribution for the quantity, Q, of fibres recovered, given contact, C, and the prosecution proposition, H_p, that Mr B is the man who struggled with Mr A. Values are adopted from Cook et al. (1999).

	C:	*light*	*medium*	*heavy*	*no contact*
Q:	*none*	0.2	0.1	0.02	1
	few	0.6	0.4	0.28	0
	many	0.2	0.5	0.70	0

the control fibres. This rarity may be assessed in various ways. The scientist may, for example, consider a target fibre study. However, as noted earlier in this chapter, they should be aware that such studies offer different probability estimates, notably probabilities of observing by chance no or one (small or large) target group of fibres that match the control.

Q_B : This variable has the same definition as Q_A with the sole difference that the matching fibres are found on Mr B's garment.

8.4.3 The expected weight of evidence

The scientist can offer a phased approach to the customer. The case can be re-assessed once one of the garments has been examined. For illustration, imagine the following. The scientist starts by considering the black sweater of Mr A. There could be red fibres on this garment, which might match the red fibres of Mr B's sweater. The scenario is simplified in the sense that it is assumed that at most one group of foreign fibres may be recovered. Thus, the scientist focuses on red fibres without interest in any other possible background fibres.

For whatever quantity Q_A of fibres that may be found, the scientist will evaluate a likelihood ratio, that is,

$$\frac{Pr(Q_A \mid H_p)}{Pr(Q_A \mid H_d)}.$$

The outputs of the various scenarios are given in the Bayesian networks shown in Figure 8.2. The model shown in Figure 8.2(i) indicates the conditional probabilities of finding quantities Q_A of fibres given H_p is true: $Pr(Q_A = none \mid H_p) = 0.062$, $Pr(Q_A = few \mid H_p) = 0.348$ and $Pr(Q_A = many \mid H_p) = 0.590$. Figure 8.2(ii) displays the expected quantities Q_A of fibres when H_d is true: $Pr(Q_A = none \mid H_d) = 0.970$, $Pr(Q_A = few \mid H_d) = 0.020$ and $Pr(Q_A = many \mid H_d) = 0.010$

Using these values, likelihood ratios for finding none, few and many fibres on Mr A's garment can be calculated as follows:

1. $V_{Q_A=none} = \frac{0.062}{0.970} = 0.064$. This scenario provides moderate support for the defence's case, that is, $1/0.064 \approx 16$.

2. $V_{Q_A=few} = \frac{0.348}{0.020} = 17.4$. In this setting, there is moderate support for the prosecution's case.

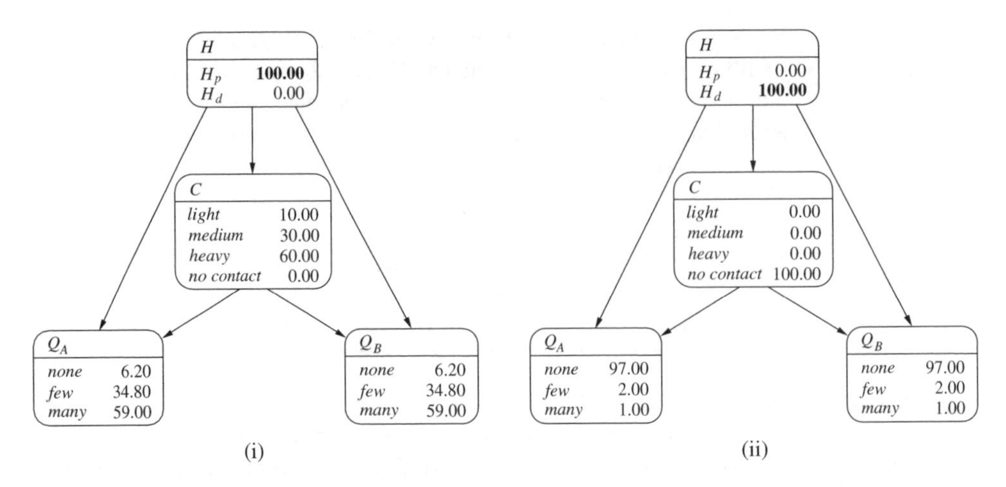

Figure 8.2 Assessment of the expected quantities Q of recovered fibres given (i) H_p is true, and (ii) H_d is true. Node C represents the level of contact. Node H represents the propositions, the prosecution proposition, H_p, that Mr B is the man who struggled with Mr A and the defence proposition, H_d, that there has been no physical contact between Mr B and Mr A. Nodes Q_A and Q_B represent the quantity of fibres recovered on, respectively, Mr A's and Mr B's garments.

3. $V_{Q_A=many} = \frac{0.590}{0.010} = 59.0$. This result also provides moderate support for the prosecution's case.

This analysis enables the scientist, even before examining the first garment, that of Mr A, to inform the customer of the following. If the suspect, Mr B, is the man who struggled with Mr A (H_p), then

- there is a probability of less than 10% (6.2%) that a result will be obtained that will provide moderate support for the innocence of Mr B,
- there is a probability of about 35% (34.8%) that a result will be obtained that will provide moderate support for the prosecution's case, that is, through a likelihood ratio of approximately 17, and
- it is more probable than not (59%) that a result will be obtained that will provide moderate support for the prosecution's case through a likelihood ratio of 59.

If, on the other hand, Mr B is truly innocent (H_d), then

- there is a 97% probability that the examination will result in moderate support for the defence case, and
- there is a 3% chance of obtaining evidence that will, falsely, incriminate Mr B.

Imagine now that the scientist proceeds with examining Mr A's garment but does not find any matching fibres. At this stage, the scientist may contact his customer in order to discuss whether to proceed with the examination of Mr B's red sweater. Evaluation of $Pr(Q_B \mid Q_A = none, H_p)$ and $Pr(Q_B \mid Q_A = none, H_d)$ then becomes relevant as it allows the scientists to assess the expected weight of evidence for different outcomes on Mr B's sweater.

By instantiating node H to H_p and node Q_A to *none*, the probabilities of the states of the variable Q_B change to 0.117 (previously 0.062), 0.441 (previously 0.348) and 0.442 (previously 0.59) for, respectively, none, few and many matching fibres recovered on Mr B's garment. Figures 8.3(i) and (ii) provide graphical representations of these evaluations. The likelihood ratios for finding quantities Q_B of fibres on Mr B's garment, given that no fibres have been found on Mr A's garment ($Q_A = none$), can then be calculated as follows:

1. $V_{Q_B=none|Q_A=none} = \frac{0.117}{0.970} = 0.121.$
2. $V_{Q_B=few|Q_A=none} = \frac{0.441}{0.020} = 22.05.$
3. $V_{Q_B=many|Q_A=none} = \frac{0.441}{0.010} = 44.1.$

On the basis of this assessment, the scientist may now inform his client of the following. If Mr B is the man who struggled with Mr A (H_p), then

- there is approximately an 11% probability that the findings on Mr B's garment will tend to support his innocence. Since the likelihood ratio is smaller than ten, this support is considered as weak ($V = 1/0.121 \approx 8$). Thus, knowledge of $Q_A = none$ has approximately doubled the probability of obtaining evidence supporting Mr B's innocence, that is, from 6.2% to 11.7%. However, the degree of support decreases from a likelihood ratio of approximately 16 to one of about 8.

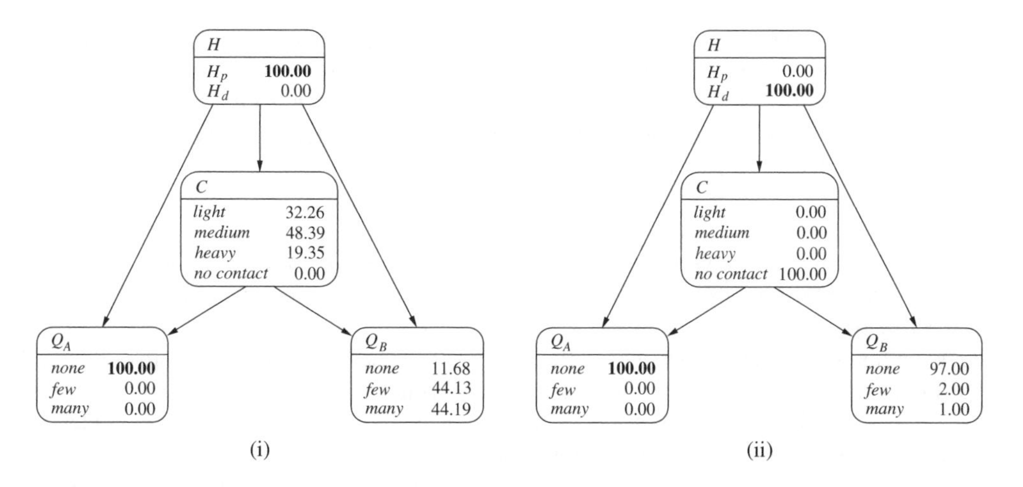

Figure 8.3 Assessment of expected quantities Q_B of fibres on Mr B's garment, given (i) H_p is true, and (ii) H_d is true. It is further assumed that no fibres have been found on Mr A's garment. Node C represents the level of contact. Node H represents the propositions, the prosecution proposition, H_p, that Mr B is the man who struggled with Mr A and the defence proposition, H_d, that there has been no physical contact between Mr B and Mr A. Nodes Q_A and Q_B represent the quantity of fibres recovered on, respectively, Mr A's and Mr B's garments.

- likelihood ratios of approximately 22 and 44 can be obtained, each with a probability of about 44%. These likelihood ratios provide moderate support for the prosecution's case. Thus, the effect of knowing $Q_A = none$ is to reduce the overall probability of obtaining evidence incriminating Mr B.

However, the client should also be advised that, if the suspect is truly innocent (H_d), then

- there is a 97% probability of obtaining evidence that provides moderate support for the defence proposition, and that

- there is also a 3% probability of obtaining evidence that will falsely incriminate the suspect.

It is also possible to allow for uncertainty about the truthstate of H and assess the probabilities of the various outcomes in node Q_B, given knowledge about an observed quantity Q_A. For example, when assuming $Pr(Q_A = none) = 1$, the probabilities of finding quantities Q_B of fibres are, respectively, 0.919, 0.045 and 0.036. This would imply, for instance, that the probability of finding no matching fibres on Mr B's garment, given that no matching fibres have been found on Mr A's garment is quite high (0.919). This can provide the customer with valuable information to support a decision.

8.5 Pre-assessment with multiple propositions

8.5.1 Preliminaries

During the pre-assessment of a case, it may not only be of interest to assess the value of different outcomes of the evidence. It may also be of interest to assess the value of the evidence given different propositions of interest. Evaluations of this kind may be required in order to help a scientist's client in determining an optimal argumentative strategy. Such an example is discussed in further detail below.

Imagine the following scenario. A condom is found in a public place, such as a car park. This item of evidence could be associated with a violent crime, for example, a rape followed by murder. Analysis reveals the presence of semen inside the condom. A suspect is apprehended for reasons unconnected with the biological fluid found in the condom. The suspect denies being involved in the crime. However, he does not deny that the semen could be his, so no DNA analysis is performed for the time being. In other words, it is not felt to be of primary interest to gather scientific evidence to characterise the link between the suspect and the condom. What is of interest here is to see whether there is evidence associating the condom, and by extension the suspect, with the rape of the victim.

Scientists may suggest analyses of the external surface of the condom in order to detect eventual vaginal cells of the victim. The possible outcomes of such analyses together with their evidential value may be studied in a case pre-assessment. As will be seen shortly, the kind of results that will be considered in the pre-assessment depend in part on the propositions specified initially. A particular outcome worth noting is, however, the detection of no vaginal cells. This is a kind of evidence that may be termed *missing*, as it is expected, but not found. Note that missing evidence has already been discussed at some length in Section 6.5.

8.5.2 Construction of a Bayesian network

A Bayesian network to deal with the problem of missing evidence is illustrated in Figure 8.4. The nodes are defined as follows:

- H: This node represents the respective positions of the prosecution (H_p) and the defence (H_d). A priori, the states of node H do not refer to any particular propositions. Node H merely specifies that two competing propositions exist. Association of specific scenarios to H_p and H_d is realised through the use of a node named Exp as explained below.

- Exp: A node Exp, short for 'explanations', defines the set of scenarios that may be proposed by the prosecution and the defence. Node Exp contains a list of alternative scenarios that offer explanations for possible findings. For the purpose of the scenario considered here, Exp may be assumed to cover the following: the suspect had no sexual intercourse (exp_1), the suspect had sexual intercourse with the victim (exp_2), and the suspect had sexual intercourse with his girlfriend (exp_3). Depending on the information offered by the suspect, one could also add the proposition according to which the suspect had intercourse with a third person. This would result in an increase in the number of node probability tables. For the clarity of the present discussion, only scenarios exp_i ($i = 1, 2, 3$) are considered here.

  A direct link between nodes H and Exp is adopted. This allows users to define individual strategies for both the prosecution and the defence. For example, Table 8.8 specifies that the prosecution's position is that the suspect had sexual intercourse with the victim, whereas the defence case is that the suspect had no sexual intercourse. If, on the other hand, the defence case is that the suspect had intercourse with his girlfriend, then Table 8.9 could be used. Notice that it is also possible, theoretically, for a party to propose more than one explanation. For example, the defence may argue that the suspect probably had no intercourse, but if he had, then the intercourse occurred with his girlfriend. In such a setting, conditional probabilities need to be distributed between these two scenarios. Notice that similar, 'composed' alternative propositions have already been encountered in the context of DNA evidence; for instance, the suspect population under H_d may be one that consists of both individuals related and individuals unrelated to the suspect.

Figure 8.4 Bayesian network for the pre-assessment of cases in which missing evidence may be expected. Node H is a binary node with two states, H_p and H_d, the prosecution and defence propositions, respectively. Node Exp (explanations) defines the set of scenarios that may be proposed by the prosecution and defence. Node TP (true presence) provides a distinction between the results of the analyses and the true state of the samples. Node E defines the various expected outcomes of the analyses.

Table 8.8 Conditional probabilities for node Exp (explanations) given H_p and H_d, the prosecution and defence propositions, respectively. The prosecution proposition is that the suspect had sexual intercourse with the victim. The defence proposition is that the suspect had no sexual intercourse.

H :		H_p	H_d
Exp :	No sexual intercourse (exp_1)	0	1
	Intercourse with victim (exp_2)	1	0
	Intercourse with girlfriend (exp_3)	0	0

Table 8.9 Conditional probabilities for node Exp (explanations) given H_p and H_d, the prosecution and defence propositions, respectively. The prosecution proposition is that the suspect had sexual intercourse with the victim. The defence proposition is that the suspect had intercourse with his girlfriend.

H:		H_p	H_d
Exp:	No sexual intercourse (exp_1)	0	0
	Intercourse with victim (exp_2)	1	0
	Intercourse with girlfriend (exp_3)	0	1

- TP: This node is used to introduce a distinction between the actual results of the analyses, that is, an indication of the presence or absence of vaginal cells (as represented by a node E defined below), and the 'true' presence or absence of such cells. This node is named TP, short for 'true presence', and covers the following states: true presence of vaginal cells of the victim (TP_v), true presence of vaginal cells of the suspect's girlfriend (TP_g), and absence of vaginal cells (TP_a). Notice that the number and definitions of the states of the node TP is directly dependent on the explanations implied by the states of the node Exp. For example, no possibility for the presence of vaginal cells of a third woman is included because node Exp does not contain a scenario that could explain such an event.

- E: This variable defines the various expected outcomes of analyses performed on the external surface of the condom. Three states are proposed, namely, E_m, E_v and E_g. The state E_m denotes the event that evidence is missing, that is, no vaginal cells are detected. The state E_v refers to an event where vaginal cells are found and the DNA profile matches that of a sample obtained from the victim. The state E_g also stands for a case in which vaginal cells are found, the difference being, however, that the DNA profile corresponds to a sample provided by the suspect's girlfriend. Incorporation of these states within a single node E means that they are assumed to be mutually exclusive. This signifies that one assumes that there cannot be a match with more than one individual. Values for the random-match probability thus need not be considered. In case one wishes to avoid this assumption, one may adopt distinct nodes for observed correspondences with, respectively, the victim and the girlfriend. See Section 8.5.4 (Figure 8.5) for a possible Bayesian network that allows such issues to be handled.

Note that node E depends on the node TP. The conditional probabilities associated with node E allow for the accounting of a number of issues. For example, if it is assumed that the analyses will correctly indicate the presence and absence of cells, then probabilities could be specified as shown in Table 8.11. However, there may eventually be special circumstances so that analyses may fail to detect the presence of vaginal cells. In such cases, probabilities for the states $Pr(E_m \mid TP \neq TP_a)$ should be chosen which are different from zero.

8.5.3 Evaluation of different scenarios

Consider the prosecution's proposition that the suspect had sexual intercourse with the victim. The defence suggests that the suspect had no sexual intercourse. The conditional probabilities associated with the nodes of interest could then be as follows:

- Node *Exp*: If the prosecution's case is true, then the suspect had sexual intercourse with the victim: $Pr(exp_2 \mid H_p) = 1$. The suspect did not have any sexual intercourse if the defence's case is true: $Pr(exp_1 \mid H_d) = 1$. See Table 8.8 for a summary of the conditional probabilities assigned to the node Exp.

- Node *TP*: It is assumed that if the suspect had intercourse with either the victim or his girlfriend, then vaginal cells of the woman with whom he had intercourse would be present with probability 0.95. Formally, $Pr(TP_v \mid exp_2) = Pr(TP_g \mid exp_3) = 0.95$. Note that the term 'probability of presence' incorporates transfer and persistence. It is also assumed that there is a 0.05 probability that sexual intercourse does not result in the presence of vaginal cells, for example, $Pr(TP_a \mid exp_2) = 1 - Pr(TP_v \mid exp_2) = 1 - 0.95 = 0.05$. The absence of vaginal cells is taken to be certain if there was no sexual intercourse: $Pr(TP_a \mid exp_1) = 1$. A summary of all conditional probabilities is provided by Table 8.10.

- Node E: The probabilities assigned to this variable depend crucially on the skills of the examining scientist as well as on the performance of the methods and techniques employed. It may be assumed, for example, that if vaginal cells are truly present, then the probability that no evidence will be produced, that is, that evidence is missing, should be low or close to zero. Therefore, let $Pr(E_m \mid TP_v) = Pr(E_m \mid TP_g) = 0$. A further assumption may be that if cells from a woman are truly present, then the probability that the scientist will find and declare a match with that individual is 1. This would be an expression of the belief that the scientist is always able to detect a match when in fact a match does exist. Remember that the node table of E allows one to account for potential false positives: for example, values different from zero for $Pr(E_g \mid TP_v)$ signify that there is a possibility that vaginal cells from the victim are erroneously associated with the suspect's girlfriend.

Table 8.10 Conditional probabilities for the states of the node TP (true presence) given the scenarios specified by the node Exp (explanations). States of TP are TP_a, the absence of vaginal cells, TP_v, true presence of vaginal cells of the victim, TP_g, true presence of vaginal cells of the girlfriend. States of Exp are exp_1, the suspect had no sexual intercourse, exp_2, the suspect had sexual intercourse with the victim and exp_3, the suspect had sexual intercourse with his girlfriend.

	Exp:	exp_1	exp_2	exp_3
TP:	TP_a	1	0.05	0.05
	TP_v	0	0.95	0
	TP_g	0	0	0.95

Table 8.11 Conditional probabilities for the states of the node E (expected outcomes) given the state of TP (true presence). States of E are E_m, evidence is missing, no vaginal cells are detected, E_v, vaginal cells are found and the DNA profile matches that of the victim, E_g, vaginal cells are found and the DNA profile matches that of the girlfriend. States of TP are TP_a, the absence of vaginal cells, TP_v, true presence of vaginal cells of the victim, TP_g, true presence of vaginal cells of the girlfriend.

	TP:	TP_v	TP_g	TP_a
E:	E_m	0	0	1
	E_v	1	0	0
	E_g	0	1	0

Using a Bayesian network set up with these assumptions, the expected weight of the various findings can be assessed as follows. If the prosecution's case is true, then

- there is a 0.05 probability of finding no vaginal cells, that is, that the evidence is missing (E_m). Missing evidence supports the defence's proposition by a likelihood ratio of $(0.05/1)^{-1} = 20$.

- there is a 0.95 probability of detecting vaginal cells that are found to 'match' the victim. Such evidence supports H_p through a likelihood ratio whose value is infinite. This result is due to the particular assumptions incorporated in the model, notably, that $Pr(E_v)$ is zero under H_d.

In the case in which the suspect is truly innocent, with certainty no vaginal cells will be found. This is a direct result of the way in which the probability tables have been completed. Although, given H_d, no cells will be found with certainty, this outcome provides only moderate support for the defence case, that is, through a likelihood ratio of 20.

Imagine now a change in the defence's strategy. The defendant states that he had sexual intercourse with his girlfriend and the recovered condom is associated with this event. Therefore, the conditional probabilities for the node Exp should be chosen so as to agree with Table 8.9. The definition of the nodes TP and E is not changed. In this setting, an analysis using an appropriately specified Bayesian network indicates that both the prosecution and the defence have a 0.95 probability of finding evidence supporting their proposition infinitely, if their proposition is in fact true.

8.5.4 An alternative graphical structure

Figure 8.5 proposes a method by which the assumption of having at most one matching individual (*i.e.*, victim or girlfriend) may be avoided. The node previously termed E is substituted, for each state of E, by two nodes defined as follows:

- State E_m of E is represented by two nodes TA and M, denoting 'true absence' and 'missing', respectively;
- State E_v is turned into the nodes TM_v, denoting a true correspondence with the victim's profile, and RM_v, denoting the scientist's report of a correspondence between

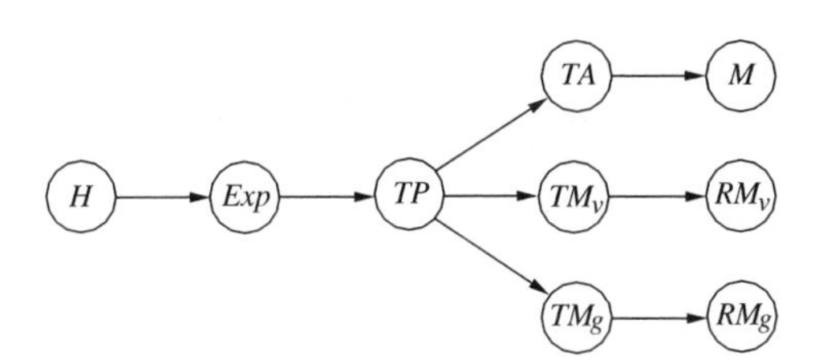

Figure 8.5 An alternative Bayesian network for the pre-assessment of cases involving potentially missing evidence. Node H is a binary node with two states, H_p and H_d, the prosecution and the defence propositions, respectively. Node Exp (explanations) defines the set of scenarios that may be proposed by the prosecution and defence. States of Exp are exp_1, the suspect had no sexual intercourse, exp_2, the suspect had sexual intercourse with the victim and exp_3, the suspect had sexual intercourse with his girlfriend. Node TP (true presence), has three states, TP_a, the absence of vaginal cells, TP_v, true presence of vaginal cells of the victim and TP_g, true presence of vaginal cells of the girlfriend. Node TA denotes 'true absence' and node M denotes 'missing'. Node TM_v denotes 'true correspondence with the victim's profile' and node RM_v denotes 'reported correspondence of the victim's profile with that extracted from material collected from outside the condom'. Node TM_g denotes 'true correspondence with the girlfriend's profile' and node RM_g denotes 'reported correspondence of the girlfriend's profile with that extracted from material collected from outside the condom'. These last six nodes are binary.

the victim's profile and that extracted from material collected from the outside of the condom;

- State E_g becomes TM_g, denoting a true correspondence with the girlfriend's profile, and RM_g, denoting the scientist's report of a correspondence between the girlfriend's profile and that extracted from material collected from the outside of the condom.

All of these newly defined nodes are Boolean. They represent, respectively, factors that are not observable (TA, TM_v and TM_g) and factors that are observable (M, RM_v and RM_g). This allows users to incorporate probabilities for a random match, for example, $Pr(TM_v \mid TP_g)$, and false positives, for example, $Pr(RM_g \mid \overline{TM}_g)$. This feature arises in part because of the structural analogies between the currently discussed Bayesian network (Figure 8.4) and the model shown in Figure 5.20, which has been constructed specifically for the evaluation of such issues.

For the purpose of illustration, consider a scenario in which propositions put forward by, respectively, the prosecution and the defence are defined according to Table 8.9. It is assumed further that sexual intercourse with either the girlfriend or the victim will lead with certainty to the true presence of their vaginal cells. When setting $Pr(TM_v \mid TP_g) = Pr(TM_g \mid TP_v) = \gamma$, and, $Pr(RM_g \mid \overline{TM}_g) = Pr(RM_v \mid \overline{TM}_v) = 0$, the following likelihood ratios are obtained:

- If, for example, the only available information is a reported match with the victim, then the prosecution's case is supported by a likelihood ratio of $1/\gamma$.

- If, in addition to a reported match with the victim, information is available on a match with the suspect's girlfriend, then the latter provides evidential support of $1/\gamma$ in favour of the defence case. As a consequence, the likelihood ratio associated with the joint occurrence of matches between the victim and the girlfriend equals 1. On the other hand, if a non-match is reported with the girlfriend, then, under the stated assumptions, the posterior probability of the prosecution's case reaches certainty.

8.6 Remarks

All the Bayesian networks developed in Chapters 5 and 6 can be easily translated into Bayesian networks to pre-assess a case.

Models for the pre-assessment of evidence play an important role in an investigation because they enable the scientist to understand the principal issues in a case and to discuss them with colleagues, prosecutors and lawyers to develop strategies. They provide a logical perspective that might otherwise be lacking, as pointed out by Evett et al. (2002a, p. 530):

> The idea of pre-assessment of a case also contributes to a balanced view. It directs the scientist to consider his/her expectations *before* collecting scientific evidence. To say, after making a particular observation 'this is what I would have expected to find' will always invite the suggestion that this is a *post hoc* rationalisation.

In such an approach, the notion of a hierarchy of propositions (*see* Section 3.3) and the choice of two (or more) competing propositions are essential. This is more fundamental than the choice of probability values as will be discussed in Chapter 9. Pre-assessment represents a first step in a more general decision-making process as illustrated in Section 11.2.

9

Qualitative and sensitivity analyses

So far, Bayesian networks have been presented as a general framework for the representation and sensible interpretation of uncertainty in knowledge. It has been pointed out that the graphical structure is a network's most robust constituent: it conveys, on a rather general level of detail, a transparent and clear idea of the parameters considered as well as their assumed dependencies. When information is available on the strength of the relationships amongst the variables, models may be specified and used for inference on a quantitative level.

However, it may be felt that the quantitative specification is too problematic a task for the use of Bayesian networks in forensic contexts. Even though data may be available from various sources such as literature, surveys, databases or from experts, forensic scientists may be reluctant to provide specific numerical values for all parameters of interest. One of the reasons for this is that forensically relevant items of evidence usually come into existence under specific sets of real-world circumstances. Because of the unavailability of appropriate experimental data, it may be hard to justify quantitative assessments of factors relevant to a given scenario. Moreover, Bayesian networks are sometimes thought of as a numerical method that requires 'exact' numbers with 'high' accuracy for their implementation.

It is in this context that the present chapter describes formalisms that can help with scenarios in which there is a lack of numerical data, namely qualitative probabilistic networks (QPNs) and sensitivity analyses. Informally, qualitative probabilistic networks are abstractions of Bayesian networks in which numerical relationships are replaced by qualitative probabilistic relationships. In such models, inferences can still be made according to the laws of probability theory. The result is an indication of the change in the direction of belief. Sensitivity analyses consist of varying the probabilities assigned to one or more variables of interest and comparing the results. Such operations provide information on the effect that uncertainties about one parameter may have on another. They are thus helpful in determining the most influential parameters; an important aspect in the allocation of elicitation efforts to target variables.

Bayesian Networks and Probabilistic Inference in Forensic Science F. Taroni, C. Aitken, P. Garbolino and A. Biedermann

9.1 Qualitative probability models

Wellman (1990a) formally defined a QPN as a pair $G = (V, Q)$, where V is a set of variables, also called vertices, of the graph and Q is a set of qualitative relationships. In analogy to the definition of Bayesian networks (Section 2.1.2), a QPN G is required to be acyclic. The vertices of G are discrete and binary or multiple-valued. The qualitative relationships Q cover qualitative influences and synergies. A qualitative influence defines the sign of the direct influence one variable has on another, whereas qualitative synergies are used to describe interactions among influences.

Throughout the next few sections these concepts will be developed in further detail with their relevance for reasoning in forensic contexts illustrated through various examples.

9.1.1 Qualitative influence

9.1.1.1 The binary case

The notion of qualitative influence between two variables of a graph is an expression of how the state of one variable influences the state of the other variable. Imagine two binary variables A and B which are connected so that $A \rightarrow B$. Then, A is said to have a positive influence on B if and only if

$$Pr(B \mid A, x) \geq Pr(B \mid \bar{A}, x), \tag{9.1}$$

where x represents an assignment to other ancestors of B. Negative and zero qualitative influences can be defined analogously by substituting $\geq$ in (9.1) by $\leq$ and $=$, respectively. A fourth sign of qualitative influence is '?' and denotes ambiguity. It applies when the effect of the influence is not known.

Verbally expressed, a positive qualitative influence between two variables A and B is, for example, a statement of the form 'A makes B more likely' (Wellman 1990a). Notice that qualitative influences adhere to the property of symmetry so that one is also allowed to state that 'B makes A more likely'. Another way to express this would be to say 'the occurrence of event B favours hypothesis A over $\bar{A}$' (Schum 1994). See Garbolino (2001) for a discussion on the use of such expressions in judicial contexts.

In forensic science, qualitative reasoning is quite common. For example, forensic scientists often follow a procedure that goes from the general to the particular. In the context of evaluating evidence, this means that scientists would start by asking questions of the kind 'Is that evidence what I would expect to observe if the prosecution's case were true'?, and, 'To what degree would I expect that evidence to occur if the defence case were true'?. Initially, the scientist would frame these questions on a quite general level and think about the possible outcomes in essentially qualitative terms. In a second stage, more detailed data may then be sought and used to sustain a particular conclusion.

As an example, consider again the evaluation of a distance of firing, an inferential problem discussed at some length in Section 7.4.4. For a given distance D, of firing, the expected number of gunshot residues (GSRs) Y has been approximated by a Normal distribution. Note that this approach has solely been concerned with the amount of visualised GSR. The main aim was to demonstrate that GSR evidence is amenable, principally, to quantification and probabilistic evaluation. In practice, however, scientists may also use information other than GSR. For example, experts may assess the presence or absence of certain characteristics of wounds, damage to textile clothing in the area of the entrance hole (*e.g.*, melted fibres), and so on. When the scientist considers these items of information as a whole, then the general pattern of reasoning involved can be described on a purely qualitative level. Basically, the scientist would need to evaluate whether the observations are more, less or equally likely given the respective propositions of interest. Such judgements may be considered as the most general level at which scientists may be required to characterise their findings. If the scientist is unable to tell whether he believes the findings to be more, less or equally likely under each of the specified propositions, then his evidence is not suitable for discrimination amongst those propositions. In the latter situation, the uninformative '?'-sign is applicable, the notation chosen to illustrate its intuitive meaning.

Notice that the concept of qualitative influence allows one to describe the relation between two variables on the basis of two conditional probabilities. The revision of belief can be made according to the laws of probability even though no specific numerical probabilities have been specified.

9.1.1.2 The non-binary case

Let a and b denote random variables. Their outcomes are indexed such as a_i, for example. Also, an ordering from the highest to lowest value is assumed so that for all $i < j$, one has, for example, $a_i \geq a_j$. Then, a variable a may be said to have a positive qualitative influence on a variable b, if and only if for all values a_j and a_k of a, with $a_j > a_k$, all values b_i of b, and x,

$$Pr(b \geq b_i \mid a_j, x) \geq Pr(b \geq b_i \mid a_k, x). \tag{9.2}$$

Again, x represents an assignment to other ancestors of b and the relationship $\geq$ in the centre of (9.2) can be substituted with $\leq$ and $=$ to define negative and zero qualitative influences, respectively. A verbal expression of a qualitative influence between two non-binary variables may be, for example, 'higher values of a make higher values of b more likely', assuming the influence under investigation to be a positive one.

The direction of influence between two variables can be evaluated by comparing cumulative probability distribution functions (CDFs) (Wellman 1990a). The cumulative distribution function CDF of a variable b with states $b_1 < ... < b_n$, and $n \geq 1$, is given by the function $F_b(b_i) = Pr(b_1 \vee b_2 \vee ... \vee b_i)$, noted $Pr(b \leq b_i)$ for short. Cumulative conditional probability distribution functions are defined analogously. For example, $F_{b|a_j}$ denotes the CDF of b given a_j.

Consider the comparison of CDFs in the context of a positive qualitative influence between the non-binary variables a and b, connected so that $a \rightarrow b$. In order that the statement 'higher values of a make higher values of b more likely' to be valid, the following must hold for all values a_i, a_j with $a_i > a_j$:

$$F_{b|a_i}(b_i) \leq F_{b|a_j}(b_i) \text{ for all values } b_i \text{ of } b.$$

As an example, consider the relation between the variables c, for 'degree of contact', and q, for 'amount of transferred fibres recovered from a garment' as described by Cook et al. (1999). The variable c has the three states 'heavy', 'medium' and 'light', abbreviated

Table 9.1 Conditional probabilities for the quantity of transferred fibres recovered from a garment given different degrees of contact. Values are adopted from Cook et al. (1999).

Degree of contact (c):		*heavy* (c_1)	*medium* (c_2)	*light* (c_3)
Quantity (q):	*none* (q_1)	0.02	0.1	0.2
	few (q_2)	0.28	0.4	0.6
	many (q_3)	0.70	0.5	0.2

c_1, c_2 and c_3, respectively. Three states are also proposed for the variable q, namely *none*, *few* and *many*, abbreviated q_1, q_2 and q_3.

Table 9.1 summarises the conditional probabilities for q given c proposed by Cook et al. (1999). The corresponding cumulative conditional probability distributions thus are as follows:

$$Pr(q \leq q_1 \mid c_1) = 0.02 \quad Pr(q \leq q_1 \mid c_2) = 0.10 \quad Pr(q \leq q_1 \mid c_3) = 0.20$$

$$Pr(q \leq q_2 \mid c_1) = 0.30 \quad Pr(q \leq q_2 \mid c_2) = 0.50 \quad Pr(q \leq q_2 \mid c_3) = 0.80$$

$$Pr(q \leq q_3 \mid c_1) = 1.00 \quad Pr(q \leq q_3 \mid c_2) = 1.00 \quad Pr(q \leq q_3 \mid c_3) = 1.00$$

From this listing of values it may be seen that for each c_i of c, with $c_1 > c_2 > c_3$, the cumulative conditional probability distributions of q given c_i obey to:

$$F_{q|c_1}(q_i) \leq F_{q|c_2}(q_i) \leq F_{q|c_3}(q_i) \text{ for all } q_i \text{ of } q.$$

It may be concluded from this analysis that higher values of q are more probable for higher values of c. In the context of fibres transfer, this means that the higher the degree of contact, the greater the amount of transferred fibres recovered from a garment. An important aspect of this result is that the assumed relationship between the extent of contact and the quantity of recovered fibres is handled in a probabilistically rigorous way while specific numerical values need not necessarily be available.

Consider yet another example for a qualitative influence. In Section 7.4.1, a handwriting scenario with two variables S and E has been described. S is binary, defined as 'Mr Adams wrote the signature', and E is non-binary, referring to the results of comparative handwriting analyses. The possible states for E are few (e_1), some (e_2) and many (e_3) similarities. Recall, from Table 7.3, the conditional probabilities specified for each outcome of E given S_p and S_d, Mr Adams being or not being the source of the signature:

$$Pr(e_1 \mid S_p) = 0.60 \quad Pr(e_1 \mid S_d) = 0.78$$

$$Pr(e_2 \mid S_p) = 0.27 \quad Pr(e_2 \mid S_d) = 0.20$$

$$Pr(e_3 \mid S_p) = 0.13 \quad Pr(e_3 \mid S_d) = 0.02$$

The respective cumulative conditional probability distributions are as follows:

$$Pr(e \leq e_1 \mid S_p) = 0.60 \quad Pr(e \leq e_1 \mid S_d) = 0.78$$

$$Pr(e \le e_2 \mid S_p) = 0.87 \qquad Pr(e \le e_2 \mid S_d) = 0.98$$

$$Pr(e \le e_3 \mid S_p) = 1.00 \qquad Pr(e \le e_3 \mid S_d) = 1.00$$

For each e_i of e, the cumulative conditional probability distribution is smaller given S_p than given S_d. The underlying qualitative probabilistic statement thus is that higher values for e are more probable under S_p than under S_d. Another way to express this would be to say that more similarities are more likely to be found when Mr Adams wrote the signature than if someone else wrote the signature.

9.1.2 Additive synergy

'Synergy' is a term used when studying interactions among influences. This is typically the case when two nodes share a common child. Additive synergy is a concept applicable for evaluating the direction of interaction between two variables in their influence on a third (Wellman 1990b). Note that this property of additive synergy holds regardless of any other direct influence on that third variable.

Additive synergy is described here for the binary case. Let A, B and C be binary variables, connected in a converging connection: $A \rightarrow C \leftarrow B$. A and B are said to exhibit a positive additive synergy on their common descendant C if

$$Pr(C \mid A, B, x) + Pr(C \mid \bar{A}, \bar{B}, x) \ge Pr(C \mid A, \bar{B}, x) + Pr(C \mid \bar{A}, B, x), \qquad (9.3)$$

where x represents an assignment to other ancestors of C. Again, negative and zero additive synergies are defined by substituting $\ge$ by $\le$ and $=$ respectively.

Another way to express an additive synergy, for example, a positive one, is to say 'the effect of A on C is greater when B is true' (Wellman 1990b). This can be illustrated by rearranging the terms in (9.3) to obtain the following:

$$Pr(C \mid A, B, x) - Pr(C \mid \bar{A}, B, x) \ge Pr(C \mid A, \bar{B}) - Pr(C \mid \bar{A}, \bar{B}, x). \qquad (9.4)$$

The left-hand side describes the changes in C due to changes in A while B is true. The right-hand side describes the changes in C due to changes in A while $\bar{B}$ is true. The relative magnitude of these changes is compared and expressed in terms of the signs '+', '−', '0' and '?'.

In forensic contexts, patterns of reasoning exist that comply with the definition of additive synergy. Imagine the use of some sort of detection apparatus, such as an ion mobility spectrometer (IMS). Such devices have been developed and marketed for the detection of explosives and illicit drugs. Surfaces of interest such as tables, luggage, clothing and so on may be vacuumed and particles thus collected analysed. When a detectable quantity of a target substance is present, then the IMS would indicate that event by a sonar signal which is given with a certain reliability.

Consider the three variables D, 'presence of a detectable quantity of drugs', P, 'performance of the device' and S, 'sonar signal given by the device'. It is thought that the probability of obtaining a sonar signal is dependent on whether a detectable quantity of drugs is present, and on the apparatus' performance; an appropriate representation thus being $P \rightarrow S \leftarrow D$. The variables are binary and their states are defined as follows: 'present' and 'absent', (variable D), 'good' and 'poor' (variable P), 'yes' and 'no' (variable S). The

Table 9.2 Conditional probabilities for the variable S given all potential realisations of the variables P and D. Factor S has two states, S_1, sonar signal given by the device, S_2, sonar signal not given by the device. Factor P, the performance of the device has two states, P_1, good, P_2, poor. Factor D has two states, D_1, a detectable quantity of drugs is present, D_2, a detectable quantity of drugs is absent.

	P :	P_1		P_2	
	D :	D_1	D_2	D_1	D_2
S :	S_1	0.99	0.01	0.5	0.5
	S_2	0.01	0.99	0.5	0.5

descriptor 'performance' is used here, roughly speaking, as an expression of one's belief in the apparatus' capacity to do what it is supposed to do. So, if the apparatus' performance is good and a detectable quantity of drugs is present, then a sonar signal may be expected with a certain degree of probability. In the absence of a detectable quantity of drugs, but still assuming a good performance, a sonar signal may be expected with a reasonably low probability. Notice that whatever numerical values these two parameters may in fact assume, one's beliefs may qualitatively be expressed as $Pr(S_1 \mid P_1, D_1) > Pr(S_1 \mid P_1, D_2)$. Sample values are given in Table 9.2. When the performance of the detection apparatus is poor, then the probability of obtaining a sonar signal may be less influenced by the actual presence or absence of drugs. Qualitatively speaking, $Pr(S_1 \mid P_2, D_1) \approx Pr(S_1 \mid P_2, D_2)$, or, even $Pr(S_1 \mid P_2, D_1) = Pr(S_1 \mid P_2, D_2)$. Again, sample values are given in Table 9.2.

All the necessary conditional probabilities of the variable S are now defined. They can be used to examine the interaction among the influences that both P and D have on S. The relative magnitude of the target probabilities is such that the following holds:

$$Pr(S_1 \mid D_1, P_1) - Pr(S_1 \mid D_2, P_1) > Pr(S_1 \mid D_1, P_2) - Pr(S_1 \mid D_2, P_2).$$

The two variables P and D thus exhibit a positive additive synergy on S. Verbally expressed, the presence of drugs is thought to have a stronger effect on the probability of obtaining a sonar signal when the apparatus is a viable device rather than when its performance is poor.

9.1.3 Product synergy

One can distinguish between two types of synergies, one of which, additive synergy, has been introduced in the previous section. A second type of interaction is product synergy. It regroups the sign of conditional dependence between a pair of immediate predecessors of a node that has been observed or has indirect evidential support (Druzdzel and Henrion 1993a). As with additive synergy, the notion of product synergy applies to scenarios involving converging connections. Product synergy has been proposed by Henrion and Druzdzel (1991) and has been further studied by Wellman and Henrion (1993).

Consider a converging connection involving two variables A and B with direct influences on a third variable C: $A \rightarrow C \leftarrow B$. The variables A and B are said to exhibit a negative product synergy with respect to the outcome C of C, if and only if

$$Pr(C \mid A, B) \times Pr(C \mid \bar{A}, \bar{B}) \leq Pr(C \mid A, \bar{B}) \times Pr(C \mid \bar{A}, B). \tag{9.5}$$

Again, positive and zero product synergies are defined by replacing $\leq$ by $\geq$ and $=$, respectively. Notice that there are as many product synergies as there are possible outcomes for C. So, for the binary case, there would be a product synergy for both C and $\bar{C}$. Also note that the above definition of product synergy assumes that C either has no parents other than A and B, or, if there are parents, they are all instantiated. See Druzdzel and Henrion (1993c) for an extension to situations involving additional, unobserved predecessors of C.

The negative product synergy between A and B as expressed by (9.5) means that, given C, confirmation of A tends to makes B less likely. It is also useful to note that given C, A and B are no longer d-separated. Product synergy can thus be taken as a concept describing the interactions between two parental variables who become dependent upon receiving information on the truth state of their common child.

Hereafter, product synergy will be used to study two particular aspects of reasoning among variables sharing a common descendant (Wellman and Henrion 1993):

1. A frequently encountered pattern of reasoning that became known as 'explaining away', which applies when the confirmation of a believed event reduces the need to invoke alternative events.

2. The opposite of 'explaining away', which is a mode of reasoning applied when the confirmation of one event increases the belief in another.

9.1.3.1 Explaining away

Discussions on explaining away may be found, for example, in Kim and Pearl (1983), Henrion (1986) or Pearl (1988). More general conditions for this particular form of reasoning have been formulated by Henrion and Druzdzel (1991) and Wellman and Henrion (1993). The latter authors have shown that negative product synergy is a general probabilistic criterion that precisely justifies explaining away.

Consider this through an example with a forensic connotation. Imagine that an individual suspected of arson is arrested a few hours after a house had been set on fire, aided by spilt inflammable liquid. Forensic scientists examine the suspect's hands and clothing for the presence of residual quantities of a flammable liquid, that is, its vapours. Assume that analyses indicate the presence of considerable quantities of such a substance. Represent this event by P. When evaluating this finding, the scientist may consider various, not necessarily mutually exclusive, events that may lead to the detection of a flammable substance on the hands and the clothing of a suspect. For the purpose of illustration, consider two potential sources defined as follows:

- A variable B, termed *background contamination*, refers to the accidental presence of a flammable liquid, acquired, for example, while visiting or working in an environment that is contaminated with such a substance.

- A variable T refers to a recent manipulation of a flammable substance, eventually resulting in a primary transfer. It shall be ignored here, however, whether that transfer (or spill) is crime-related or not.

Both B and T may be realised simultaneously; so they are modelled as distinct parental nodes of P. Conditional probabilities for P could be as shown in Table 9.3. These values reflect a scientist's point of view according to which a positive analytical result

- is highly likely whenever a transfer occurred (T) and a background contamination (B) is present,
- is still likely to occur even though only T is true and B is not, and vice versa,
- is unlikely if neither a transfer (T) occurred nor a background contamination (B) is present.

Note that the absolute values displayed in Table 9.3 are not of primary interest here. It is sufficient here to consider them as rough indications of the beliefs held by a reasoner. A closer examination of these beliefs shows that they satisfy the following:

$$\frac{Pr(P \mid T, B)}{Pr(P \mid T, \bar{B})} \leq \frac{Pr(P \mid \bar{T}, B)}{Pr(P \mid \bar{T}, \bar{B})} . \tag{9.6}$$

Table 9.3 Conditional probabilities for the factor P given all potential realisations of the factors T and B. Factor P, the analysis of the suspect's hands and clothing for the presence of residual quantities of a flammable liquid, has two states, P, the presence, and $\bar{P}$, the absence of such residual quantities. Factor T, the recent manipulation of a flammable substance with a resultant primary transfer, has two states, T, the occurrence of such a manipulation and $\bar{T}$, the absence of such a manipulation. Factor B, background contamination, has two states, B and $\bar{B}$, the presence and absence, respectively, of such a contamination.

	T:	T		$\bar{T}$	
	B:	B	$\bar{B}$	B	$\bar{B}$
P:	P	0.99	0.97	0.7	0.01
	$\bar{P}$	0.01	0.03	0.3	0.99

Verbally, this relation can be taken to mean that the proportional increase of the probability of P, the occurrence of a positive analytical result, due to B, the accidental background presence of a flammable substance, is smaller given T, the presence of a transferred flammable liquid, than $\bar{T}$, the absence of such substance. Alternatively, one could also say that the incremental effect of one of the two possible sources, B for example, is less than it would be if the other, T, were absent.

For the above statements to be valid, one's beliefs must comply with the formal requirement of negative product synergy. This can be explained by the fact that (9.6) is obtained from (9.5) by rearranging the terms. A practical consequence of this is that, given a positive analytical result, the confirmation of one potential 'cause' would decrease one's belief in the alternative event that may lead to the observed result. It is modes of reasoning of this kind that are referred to as *explaining away*. Instances where explaining away applies can be found, for example, in Chapter 6. Further applications of Bayesian networks for evaluating the evidence of residual quantities of flammable liquids are described by Biedermann et al. (2005a,b).

9.1.3.2 The opposite of explaining away

When there are two events that may both lead to the same consequence, then these events may not only compete with one another in explaining the observed effect but they might also complement each other (Wellman and Henrion 1993). The latter situation is one in which 'explaining away' does not appear to be appropriate, that is, when the confirmation of one event, given its consequence, tends to increase the probability of another event that may lead to the same effect.

As an example, consider once again a scenario discussed earlier in Section 9.1.2. An analytical device is used in order to detect the eventual presence of traces of an illicit drug. The pattern of reasoning involved when obtaining a positive sonar signal could be as follows: given the apparatus' sonar signal (S_1), one's suspicion of traces of drugs effectively being present (D_1) would be higher the more one believes the apparatus' performance to be good (P_1), that is, the apparatus is sensitive to drugs. This sort of reasoning applies when the beliefs about the various outcomes of S given P and D satisfy:

$$\frac{Pr(S_1 \mid P_1, D_1)}{Pr(S_1 \mid P_1, D_2)} \geq \frac{Pr(S_1 \mid P_2, D_1)}{Pr(S_1 \mid P_2, D_2)} . \tag{9.7}$$

The opposite of explaining away can thus be considered in situations where a positive product synergy applies. The condition defined by (9.7) states that the proportional increase in the probability of S_1 (the apparatus giving a sonar signal) due to D_1 (the presence of a detectable quantity of drugs), is greater given P_1 (the apparatus' performance being good) than P_2 (the apparatus' performance being poor).

9.1.4 Properties of qualitative relationships

So far, qualitative influences have been considered as direct relationships between two adjacent variables of a network, that is, as a concept associated with each of a network's arcs.

However, influences among nodes may not only be evaluated along arcs. It is also possible to consider indirect influences between separated variables. To this end, Wellman

Table 9.4 Sign product ($\otimes$) and sign addition ($\oplus$) operators as defined by Wellman (1990a).

$\otimes$	+	−	0	?
+	+	−	0	?
−	−	+	0	?
0	0	0	0	0
?	?	?	0	?

$\oplus$	+	−	0	?
+	+	?	+	?
−	?	−	−	?
0	+	−	0	?
?	?	?	?	?

(1990a) has proposed an approach based on network transformations where all trails – alternating sequences of directed links and nodes such that each link starts with the node ending the preceding link – between two nodes of interest are collapsed into a single arc. The sign of this resulting arc is determined from the signs associated with the collapsed arcs. Two operators are available for computing the sign of a resulting arc: the sign product operator and the sign addition operator (Table 9.4).

The sign product operator $\otimes$ is used for computing the sign of qualitative influence between two separated variables based on the signs of the qualitative influences that are associated with each arc of the trail connecting two variables (*transitivity*). Suppose a QPN involving the three variables A, B and C with $A \rightarrow B$ and $B \rightarrow C$ forming a trail from A to C Figure 9.1(i). Let δ_1 and δ_2 denote the signs of qualitative influences associated to $A \rightarrow B$ and $B \rightarrow C$, respectively. The sign of qualitative influence of A on C along the trail $A \rightarrow B \rightarrow C$ can be evaluated by collapsing the two arcs $A \rightarrow B$ and $B \rightarrow C$ into a single arc $A \rightarrow C$. Denote the sign of that single arc by δ_3. It is obtained by applying the sign product operator to δ_1 and δ_2. For example, if $\delta_1 = +$ and $\delta_2 = -$, then δ_3 is given by $\delta_1 \otimes \delta_2 = + \otimes - = -$. It is assumed here that no direct influence exists between A and C. However, as noted by Wellman (1990b), even if there were such a direct link, it still makes sense to say that A influences C with the sign δ_i along the particular trail considered, that is, $A \rightarrow B \rightarrow C$. Thus, it is possible to maintain the original graph and to use the sign product operator for defining the sign of qualitative influence along a trail in a graph.

The sign addition operator $\oplus$ is used to combine parallel influences (*composition*). Consider again a QPN with a trail from A to C via B. In addition, assume a direct link from A to C as shown in Figure 9.1(ii). A sign δ_i of qualitative influence is associated with each link. The sign of qualitative influence that A exerts on C can be evaluated in two steps. First, the sign of qualitative influence along the trail $A \rightarrow B \rightarrow C$ is evaluated. For this purpose, the sign product operator is applicable as mentioned above: $\delta_1 \otimes \delta_2$. Second, the sign associated with this trail is combined with the sign associated with the direct link

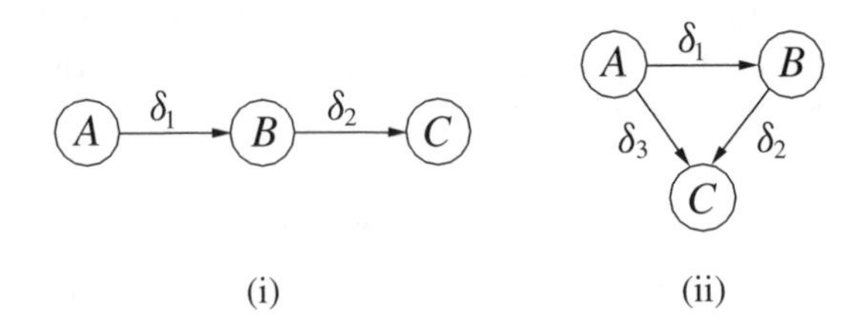

Figure 9.1 Qualitative probabilistic networks over three variables, A, B and C. The δ_i designate the signs of qualitative influence ($\delta_i \in \{+, -, 0, ?\}$).

$A \rightarrow C$. This step involves a combination of parallel influences that requires the application of the sign addition operator. For the network shown in Figure 9.1(ii), the sign addition operator is applied as follows: $(\delta_1 \otimes \delta_2) \oplus \delta_3$.

Generally, note that the operators $\otimes$ and $\oplus$ commute, associate, and distribute as ordinary multiplication and addition do. Besides transitivity and composition, qualitative relationships also adhere to the property of *symmetry*. In the case of a qualitative influence (Section 9.1.1), the property of symmetry states that if a node A exerts a qualitative influence on another node B, then, inversely, B exerts a qualitative influence of the same sign on A. Notice, however, that the term symmetry is restricted to the sign of influence. The magnitude of the influence of a variable A on a variable B can be arbitrarily different from the magnitude of the influence of B on A (Druzdzel and Henrion 1993b).

The property of symmetry also applies to additive and product synergy respectively. In this context, symmetry refers to the property that the two nodes exhibiting a synergy are interchangeable.

9.1.5 Evaluation of indirect influences between separated nodes: a forensic example

This section aims to illustrate potential applications of some of the previously introduced qualitative probabilistic concepts. To this end, consider again a scenario of the suspected trafficking of illegal drugs, discussed at some length in Section 6.2.6. Let H be the major proposition with H_p defined as 'the suspect is involved in illegal drug trafficking, reselling and activities related thereto' and H_d defined as 'the suspect is not involved in illegal drug trafficking, reselling and activities related thereto'. Material collected from the surface of the suspect's clothing is analysed by an IMS. A positive result is obtained for cocaine. This result is confirmed by gas chromatography (GC)/mass spectrometry (MS). Let these findings be denoted by a variable E. This is a binary variable that relates to the proposition that the instrumental analysis yields a positive result for cocaine. The two states are E and $\bar{E}$, the truth and falsity of that proposition. An argument is constructed to the variable H by defining the following propositions:

- C: the suspect has been in contact with a primary source of illegal drugs (*i.e.*, cocaine);
- $\bar{C}$: the suspect has never been in contact with a primary source of illegal drugs;
- T: there has been some sort of transfer (primary, secondary, tertiary or a combination of these) of a certain quantity of illegal drugs;
- $\bar{T}$: no transfer has occurred.

Recall that an illegal drug, in powder form or as a compressed bloc, has been referred to as a *primary source*. It provides an instance for a possible (primary) transfer to an individual, which, in turn, may initiate secondary and tertiary transfers as explained in Section 6.2.6.

Figure 9.2 reflects the assumed relationships among the variables. The δ_i indicate the sign of qualitative influence. Hereafter, different scenarios will be discussed where the signs of qualitative influence may change because of varying case-specific circumstantial information, or the strategy chosen by the defence. Each time assessments change, the aim will be to re-evaluate the direction of influence that knowledge of E exerts on H.

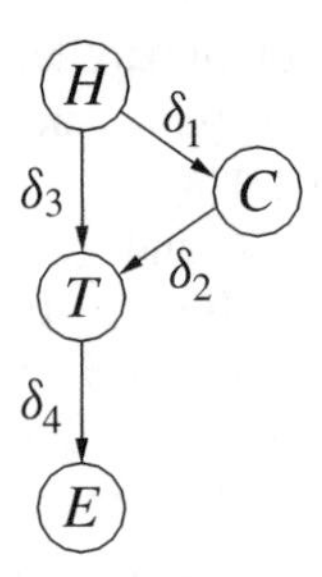

Figure 9.2 QPN for evaluating transfer evidence under activity level propositions. Node C has two states, C, the suspect has been in contact with a primary source of illegal drugs, $\bar{C}$: the suspect has never been in contact with a primary source of illegal drugs. Node T has two states, T, there has been some sort of transfer of a certain quantity of illegal drugs, $\bar{T}$, no transfer has occurred. Node H has two states, H_p and H_d, the suspect is (is not) involved in illegal drug trafficking, reselling and activities related thereto. Node E represents the findings of the IMS and GC/MS analyses.

9.1.5.1 Scenario 1

Imagine that a fairly high intensity signal is recorded by the IMS during analysis of the trace material collected on the suspect. Usually, confirmative analyses by GC/MS would not involve a quantification. Nevertheless, the intensity of the signal obtained through IMS can provide a rough idea as to the extent of the contamination.

Let the proposition proposed by the defence, H_d, be that the suspect has never been involved in any way in activities of which he is accused by the prosecution, that is, the trafficking of illegal drugs, reselling and activities related thereto. An inference from E to H requires a series of assessments to be made. They may be a consideration of the following:

- The probability of contact C: If the suspect engages in trafficking of illegal drugs, their reselling and activities such as their packaging (H_p), then the probability of a direct contact with a primary source of illicit drugs appears to be quite likely: Let $Pr(C \mid H_p)$ assume a value of 0.9, for example. If the suspect is innocent (H_d), then the probability of a contact with a primary source can be considered impossible: $Pr(C \mid H_d) = 0$. Therefore, it is assumed that there is no possibility for a legitimate contact with a primary source of illegal drugs.

- The probability of transfer T: If the suspect has been in contact with a primary source (C), then it is taken to be certain that a transfer has occurred, independently on H_p and H_d: $Pr(T \mid C, H_p) = Pr(T \mid C, H_d) = 1$. If the suspect was not in contact with a primary source ($\bar{C}$), then the probability of transfer is reduced, but not necessarily zero. For example, if the suspect is involved in drug trafficking (H_p), but did not himself handle any drugs, then secondary or tertiary transfers may eventually occur because of exposure to a contaminated environment, for example, a car used for the transportation of drugs, or a room used for the storage of drugs. For the purpose of illustration, consider a value of 0.4 for $Pr(T \mid \bar{C}, H_p)$. If the suspect is not involved in any of the alleged activities (H_d), and he did not have contact with a primary source, then the probability of a transfer can be set to zero or close to zero: $Pr(T \mid \bar{C}, H_d) = 0$. The absolute value of this parameter is not of primary importance here; rather its relative value when compared to $Pr(T \mid \bar{C}, H_p)$: $Pr(T \mid \bar{C}, H_p) > Pr(T \mid \bar{C}, H_d)$.

- The probability of the evidence E: It is assumed that the positive analytical results are much more likely to occur when there was a transfer of micro-traces of drugs than when there was no such transfer: $Pr(E \mid T) >> Pr(E \mid \bar{T})$. Sample values could be, for instance, 0.99 versus 0.01.

These qualitative expressions of beliefs about the various probabilistic relationships can now be used to extract the signs of qualitative influence associated with each arc of the QPN shown in Figure 9.2.

- Relationship between H and C (δ_1): The assumption $Pr(C \mid H_p) > Pr(C \mid H_d)$ implies that H positively influences C along $H \rightarrow C$.

- Relationship between C and T (δ_2): Given H_p or H_d, the probability of transfer given contact is considered to be higher than transfer given no contact. Formally, this may be written $Pr(T \mid C, H_p) > Pr(T \mid \bar{C}, H_p)$ and $Pr(T \mid C, H_d) > Pr(T \mid \bar{C}, H_d)$. This implies that $Pr(T \mid C, x) > Pr(T \mid \bar{C}, x)$ for any value of x of other parents of T, that is, other than C. The conclusion thus is that the variable C positively influences T along $C \rightarrow T$.

- Relationship between H and T (δ_3): Here, the changes in the probability of T due to varying H need to be considered for each state of C. If C is true, one has $Pr(T \mid H_p, C) = Pr(T \mid H_d, C)$. If $\bar{C}$ is true, one has $Pr(T \mid H_p, \bar{C}) > Pr(T \mid H_d, \bar{C})$. The overall relationship is thus described by $Pr(T \mid H_p, x) \geq Pr(T \mid H_d, x)$. Thus H positively influences T along $H \rightarrow T$.

- Relationship between T and E (δ_4): The above assumption $Pr(E \mid T) > Pr(E \mid \bar{T})$ implies a positive qualitative influence of T on E.

The signs of qualitative influence in the QPN are now defined and can be used for inference. Attention is drawn here to the effect that knowledge about E may have on the truth state of H. Let δ_5 denote the indirect influence between H and E. The direction of this influence depends, on the one hand, on the graphical structure of the QPN, that is, the various trails between H and E, and, on the other hand, on the signs associated with each arc of the trails connecting H and E. In the case at hand, δ_5 derived from the QPN shown in Figure 9.2 is

$$\delta_5 = [(\delta_1 \otimes \delta_2) \oplus \delta_3] \otimes \delta_4 \ . \tag{9.8}$$

Following the definition of the sign product and sign addition operators (Table 9.4), one can find that $\delta_5 = +$ for $\delta_1 = \delta_2 = \delta_3 = \delta_4 = +$.

The result of the qualitative analysis is that, under the stated assumptions, knowledge of E tends to increase the probability of H. The computation of the sign of overall influence between H and E, based on (9.8), is more attractive, in some way, than the development of a formal likelihood ratio: The contribution of each of the underlying qualitative influences is shown in (9.8). In addition, whenever one or more of the δ_i ($i = 1, ..., 4$) changes, (9.8) can readily be updated. This is illustrated in the scenarios discussed in Sections 9.1.5.2 and 9.1.5.3.

9.1.5.2 Scenario 2

Consider a scenario slightly different from the one discussed in the previous section. Imagine, as was done before, that the position of the defence is to argue that the suspect is not involved in drug trafficking. However, information is offered by the defence according to which the suspect has frequently been exposed to a potentially contaminated environment. This may be the case, for example, if accomplices of the suspect have already been proven to be involved in drug trafficking. It may then be a question for consideration whether the newly acquired circumstantial information should change the previously derived direction of inference between E and H. The additional information offered by the defence seems relevant for assessing the probabilities of transfer. Consider Table 9.5 for this purpose.

If there were a contact with a primary source (C), then a transfer is assumed to occur with certainty: $Pr(T \mid C, H_p) = Pr(T \mid C, H_d) = 1$. If there were no contact but the suspect is involved in drug trafficking, a non-zero transfer probability, denoted as t, may be appropriate. As noted earlier, secondary or tertiary transfers may occur in such circumstances. These assessments are all analogous to Scenario 1 (Section 9.1.5.1).

In Scenario 1 it was assumed that, if there was no contact, and the suspect was not involved in drug trafficking, there was no possibility for a transfer: $Pr(T \mid \bar{C}, H_d) = 0$. In the scenario considered here, information offered by the defence sheds new light on this assessment. It may appear reasonable to argue that there could be a secondary or tertiary transfer in the same way as that for someone who is involved in drug trafficking. So, if $Pr(T \mid \bar{C}, H_d)$ is to be abbreviated by t', probabilities could be assessed so that $t \approx t'$, or, $t = t'$.

Because of changes in the probabilities associated with the node T, the signs δ_2 and δ_3 (Figure 9.2) need to be reviewed.

Table 9.5 Conditional probabilities for the variable transfer (T), with two states, T and $\bar{T}$, transfer or no transfer, respectively. Factor C has two states, C and $\bar{C}$, contact or no contact, respectively, with a primary source of drugs. Factor H has two states, H_p and H_d, the suspect is, or is not, involved in drug trafficking. Probabilities t and t' are the transfer probabilities when there is no primary contact, $\bar{C}$, under propositions H_p and H_d, respectively.

	H :	H_p		H_d	
	C:	C	$\bar{C}$	C	$\bar{C}$
T :	T	1	t	1	t'
	$\bar{T}$	0	$1-t$	0	$1-t'$

- Influence associated to $C \rightarrow T$ (δ_2): $Pr(T \mid C, H_p) > Pr(T \mid \bar{C}, H_p)$ and $Pr(T \mid C, H_d) > Pr(T \mid \bar{C}, H_d)$ implies that C exerts a positive qualitative influence on T: $\delta_2 = +$.

- Influence associated to $H \rightarrow T$ (δ_3): From $Pr(T \mid H_p, C) = Pr(T \mid H_d, C)$ and $t = t'$, one can find that $\delta_3 = 0$.

The signs thus determined can now be entered in (9.8). The sign of overall influence between H and E is given by

$$\delta_5 = \underbrace{[\underbrace{\underbrace{(+ \otimes +)}_{+} \oplus 0}_{+}] \otimes +}_{+} = + \, .$$

The result of the qualitative analysis is that the evidence E still supports H_p over H_d despite the circumstantial information put forward by the defence. Does this mean that the additional information is without effect? The answer to this question depends on one's point of view. As far as the effect of E on H is concerned, the *direction* of inference remains unchanged. Effects may, however, be observed more locally, notably at the arc $H \rightarrow T$. Here, information relating to a possible secondary or tertiary transfer has 'neutralised' the qualitative influence that H exerts on T: the '+'-sign has changed to a '0'-sign. Notice that the assumption $Pr(T \mid H_p, C) = Pr(T \mid H_d, C)$ and $t = t'$ is equivalent to say that the edge $H \rightarrow T$ is uninformative and could be omitted, as long as the assumption holds.

The qualitative analysis is also capable of illustrating the reasons why there is still a positive effect of E on H. Assistance in determining assessments can again be found in the probability table of the node T. Notably, it has been assumed that, for H_p and H_d, $Pr(T \mid C) > Pr(T \mid \bar{C})$. So, even though there may be a possibility for a secondary or tertiary transfer, a contact with a primary source (C) is still better in explaining a transfer (T). Consequently, evidential support for T due to E favours C over $\bar{C}$, which, in turn is evidence supporting H_p over H_d. This scenario is considered further in Section 9.2.3.

9.1.5.3 Scenario 3

Imagine a scenario where the analytical result is negative. The definition of the variable E could then be changed in order to refer to that event. Doing so, the conditional probabilities of E given T need to be reviewed. Clearly, a negative result is less likely given T than $\bar{T}$. Consequently, $Pr(E \mid T) < Pr(E \mid \bar{T})$ and $\delta_4 = -$. Assuming the remaining signs to be the same as in Scenario 1, the sign of overall influence between E and H is given by

$$\delta_5 = \underbrace{[\underbrace{\underbrace{(+ \otimes +)}_{+} \oplus +}_{+}] \otimes -}_{-} = - \, .$$

The result of this analysis is that E, a negative result, is evidence for the non-occurrence of T, that is, $\bar{T}$. An increased probability of $\bar{T}$ is better explained if no contact occurred ($\bar{C}$) and the suspect is not the offender (H_d), a result that appears perfectly reasonable.

9.1.6 Implications of qualitative graphical models

Notice that in the scenarios discussed above, no probabilities have been assigned to the root variable H. In fact, they are irrelevant for the kind of qualitative analyses studied here. In some way, this seems a convenient property as it can help forensic scientists to focus on the probability of the evidence while avoiding the provision of an opinion on an issue, that is, H, that lies outside their area of competence.

The qualitative notions presented here are not intended for use in written reports, or for presentation before trial. The qualitative framework is primarily intended as an aid for probabilistic reasoning in situations where there is a lack of numerical data. The aim here is to show that such models can offer logical guidance in evaluative processes. Scientists can be provided with an approach that allows them to discuss, lay down and articulate the foundations of their inferences and beliefs in order to justify the coherence and credibility of their conclusions.

The elegance of the qualitative framework lies in its abstraction from numbers. The relative magnitude of probabilistic assessments can be sufficient to process uncertain knowledge according to accepted rules for inference. Also, literature (Druzdzel 1996, 1997; Druzdzel and Henrion 1993a; Wellman and Henrion 1993) has noted the following:

- It may be advantageous to maintain qualitative distinctions even when numerical information is available.
- A qualitative specification might often be obtained with much less effort than a quantitative one. Qualitative probabilistic networks can thus supplement Bayesian networks where numerical values are either not available or not necessary for the questions of interest.
- If a numerically specified network is already available, the qualitative dependencies may readily be determined from the formal definitions of the various QPN-properties.
- Along with Bayesian networks, QPNs extend the spectrum of possible levels of precision for specifying probabilistic inference models. Notably, the degree of specificity can be made dependent on the extent of available information, allowing maximum efficiency with the least possible effort.
- The results of qualitative analyses provide a workable basis for the generation of verbal explanations of probabilistic reasoning. Explanations of this kind appear to be easier to follow than explanations using numerical probabilities.

Generally, the understanding of both the assumptions, and the reasoning of probabilistic expert systems, are prerequisites for a successful collaboration between these systems and their user (Henrion and Druzdzel 1991). In this context, qualitative verbal explanations extracted from QPNs are of particular interest. A distinction may be made between the explanation of assumptions and the explanations of reasoning (Druzdzel 1996; Henrion and Druzdzel 1991). The explanation of assumptions is concerned with the communication of static knowledge, for example, a model of a graph structure as a representation of a real-world problem. The explanation of reasoning is a more dynamic property as it focuses on extracting conclusions from assumptions encoded in the original model and the observed evidence. For the purpose of inference in QPNs, the interested reader may find it useful to consider further readings in a topic concerned with a method known as *qualitative belief propagation* (Druzdzel and Henrion 1993b; Henrion and Druzdzel 1991). Its underlying algorithm traces the qualitative effect of evidence from one variable to the next and thus provides an illustrative basis for evaluating the effect of newly acquired knowledge and its propagation through the network.

9.2 Sensitivity analyses

It is possible that a scientist has set up a graphical model for the structuring of some sort of inferential task. Such a model would usually include certain structural assumptions, that is, relationships believed to be true among specific variables of interest. Within Bayesian networks, probability theory is used as a concept to characterise the nature of the relationships that are assumed to hold among the variables. Rarely, however, is a model considered a 'definite' solution for a given problem. Rather, a model is a starting point for questions of the kind 'How does knowledge about one variable affect the truth state of another variable'?, or, 'How is a probabilistic analysis affected by changes in one or more parameters of interest'?. The provision for answers to such questions is a crucial element in the process of deciding whether a model is acceptable for the purpose for which it has been designed. In this context, it has been pointed out that

> [...] we should be able to obtain useful insights about the relative importance to our conclusions of the various assumptions, decisions, uncertainties and disagreements in the inputs. These can help us decide whether it is likely to be worthwhile gathering more information, making more careful uncertainty assessments, or refining the model, and which of these could most reduce the uncertainty in the conclusions. (Morgan and Henrion 1990, p. 172)

Sensitivity analysis is a general technique that allows one to investigate the properties of a mathematical model. For example, when information is available on a set of variables, one may wish to evaluate the way in which such evidence affects other nodes of interest. This is what sometimes is termed *sensitivity analysis applied to evidence*. A common measure for sensitivity to evidence is the likelihood ratio. So, when H denotes a proposition of interest, and E is the evidence, then

$$\frac{Pr(E \mid H_p)}{Pr(E \mid H_d)}$$

measures the impact that evidence E has on the proposition H, or, alternatively, the degree to which evidence E discriminates H_p from H_d. The reader is already well acquainted with this measure, as it plays a central role in many of the examples studied so far in this book.

Sensitivity analysis, as applied to parameters, amounts to a study of the effects of the uncertainties in the parameters of a model on its outcome. In a Bayesian network, a parameter is typically an assessment of a node's (conditional) probability. In Section 9.2.1, the changes in the truth state of a target node arising from uncertainties in the assessment of a single parameter are studied. In Section 9.2.4, analyses are extended to uncertainties in two parameters.

9.2.1 Sensitivity to a single parameter

Consider a Bayesian network described in Section 5.11.2 (Figure 5.20), which can be used for the evaluation of the potential for error associated with DNA evidence (Biedermann and Taroni 2004). There are three nodes H, M and R that are serially connected: $H \rightarrow M \rightarrow R$. The network is chosen here for purely illustrative purposes because it contains, on the one hand, logical probabilities of zero and one, and, on the other hand, epistemic probabilities such as γ and fpp. The model involves only a few variables; however, it may be thought of as a local part of a larger network.

Proposition H is the proposition according to which a crime stain comes from the suspect. The variable M refers to the event that there is a true match between the profile of the crime stain and the profile of a sample taken from the suspect. Recall that it was assumed that when the suspect is in fact the source of the crime stain, that is, H_p is true, then certainly there is a true match: $Pr(M \mid H_p) = 1$. Notice that M may also be true even though H may be false. The probability of such an event has been subsumed within γ, short for the random match probability. The parameter R is the scientist's report of a match between the profile of the crime stain and that of the sample known to originate from the suspect. It is assumed that if there actually is a match, that is, M is true, then the scientist would report a match with certainty: $Pr(R \mid M) = 1$. The probability with which the scientist would falsely declare a match, $Pr(R \mid \bar{M})$, is abbreviated by fpp.

Imagine now a scientist's report of a match. Such information would be communicated to the network by instantiating the node R. The target probability is $y = Pr(H_p \mid R)$, the suspect being the source of the crime stain, given the evidence provided by the scientist. Let $x = Pr(R \mid \bar{M})$, the false-positive probability, denote the parameter whose actual value is uncertain.

Here, sensitivity analysis aims to investigate the functional relation between y and x. Start by rewriting y as:

$$y = Pr(H_p \mid R) = \frac{Pr(R \mid H_p)Pr(H_p)}{Pr(R)} .$$

In its extended form, the numerator is

$$[Pr(R \mid M)Pr(M \mid H_p) + Pr(R \mid \bar{M})Pr(\bar{M} \mid H_p)]Pr(H_p) .$$

Consider this as a linear function in x:

$$ax + b ,$$

where $a = Pr(\bar{M} \mid H_p)Pr(H_p)$ and $b = Pr(R \mid M)Pr(M \mid H_p)Pr(H_p)$. For the denominator,

$$Pr(R) = \sum_M \sum_H Pr(R \mid M)Pr(M \mid H)Pr(H) .$$

The denominator can also be considered as a linear function in x, notably,

$$Pr(R) = cx + d .$$

The factor c is given by $Pr(\bar{M} \mid H_p)Pr(H_p) + Pr(\bar{M} \mid H_d)Pr(H_d)$ and d is given by $Pr(R \mid M)[Pr(M \mid H_p)Pr(H_p) + Pr(M \mid H_d)Pr(H_d)]$.

As may be seen from these transformations, the numerator and denominator of y relate linearly to the conditional probability x. The probability of interest, y, is thus expressed as a quotient of two linear functions of x. A more formal statement of this property can be found, for example, in Coupé and Van der Gaag (1998).

Recall that earlier in this section, $Pr(M \mid H_p)$ and $Pr(R \mid M)$ have both been set to 1. For the purpose of illustration, let γ be 10^{-5}. For H, uninformative probabilities of $H_p = H_d = 0.5$ are used. Notice that this assignment of prior probabilities is one of a purely technical nature. In practice, it may well be informed by case-specific background information that lies outside the province of the forensic scientist.

On the basis of these assessments, it can be shown that

$$y = \frac{ax + b}{cx + d} = \frac{1}{10^{-5}(1 - x) + 1 + x} .$$

This function is shown in Figure 9.3. It shows that when the probability of a false-positive tends to zero, then the probability of interest, $Pr(H_p \mid R)$, becomes $1/[1 + Pr(M \mid H_d)] = 1/(1 + \gamma)$. On the other hand, when fpp reaches one, then the probability of interest falls back to the prior value of 0.5 as specified initially.

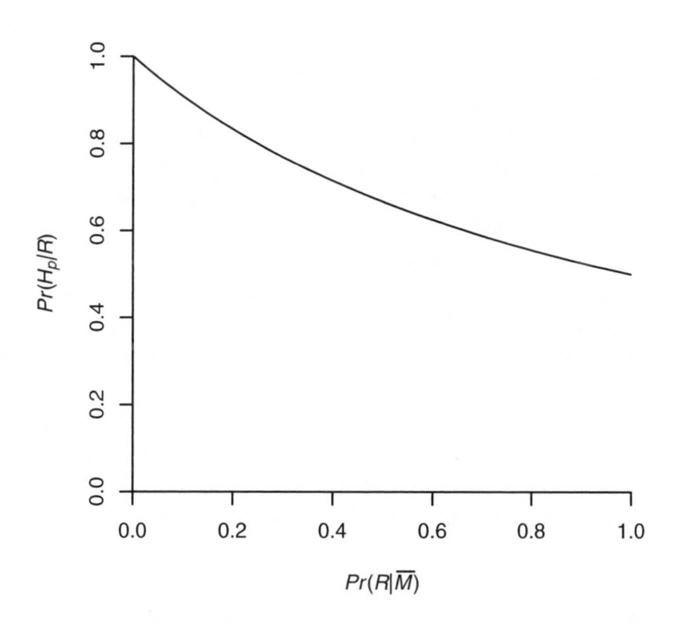

Figure 9.3 Representation of the conditional probability $Pr(H_p \mid R)$ as a function of $Pr(R \mid \bar{M})$, the false-positive probability fpp. Parameter R is the report of a match between the profile of a crime stain and that of a sample known to originate from a suspect. Parameter $\bar{M}$ is the event that there is no match between the profile of a crime stain and that of a sample known to originate from a suspect. Proposition H_p is that the suspect is the source of the crime stain.

9.2.2 One-way sensitivity analysis based on a likelihood ratio

In the previous section, consideration has been given to a one-way sensitivity analysis of a conditional probability of interest for varying assessments of a certain parameter. A graphical representation is provided in Figure 9.3. The approach requires scientists to make an assumption of prior probabilities for the proposition of interest, that is, that the suspect is the source of the crime stain ($Pr(H_p)$). There may be objections to this: One may prefer to work with as few assumptions as possible, and, as far as forensic scientists are concerned, the addressing of source-level propositions may be felt to be inappropriate.

Assumptions on prior probabilities can be avoided when working with a likelihood ratio. It draws one's attention to the value of the evidence while having the same potential for analysing uncertainties about the values of the parameters.

Consider again the Bayesian network examined in the previous section. The probability of the scientist falsely reporting a match, $Pr(R \mid \bar{M})$, is the parameter about which uncertainty exists. One wishes to evaluate the effect this uncertainty has on the evidential value of the scientist's report. A one-way sensitivity analysis could thus consist of evaluating the changes in the magnitude of the likelihood ratio due to varying assessments of $Pr(R \mid \bar{M})$. The likelihood ratio associated with the currently discussed Bayesian network, $H \rightarrow M \rightarrow R$, is (Thompson et al. 2003):

$$V = \frac{Pr(R \mid H_p)}{Pr(R \mid H_d)} = \frac{1}{\gamma + [fpp \times (1-\gamma)]} .$$

Figure 9.4 illustrates the effect that varying assessments of the false-positive probability fpp have on the magnitude of the likelihood ratio V.

Sensitivity analyses enable the scientist to evaluate a likelihood ratio even though exact point probabilities cannot be provided for all parameters. In the example under discussion, uncertainty with respect to one parameter is considered. This uncertainty can be expressed by a plausible interval that defines a range of values in which the 'true' probability is assumed to lie with reasonable certainty (Coupé et al. 2000). For example, it may be thought that the value of fpp lies somewhere in between a and b (with $b < a$), these denoting values between 0 and 1. The changes in $V(fpp)$ within that interval then provide an indication of the relative importance of the uncertainty about the actual value of the parameter fpp.

From Figure 9.4, one can see that fpp has, notably for low values, a rather drastic effect on V. Analysts can thus gain an idea of the effect that uncertainty about a specific parameter should be expected to have on the value of the likelihood ratio. Such information may be helpful for deciding whether efforts should be deployed for the elicitation of a parameter's value.

9.2.3 A further example of one-way sensitivity analysis

Consider again a context involving the analysis of trace material recovered from the surface of a suspect's clothing. In Section 9.1.5, such cases have been discussed on a purely qualitative level. It has been seen that a qualitative model specification may suffice to review the *direction* of inference whenever case circumstances change. Recall two of the scenarios studied in further detail:

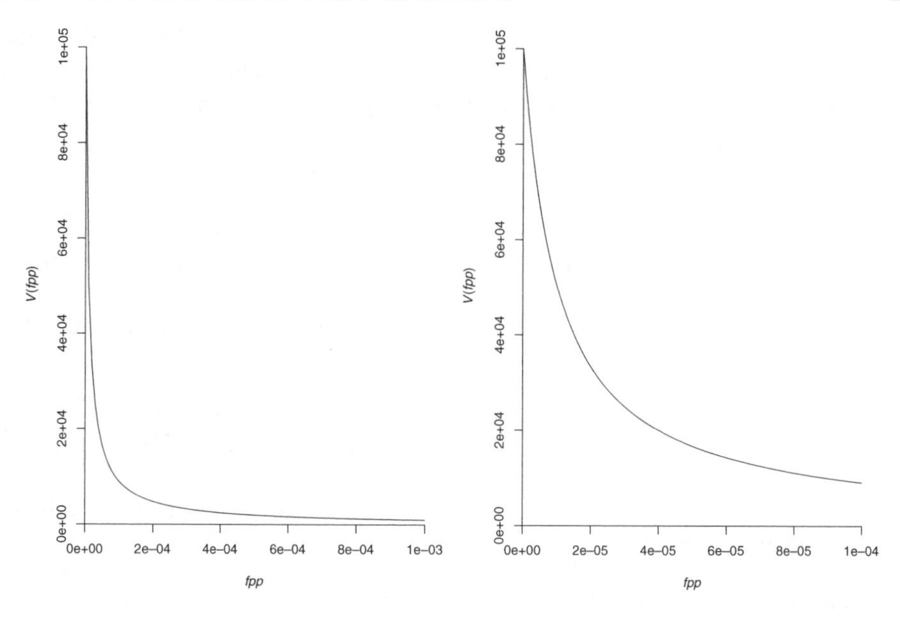

Figure 9.4 One-way sensitivity analyses of the likelihood ratio V for varying assessments of the false-positive probability fpp ($Pr(R \mid \bar{M})$, the probability of reporting a match when there is no match). The value for γ, the frequency of the profile in the relevant population, has been fixed at 10^{-5}. The right-hand graph is a magnification of the left-hand graph for $fpp < 1e^{-04}$. [Reproduced with permission from Elsevier.]

- Scenario 1 (Section 9.1.5.1): Here it was assumed that there was no possibility for a transfer for innocent reasons. Stated otherwise, if the suspect is innocent, and he was not in contact with a primary source of illegal drugs, then there was no possibility for a transfer of traces of drugs (including secondary and tertiary transfer). Thus, $Pr(T \mid \bar{C}, H_d)$ was set to zero. If the suspect is involved in drug trafficking and related activities (H_p), and no contact with a primary source occurred, then the probability of there being trace material may be different from zero, for example, because of secondary or tertiary transfer. A probability $Pr(T \mid \bar{C}, H_p) = t$ is used to denote the probability of this event.

- Scenario 2 (Section 9.1.5.2): This scenario is different from the former in the sense that there could be a transfer of trace material even though the suspect is innocent. A probability t', different from zero, has thus been adopted for $Pr(T \mid \bar{C}, H_d)$. Note, the defence has been assumed to argue that case circumstances are such that the transfer probabilities in the absence of contact with a primary source are the same for both an innocent and a guilty suspect. It has thus been accepted that $t = t'$.

Qualitative analyses indicated that the likelihood ratio in the first scenario was greater than one. The change in the defence strategy in the second scenario did not, however, change anything on that direction of inference. However, the two scenarios seem clearly different.

The reason for this result is that qualitative analyses operate on a level too general to detect such differences. As noted earlier in this chapter, qualitative analyses concentrate on the direction of inference while not discriminating between inferences described by the same sign, notably positive or negative ones.

These are typical instances where sensitivity analyses can aid in the achievement of higher levels of detail. Consider this in the context of Scenario 1. Choose the likelihood ratio as the quantity of interest and let t be the parameter about which uncertainty exists. Here, a sensitivity analysis consists of varying the value of t and evaluating the effect on the likelihood ratio V.

Formally written, V associated with the Bayesian network shown in Figure 9.2 is given by

$$V = \frac{Pr(E \mid H_p)}{Pr(E \mid H_d)} = \frac{\sum_{TC} Pr(E \mid T)Pr(T \mid C, H_p)Pr(C \mid H_p)}{\sum_{TC} Pr(E \mid T)Pr(T \mid C, H_d)Pr(C \mid H_d)}. \tag{9.9}$$

Following the discussion in Section 9.1.5.1 one can reasonably agree about the values for the following probabilities:

- Node C: $Pr(C \mid H_p) = 0.9$, $Pr(C \mid H_d) = 0$;
- Node T: $Pr(T \mid C, H_p) = Pr(T \mid C, H_d) = 1$, $Pr(T \mid \bar{C}, H_d) = t' = 0$
- Node E: $Pr(E \mid T) = 1$, $Pr(E \mid \bar{T}) = 0.01$.

A graphical representation of a one-way sensitivity analysis with these values is shown in Figure 9.5 (solid line). For $t = 0$ and $t = 1$, one can obtain likelihood ratios of, respectively, 90.01 and 100. Note that the latter is a setting in which (9.9) reduces to $1/Pr(E \mid \bar{T})$.

This analysis is now contrasted with Scenario 2, where t' is varied as well. In order to assure consistency with the scenario as described in Section 9.1.5.2, and, for the ease of discussion, the constraint $t = t'$ is accepted. This setting is represented in Figure 9.5 by a dotted line.

As may be seen, comparison of Scenarios 1 and 2 indicates that when the defence's strategy is to suggest that transfer probabilities given no contact are the same for a guilty and an innocent suspect, then the likelihood ratio may be drastically reduced, although not falling below 1.

9.2.4 Sensitivity to two parameters

Two-way sensitivity analyses may be addressed when information is needed about the joint effect of two parameters. As an example, consider again a situation in which a scientist's report of a match tends to associate a suspect with a crime stain. With reference to the previous sections, a Bayesian network with $H \rightarrow M \rightarrow R$ is used.

Also part of the evidence is that the suspect has been selected on the basis of a search of the crime stain's DNA profile against a relevant database (containing N profiles). Following the arguments presented in Section 5.10, the current Bayesian network can be extended with a binary variable D, defined as 'the other $(N - 1)$ profiles of the database do not match'. D is, besides M, a further child variable of H.

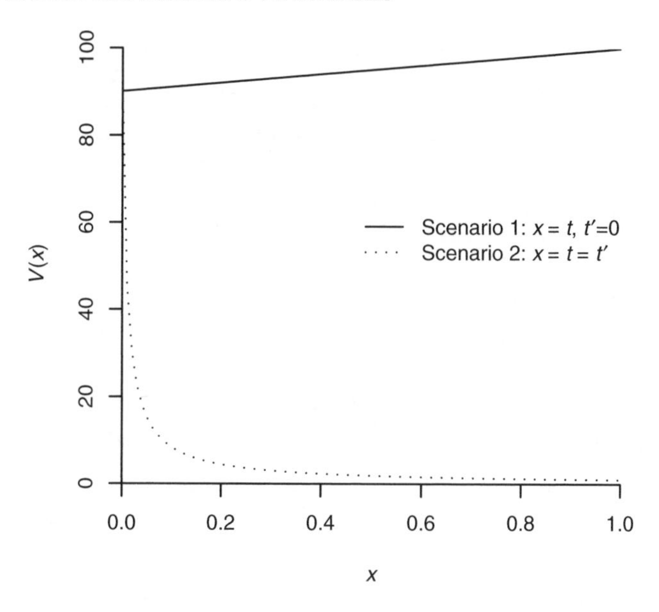

Figure 9.5 Sensitivity analyses of two scenarios involving the transfer of traces of illegal drugs. For the first scenario, Scenario 1, x represents $Pr(T \mid \bar{C}, H_p)$, denoted as t. In Scenario 1, $t' = 0$. In Scenario 2, x represents both $Pr(T \mid \bar{C}, H_p), t$, and $Pr(T \mid \bar{C}, H_d), t'$, which are set equal, $t = t'$. $V(x)$ denotes the value of the evidence T for the given value of x.

When the evidence consists of both R and D, then the likelihood ratio is given by

$$V = \frac{Pr(R, D \mid H_p)}{Pr(R, D \mid H_d)} = \frac{1}{\gamma + [fpp + (1 - \gamma)]} \times \frac{1}{1 - \phi} .$$

The parameter ϕ stands for the probability that the source of the crime stain is to be found among the other $(N - 1)$ suspects.

The current scenario will focus on uncertainties with respect to the parameters γ and ϕ, while fpp is assumed to be 0. As in the previous section, the likelihood ratio V is chosen as the quantity of interest. Here, a two-way sensitivity analysis consists of the evaluation of the changes in the magnitude of V due to the simultaneous variation of γ and ϕ. This sort of analysis can be represented graphically in the form of contour lines (*see* Figure 9.6). These connect the combinations of values for γ and ϕ that result in the same value for the likelihood ratio V.

Figure 9.6 shows contour lines for likelihood ratios of 1.5, 3, 6, 9, 15, 30, 60, 100, 250 and 1000. The largest distances between the contour lines can be observed in the lower right part of the graph. This indicates that the likelihood ratio is relatively insensitive for high values of γ and low values of ϕ. Whenever γ decreases and/or ϕ increases, the contour lines become more close, that is, changes in the values of the parameters tend to provoke greater changes of the value of V.

Imagine that a scientist is willing to express uncertainty about the values of γ and ϕ by means of plausible intervals. In Figure 9.6, dashed parallel lines are used to delimit

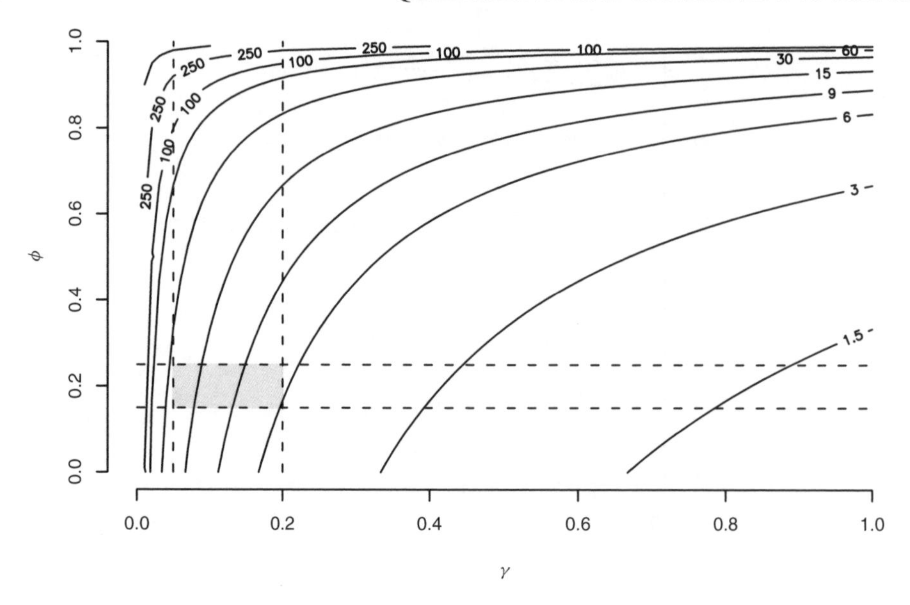

Figure 9.6 Two-way sensitivity analysis: simultaneous variation of the parameters γ and ϕ. A suspect has been selected on the basis of a search of a crime stain's DNA profile against a relevant database of N profiles. The relative frequency of the profile is γ. The probability that the source of the crime stain is to be found among the other $(N - 1)$ suspects is ϕ. The value for fpp has been set to zero. Each contour line represents a likelihood ratio of a specific value. The dashed parallel lines delimit hypothetical plausible intervals. [Reproduced with permission from Elsevier.]

hypothetical intervals. These provide an indication of the *plausible effect* that the scientist's parameter uncertainties have on the magnitude of the likelihood ratio. This is visualised by the grey shaded area: The number of contour lines together with their values allow one to appreciate the magnitude of the plausible effect.

Figure 9.6 further illustrates that, depending on the area where an uncertain parameter is thought to lie, the likelihood ratio may vary considerably even though the plausible intervals may be narrow. Conversely, configurations may exist where plausible intervals are large but the likelihood ratio varies only little.

9.2.5 Further issues in sensitivity analyses

In the previous sections, sensitivity analyses with up to two parameters have been considered. The number of parameters varied simultaneously may further be increased but results tend to become more difficult to interpret.

For clarity of exposition, the discussion has solely focused on a local Bayesian network fragment. Extending analyses to greater networks may become more time-consuming, so methods have been sought that may allow that burden to be reduced. For example, inspection of a graph's structure for the presence of independence relationships can often aid to reduce the topology that needs to be included in an analysis. Such independence relationships may exist prior to the receipt of evidence; on other occasions, owing to learning of specific evidence, one part of a network may become independent from another.

Generally, the nodes whose assessments upon variation may influence a network's probability of interest are referred to as this probability's *sensitivity set* (Coupé et al. 2000). Further details on sensitivity sets and the technicalities of their computation can be found, for example, in Coupé and Van der Gaag (1998) and Kjærulff and Van der Gaag (2000).

10

Continuous networks

10.1 Introduction

So far, events and counts and the probability of their occurrence have been discussed. These ideas may be extended to consider the measurements about which there may be some uncertainty or randomness. In certain fairly general circumstances, the way in which probability is distributed over the possible values for the measurements can be represented mathematically by functions known as probability distributions. The most well-known distribution for measurements is that of the Normal distribution. This will be described in Section 10.3.2. Before probability distributions can be discussed here, however, certain other concepts have to be introduced.

10.2 Samples and estimates

For any particular type of evidence, the distribution of the characteristic of interest is important. This is so that it may be possible to determine the rarity or otherwise of any particular observation. For the refractive index and elemental composition of glass, the distributions of the refractive index and elemental composition measurements are important. These distributions relate to variability within and between pieces of glass from the same or different populations such as window glass and car head lamp glass. Sometimes the population may be a conceptual population only (*e.g.*, the population of all possible fragments of window glass). In practice, these distributions are not known exactly. They are estimated from a sample.

A characteristic of interest from the population is known as a *parameter*. The corresponding characteristic from the corresponding sample is known as an *estimate*. It is hoped that an estimate will be a good estimate in some sense. Different samples from the same population may produce different estimates. Different results from different samples do not mean that some are wrong and others are right. They merely indicate the natural variability in the distribution of the characteristics amongst different members of the population.

Variation is accounted for by including a measure of the variation, known as the *standard deviation*, in the process. The square of the standard deviation is known as the *variance*; a sample variance may be denoted as s^2, and the corresponding population variance σ^2. These terms have been used earlier in Section 7.4.4.

A notational convention uses Roman letters for functions evaluated from measurements from samples and Greek letters for the corresponding parameters from populations. Thus, a sample mean may be denoted as $\bar{x}$ and the corresponding population mean θ. A sample standard deviation may be denoted as s and the corresponding population standard deviation σ. Another more general notation is the use of the symbol ^ to indicate an estimate of a parameter. Thus, an estimate of a population mean θ may be indicated as $\hat{\theta}$.

The concept of a *random variable* [or *random quantity* or *uncertain quantity*, Lindley (1991)] needs some explanation. There has been some discussion in Section 1.1.9 and Section 2.1.2. A *random variable*, in a rather circular definition, is something that varies at random. For example, there is variation in the elemental concentrations between glasses.

A *statistic* is a function of the data. Thus, the sample mean and the sample variance are statistics. A particular value of a statistic, which is determined to estimate the value of a parameter is known as an *estimate*. The corresponding random variable is known as an *estimator*. An estimator, say X, of a parameter, say θ, which is such that $E(X) = \theta$ is said to be unbiased. If $E(X) \neq \theta$, the estimator is said to be biased.

Notation is useful in the discussion of random variables. Rather than write out phrases such as 'the concentration of silicon in a fragment of glass' in long-hand, the phrases may be abbreviated to a single upper-case Roman letter. For example, let 'X' be short for 'the concentration of silicon in a fragment of glass' and the phrase 'the probability that the concentration of silicon in a fragment of glass is less than 30' may be written as $Pr(X < 30)$, or more generally as $Pr(X < x)$ for 'the probability that the concentration of silicon in a fragment of glass is less than x' for a general value x of the concentration of silicon.

The mean of a random variable is the corresponding population mean. In the examples given here, this would be the mean refractive index of the population of all fragments of glass or the mean concentration of silicon in glass (both conceptual populations). The mean of a random variable is given a special name, the *expectation*, and for a random variable, say X, it is denoted as $E(X)$. Similarly, the variance of a random variable is the corresponding population variance. For a random variable X, it is denoted as $Var(X)$. The mean may be thought of as a *measure of location* to indicate the size of the measurements. The standard deviation may be thought of as a *measure of dispersion* to indicate the variability in the measurements.

The applications of these concepts are now discussed in the context of probability distributions for measurements.

10.3 Measurements

10.3.1 Summary statistics

Consider a population of continuous measurements with mean μ and standard deviation σ.

Given sample data, $(x_1, x_2, \ldots, x_n)$, of measurements from this population, μ and σ may be estimated from the sample data as follows. The sample mean, denoted as $\bar{x}$, is defined by

$$\bar{x} = \sum_{i=1}^{n} x_i/n, \quad (10.1)$$

where $\sum$ denotes summation, such that $\sum_{i=1}^{n} x_i = x_1 + \cdots + x_n$. The sample standard deviation, denoted as s, is defined as the square root of the sample variance, s^2, which is itself defined by

$$s^2 = \sum_{i=1}^{n} (x_i - \bar{x})^2/(n-1). \quad (10.2)$$

Expression (10.2) can also be calculated as

$$s^2 = \left\{ \sum_{i=1}^{n} x_i^2 - \left(\sum_{i=1}^{n} x_i \right)^2 /n \right\} /(n-1). \quad (10.3)$$

As an example of the calculations, consider the following three measurements of the concentrations of silicon in a fragment of glass ($n = 3$).

x_1	x_2	x_3
27.85	27.83	28.27

Then, from (10.1) with $n = 3$,

$$\bar{x} = (27.85 + 27.83 + 28.27)/3 = \sum_{i=1}^{n} x_i/n = 83.95/3 = 27.983.$$

From (10.3),

$$\begin{aligned} s^2 &= \left\{ \sum_{i=1}^{n} x_i^2 - \left(\sum_{i=1}^{n} x_i \right)^2 /n \right\} /(n-1) \\ &= \{(27.85^2 + 27.83^2 + 28.27^2) - 83.95^2/3\}/2 \\ &= (2349.32 - 83.95^2/3)/2 = 0.062, \end{aligned}$$

and the sample standard deviation is

$$s = \sqrt{(0.062)} = 0.248.$$

Note that the sample mean and standard deviation are quoted to one more decimal place than the original measurements.

10.3.2 Normal distribution

When considering data in the form of counts, the variation in the possible outcomes is represented by a function known as a *probability function*. The variation in measurements, which are continuous, may also be represented mathematically by a function, known as a *probability density function*. Probability functions and probability density functions are both examples of *probability models*.

As an example of a probability model for a continuous measurement, consider the estimation of the quantity of alcohol in blood. From experimental results, it has been determined that there is variation in the measurements, x (in g/kg), provided by a certain procedure. The variation is such that it may be represented by a probability density function which in this case is 'unimodal', symmetric and bell-shaped. The particular function that is used here is the *Normal* or *Gaussian* probability density function (named after the German mathematician Carl Friedrich Gauss, 1777–1855).

Two characteristics of the measurement are required to define the Normal probability density function. These are the mean, or expectation, θ, and the standard deviation, σ. Given these parameters, the Normal probability density function for x, $f(x \mid \theta, \sigma^2)$, is given by

$$f(x \mid \theta, \sigma^2) = \frac{1}{\sqrt{2\pi\sigma^2}} \exp\left\{-\frac{(x-\theta)^2}{2\sigma^2}\right\}. \tag{10.4}$$

The function $f(x \mid \theta, \sigma^2)$ is a function which is symmetric about θ. It takes its maximum value when $x = \theta$, it is defined on the whole real line for $-\infty < x < \infty$ and is always positive. The area underneath the function is 1, from the law of total probability, since x has to lie between $-\infty$ and ∞.

In some countries, if the alcohol level in blood is estimated to be greater than 0.8 g/kg, a person is considered to be under the influence of alcohol. The variability inherent in a measurement, x, of alcohol quantity is known from previous experiments to be such that it is normally distributed about the true value θ with variance, σ^2, of 0.005. The variability in the measurements is taken to be constant over all instances of use of the measuring instrument. Consider a person whose actual quantity, θ, of alcohol in the blood is 0.7 g/kg. This value is taken as the mean of the measurements of the alcohol in the blood; that is, the measuring instrument is unbiased. The probability density function $f(x \mid \theta, \sigma^2)$ for the measurement of the quantity of alcohol in the blood is then obtained from (10.4) with the substitution of 0.7 for θ and 0.005 for σ^2. The function is illustrated in Figure 10.1. Note the labelling of the ordinate as 'probability density'. In particular, it is possible for the probability density function to take values greater than 1.

There is a special case of zero mean ($\theta = 0$) and unit variance ($\sigma^2 = 1$). The Normal probability density function is then

$$f(z \mid 0, 1) = \frac{1}{\sqrt{2\pi}} \exp\left(-\frac{z^2}{2}\right), \tag{10.5}$$

where z is used instead of x to denote the special nature of a normally distributed random variable with zero mean and unit variance. A random variable with this distribution is said to have a standard Normal distribution. The Normal probability density function is so common that it has special notation. If a random variable Z has a standard Normal distribution, it is denoted as

$$Z \sim N(0, 1),$$

where the conditioning on $\theta = 0$ and $\sigma^2 = 1$ on the left-hand side has been omitted for clarity. In general, a normally distributed measurement, say X, with mean θ and variance σ^2, is denoted as

$$\left(X \mid \theta, \sigma^2\right) \sim N\left(\theta, \sigma^2\right).$$

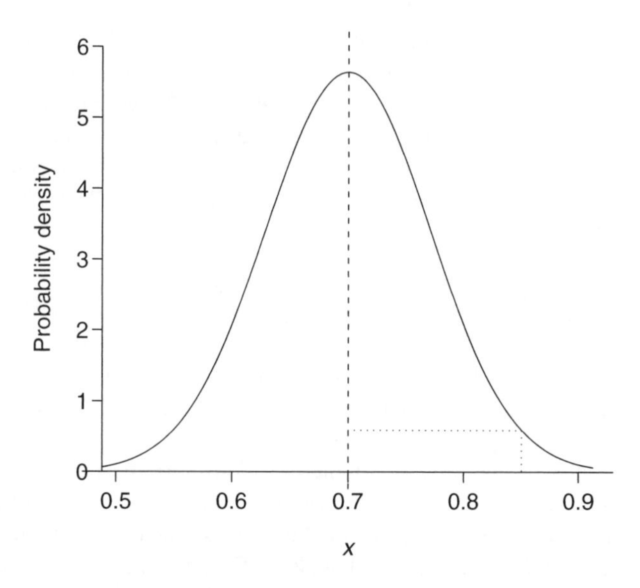

Figure 10.1 Probability density function for a Normal distribution, with mean 0.7 and standard deviation 0.005. The value 0.60 for the probability density function when $x =$ 0.85 is indicated by the dotted lines. [Reproduced by permission of John Wiley & Sons, Ltd.]

The first symbol within parentheses on the right-hand side of the expression conventionally denotes the mean, the second conventionally denotes the variance. As with Z denoted earlier, it is not always necessary for the notation to make explicit the dependence of X on θ and σ^2. The distributional statement may then be denoted as

$$X \sim N(\theta, \sigma^2),$$

and such an abbreviated notation will be used often.

The determination of probabilities associated with normally distributed random variables is made possible by a process known as *standardisation*, whereby a general normally distributed random variable is transformed into one which has a standard Normal distribution. Let

$$Z = (X - \theta)/\sigma.$$

Then, $E(Z) = 0$ and $Var(Z) = 1$ and the random variable Z has a standard Normal distribution. Notice that standardisation requires variability, as represented by σ, to be taken into account. For example, the division by σ ensures that the resulting statistic is dimensionless. Consider the measurements on blood alcohol concentration in units of g/kg. Both the numerator and the denominator of $(X - \theta)/\sigma$ have these units, so the resultant statistic is dimensionless.

Consider the following numerical example of blood alcohol measurements using the parameter values given earlier. Let X be the random variable of measurements of blood alcohol for a particular person, with x denoting the value of a particular measurement. Suppose the true, unknown, level of alcohol in the person's blood is $\theta = 0.7$ g/kg and the standard deviation in the measurement is σ, the square root of 0.005, which equals 0.07 (to two decimal places). Suppose the measurement x of the blood alcohol quantity recorded by the measuring apparatus is 0.85 g/kg, which is over the permitted limit of 0.8 g/kg. Remember that θ is unknown in this case. The variance σ^2 is assumed to be known as it has been estimated from many previous experiments with the measuring apparatus and it is assumed to be a constant, independent of θ. Substitution of $x = 0.85$ g/kg, $\theta = 0.7$ g/kg, and $\sigma^2 = 0.005$ into (10.4) gives

$$f(x) = f(0.85) = \frac{1}{\sqrt{0.01\pi}} \exp\left(-\frac{(0.85 - 0.7)^2}{0.01}\right) = 0.60, \qquad (10.6)$$

see Figure 10.1. In practice, what is of interest is the probability that the true blood alcohol level is greater than 0.8 g/kg, when the instrument provides a measurement of 0.85 g/kg, and the true mean is 0.7 g/kg with a variance of 0.005.

The probability that a normally distributed random variable lies in a certain interval cannot be determined analytically and reference has to be made to tables of probabilities of the standard Normal distribution or to statistical packages. The probability that Z is less than a particular value z, $Pr(Z < z)$, is denoted as $\Phi(z)$. Certain values of z are used commonly in the discussion of significance probabilities, particularly those values for which $1 - \Phi(z)$ is small, and some of these are tabulated in Table 10.1. Corresponding probabilities for absolute values of Z may be deduced from the tables by use of the symmetry of the Normal distribution. By symmetry,

$$\Phi(-z) = Pr(Z < -z) = 1 - Pr(Z < z) = 1 - \Phi(z).$$

Thus,

$$\begin{aligned} Pr(\mid Z \mid < z) &= Pr(-z < Z < z) \\ &= Pr(Z < z) - Pr(Z < -z) \\ &= \Phi(z) - \Phi(-z) \\ &= 2\Phi(z) - 1. \end{aligned}$$

Particular, commonly used, values of z with the corresponding probabilities for the absolute values of z are also given in Table 10.1.

Table 10.1 Values of cumulative distribution function $\Phi(z)$ and its complement $1 - \Phi(z)$ and absolute values for the standard Normal distribution for given values of z.

z	$\Phi(z)$	$1 - \Phi(z)$	$Pr(\mid Z \mid < z)$ $= 2\Phi(z) - 1$	$Pr(\mid Z \mid > z)$
1.6449	0.950	0.050	0.90	0.10
1.9600	0.975	0.025	0.95	0.05
2.3263	0.990	0.010	0.98	0.02
2.5758	0.995	0.005	0.99	0.01

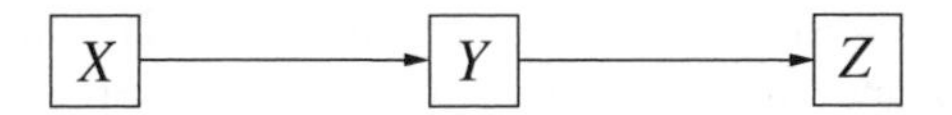

Figure 10.2 A Bayesian network with three continuous nodes, X, Y and Z such that X is a parent node for Y, Y is a child of X and a parent of Z and Z is a child of Y.

10.3.3 Propagation in a continuous Bayesian network

Propagation in discrete networks is *exact*, and as certain variables within the network become instantiated, probabilities can be updated accordingly. Distributions supported by HUGIN 6.5 include the binomial, geometric, negative binomial and Poisson distributions. However, in reality one would often wish to be able to use models in which some or all of the variables in the system of interest were actually measurements and could take values within a continuous range. This continuity is dealt with in HUGIN for several non-Normal distributions through the creation of a discrete categorical variable, splitting the continuous range into distinct subintervals and apportioning probabilities with reference to the underlying probability density function (Cowell et al. 1999). This procedure is used in version 6.5 for the beta, gamma, exponential, Weibull and uniform distributions. The accuracy of such approximations will clearly depend on the number of such subintervals the user deploys.

The Normal distribution is dealt with exactly and the methods used are described in the following text. For now, we shall consider the case in which all nodes within the network are continuous. A generalisation of the univariate Normal distribution to a multivariate Normal distribution lends itself extremely well to the field of continuous networks, since exact updates can be made in the light of evidence as a direct result of the properties of the multivariate normal (MVN) (potentials can be taken as proportional to the corresponding Gaussian probability density functions, then propagation of information does not change this proportionality structure). Indeed, under the normality assumptions, local computation is possible on the directed acyclic graph (DAG), yielding exact updates on associated means and variances under scrutiny. The potential functions that are specified in the corresponding network in the Gaussian case are actually taken as proportional to the corresponding conditional density functions (see following illustration). The MVN has advantages for modelling nodes in a continuous network (Cowell et al. 1999) in that the results are exact. If the normality assumptions appear inappropriate, the modeller may consider transforming the data [*e.g.*, lognormal, Aitken et al. (2005)].

In order to illustrate the propagation within a network containing continuous nodes, the Bayesian network in Figure 10.2 is considered (Cowell et al. 1999). Note that nodes for continuous variables are denoted with rectangles (squares) and nodes for discrete variables are denoted with ellipses (circles). Care has to be taken with the context. In Chapter 2, the flow of computations (2.3) - (2.9) was illustrated by Figure (2.10), where rectangular boxes were inputs and outputs, and oval boxes indicated the computations that were performed.

The three variables X, Y and Z are all random variables. The model is different from the more familiar regression representation where one variable is taken to be known and fixed and the other variable is unknown and random. For example, in an experiment to investigate the relationship between body temperature and time since death, the time since death of various animals may be known exactly and the body temperature is then taken to be a random variable which is dependent on time since death.

The probability density functions are specified through the following conditional distributions $X \sim N(0, 1)$, $(Y|X = x) \sim N(x, 1)$ and $(Z|Y = y) \sim N(y, 1)$. The mean of the distribution of Y, conditional on X, is given by a value x of X and the mean of the distribution of Z, conditional on Y, is given by a value y of Y.

Thus,

$$f(x) = \frac{1}{\sqrt{2\pi}} \exp\left\{-\frac{1}{2}x^2\right\},$$

$$f(y \mid X = x) = f(y \mid x) = \frac{1}{\sqrt{2\pi}} \exp\left\{-\frac{1}{2}(y - x)^2\right\},$$

and

$$f(z \mid Y = y) = f(z \mid y) = \frac{1}{\sqrt{2\pi}} \exp\left\{-\frac{1}{2}(z - y)^2\right\}.$$

In Cowell et al. (1999), results of evidence propagation on this network are given without derivation. In the network in Figure 10.2, two cliques are identified, namely, {X,Y} and {Y, Z}. It is important to know the clique potentials $\phi(x, y)$ and $\phi(y, z)$. The clique potentials are derived as follows, where $f(x)$ denotes the probability density function for variable X, and $f(x, y)$ denotes a bivariate Normal probability density function for variables X, Y. [Further details of multivariate Normal distributions are given in Aitken and Taroni (2004).] For propagation, it is sufficient to know that $f(x, y)$ may be written as $f(y \mid x)f(x)$.

$$\begin{aligned}
\phi(x, y) &\propto f(x, y) \\
&= f(y|x)f(x) \\
&= \frac{1}{\sqrt{2\pi}} \exp\left(-\frac{1}{2}x^2\right) \frac{1}{\sqrt{2\pi}} \exp\left\{-\frac{1}{2}(y - x)^2\right\} \\
&= \frac{1}{\sqrt{2\pi}} \exp\left\{-\frac{1}{2}(y^2 + 2x^2 - 2xy)\right\}.
\end{aligned}$$

Hence,

$$\begin{aligned}
\phi(x, y) &\propto \exp\left\{-\frac{1}{2}\left(y^2 + 2x^2 - 2xy\right)\right\} \\
&\propto \exp\left(-x^2 + xy - \frac{1}{2}y^2\right). \qquad (10.7)
\end{aligned}$$

The clique potential $\phi(y, z)$ may be derived similarly. By definition, the potential $\phi(y, z) \propto f(y, z) = f(z|y)f(y)$. Hence, in order to derive $\phi(y, z)$, it is first necessary to determine $\phi(y) \propto f(y)$. This potential can be obtained through *marginalisation* of the density $\phi(x, y) \propto f(x, y)$ defined here. (The process of *marginalisation* is the process by which, in a bivariate distribution, the distribution of one of the variables is determined. The term arises from the representation of bivariate categorical data in a two-dimensional table. The distribution of one of the variables is obtained by looking in the margins of the table.)

$$\begin{aligned}
\phi(y) &\propto \int_X f(x, y)\, dx \\
&\propto \int \exp\left\{-\frac{1}{2}\left(y^2 + 2x^2 - 2xy\right)\right\} dx \\
&= \exp\left(-\frac{1}{2}y^2\right) \int \exp\left\{-\left(x - \frac{1}{2}y\right)^2 + \frac{1}{4}y^2\right\} dx \\
&\propto \exp\left(-\frac{1}{2}y^2\right) \exp\left(\frac{1}{4}y^2\right) \\
&= \exp\left(-\frac{1}{4}y^2\right).
\end{aligned}$$

Now that the separator potential (*i.e.*, y separates the two cliques) $\phi(y)$ has been derived, the potential $\phi(y, z)$ may be derived.

$$\begin{aligned}
\phi(y, z) &\propto f(y, z) \\
&= f(z|y) f(y) \\
&\propto \frac{1}{\sqrt{2\pi}} \exp\left\{-\frac{1}{4}y^2\right\} \frac{1}{\sqrt{2\pi}} \exp\left\{-\frac{1}{2}(z - y)^2\right\} \\
&\propto \frac{1}{2\pi} \exp\left\{-\frac{1}{2}(1.5y^2 + z^2 - 2yz)\right\}. \qquad (10.8)
\end{aligned}$$

Now, consider the scenario in which the information is obtained that variable $Y = 1.5$. This information may be propagated through the network in order to obtain posterior potentials reflecting the new beliefs for both X and Z. (All updated, posterior potentials are marked by the superscript $*$:for example, ϕ^*.)

Firstly, $\phi(x, y)$ becomes $\phi(x, 1.5)$, which is then the posterior potential on X, $\phi^*(x)$. Then $\phi^*(x) = \phi(x, 1.5) \propto \exp(1.5x - x^2)$. Thus, the posterior potential function on X is proportional to $\exp(1.5x - x^2)$. This potential function is proportional to the corresponding posterior marginal probability density function for X, hence the posterior probability density function for X is of the form

$$f^*(x) = \frac{1}{\sigma\sqrt{2\pi}} \exp\left\{-\frac{1}{2\sigma^2}(x - 0.75)^2\right\},$$

where σ^2 is taken equal to 0.5 to ensure that $f^*(x)$ is indeed a probability density function.

Thus, $(X|Y = 1.5) \sim N(0.75, 0.5)$. It is interesting to note that the prior belief on X was $X \sim N(0, 1)$. Learning that Y has taken the value 1.5, shifts the expectation on the value of X to 0.75 and decreases the variability from 1 to 0.5, as one would expect from the dependence structure described in Figure 10.2.

Belief in the variable Z may be updated by the calculation of the posterior potential function $\phi^*(z)$ in an analogous way such that $\phi^*(z) = \phi(1.5, z) \propto \exp(1.5z - 0.5z^2)$ and

$$f^*(z) = \frac{1}{\sigma\sqrt{2\pi}} \exp\left\{-\frac{1}{2\sigma^2}(z - 1.5)^2\right\},$$

where σ^2 is taken equal to 1 to ensure that $f^*(z)$ is indeed a probability density function.

Thus, $(Z \mid Y = 1.5) \sim N(1.5, 1)$, without decreasing the variability. Therefore, the distributions of X and Z have been updated after obtaining evidence on the value of variable Y.

Alternatively, instead of instantiating the value of the clique separator Y, consider the situation in which the value of variable Z is learnt. Assume that the evidence received is that $Z = 1.5$.

This information has to be passed back through the network, exploiting the dependence structure to update beliefs on the variables X and Y accordingly. The marginal potential for Y may be updated using (10.8); $\phi^*(y) = \phi(y, 1.5) \propto \exp(1.5y - 0.75y^2)$, so

$$f^*(y) = \frac{1}{\sigma\sqrt{2\pi}} \exp\left\{-\frac{1}{2\sigma^2}(y - 1)^2\right\},$$

where σ^2 is taken as equal to $\frac{2}{3}$ to ensure that $f^*(y)$ is indeed a probability density function. Thus, $(Y|Z = 1.5) \sim N(1, \frac{2}{3})$.

The update of beliefs on X is more complex because there is no direct causal link between X and Z. The information about Z is *passed* through to X *via* Y. Variables X and Z are said to be *conditionally independent* given a value for Y. It is of interest to determine the potential function that is proportional to the conditional probability density function $f(x|z)$. Firstly, we require the updated clique potential $\phi^*(x, y)$, which is proportional to $f(x, y|z)$. The following calculations are therefore required (the notation $f(y|z)$ and $\phi^*(y)$ are essentially analogous).

$$\begin{aligned}
\phi^*(x, y) &\propto f(x, y|z) \\
&= f(x|y, z) f(y|z) \\
&= f(x|y) f(y, z) \\
&= \frac{f(x, y)}{f(y)} f(y|z).
\end{aligned}$$

Thus,

$$\phi^*(x, y) \propto \phi(x, y)\frac{\phi^*(y)}{\phi(y)}.$$

Note that this expression is the continuous version of (2.13).

Using this relation, it is possible to calculate the updated clique potential on $\{X, Y\}$, $\phi^*(x, y)$:

$$\begin{aligned}
\phi^*(x, y) &\propto \exp\left\{-\frac{1}{2}(y^2 + 2x^2 - 2xy)\right\} \frac{\exp(1.5y - 0.75y^2)}{\exp(-0.25y^2)} \\
&= \exp(-x^2 + xy - y^2 + 1.5y).
\end{aligned}$$

Then, integration over y provides $\phi^*(x)$:

$$\begin{aligned}
\phi^*(x) &\propto \int \exp(-x^2 + xy - y^2 + 1.5y)\,dy \\
&= \exp(-x^2) \int \exp\left\{-y^2 + (x+1.5)\,y\right\}\,dy \\
&= \exp(-x^2)\exp\left\{\left(\frac{x+1.5}{2}\right)^2\right\} \int \exp\left\{-\left(y - \frac{x+1.5}{2}\right)^2\right\}\,dy \\
&\propto \exp(-x^2)\exp(0.25x^2 + 0.5625 + 0.75x) \\
&\propto \exp(-0.75x^2 + 0.75x) \\
&\propto \exp\left\{-0.75(x-0.5)^2\right\}.
\end{aligned}$$

Therefore, the posterior density function for X is of the form

$$f^*(x) = \frac{1}{\sigma\sqrt{2\pi}} \exp\left\{-\frac{1}{2\sigma^2}(x - \frac{1}{2})^2\right\},$$

where σ^2 is taken to be equal to $\frac{2}{3}$ to ensure that $f^*(x)$ is indeed a density (proportional to $\phi^*(x)$). Thus, $(X|Z = 1.5) \sim N(\frac{1}{2}, \frac{2}{3})$.

Consider an example from Aitken and Taroni (2004, pp. 166–170) of the propagation of evidence in a continuous network with two nodes, as illustrated in Figure 10.3. The example concerns a drunken-driving case. The left-hand node θ represents the true level of alcohol in the blood of someone suspected of driving while under the influence of alcohol. In the Bayesian paradigm, the prior distribution for θ is taken to be a Normal distribution. Denote the mean of this distribution by ν and the variance by τ^2. The right-hand node X in the network represents the measured level of alcohol in the person suspected of driving while under the influence of alcohol. The arrow pointing from θ to X indicates that the measured level of alcohol is dependent on the true level of alcohol. It is assumed that the measured level of alcohol has a Normal distribution and that the measuring device is unbiased in that the expected value of X is θ, the true level of alcohol in the blood, and the characteristic represented by the left-hand node. The arrow in Figure 10.3 illustrates this dependency of X on θ. There is uncertainty in the measurement, X, and this is represented by the variance of X, denoted as σ^2.

The conditional distribution of $(X \mid \theta)$ is $N(\theta, \sigma^2)$. The unconditional distribution of X is $N(\theta, \sigma^2 + \tau^2)$. It can be shown [*e.g.*, Lee (2004)] that the posterior distribution of θ, given a value x for X, and given σ^2, ν and τ^2 is

$$\left(\theta \mid x, \sigma^2, \nu, \tau^2\right) \sim N\left(\theta_1, \tau_1^2\right), \tag{10.9}$$

Figure 10.3 A Bayesian network with two continuous nodes, θ and X where θ is the parent of X.

Table 10.2 Suggested values in g/kg for propagation in a two-node network of the prior mean ν, the prior variance τ^2, the variance σ^2 of the observation and the observation x.

ν	τ^2	σ^2	x
0.7	0.01	0.005	0.85

where

$$\theta_1 = \frac{\sigma^2\nu + \tau^2 x}{\sigma^2 + \tau^2} \tag{10.10}$$

and

$$\left(\tau_1^2\right)^{-1} = \left(\sigma^2\right)^{-1} + \left(\tau^2\right)^{-1} \tag{10.11}$$

or, equivalently,

$$\tau_1^2 = \frac{\sigma^2\tau^2}{\sigma^2 + \tau^2}. \tag{10.12}$$

The posterior mean, θ_1, is a weighted average of the prior mean ν and the observation x, where the weights are the variance of the observation x and the variance of the prior mean ν, respectively, such that the component (observation or prior mean) which has the smaller variance has the greater contribution to the posterior mean.

These results may be verified using HUGIN. Consider the values for X, σ^2, ν and τ^2 as given in Table 10.2.

From (10.10) and (10.12), it can be verified that the posterior mean θ_1 for the distribution of θ is 0.80 g/kg, the posterior variance τ_1^2 is 0.0033 and the posterior standard deviation τ_1 is 0.0574.

Figure 10.4 provides a schematic illustration of the prior distributions for the true and measured blood alcohol levels, and Figure 10.5 illustrates the posterior distribution for the true blood alcohol level given a measured level of 0.85 g/kg. Thus, given a reading of 0.85 g/kg, the probability that the true blood alcohol level is greater than 0.80 g/kg is $1 - \Phi(\frac{0.80-0.85}{0.0574}) = \Phi(0.87) \simeq 0.81$.

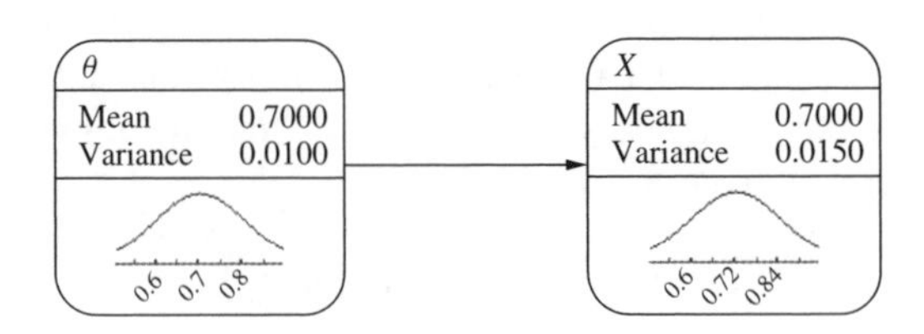

Figure 10.4 Prior distributions of the true value θ and the measured value X of blood alcohol in g/kg, where the unconditional mean of X is θ and the unconditional variance of X is $\sigma^2 + \tau^2$.

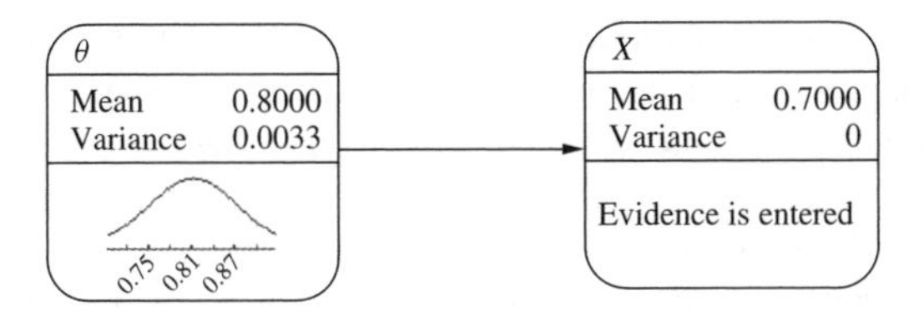

Figure 10.5 Posterior distribution of the true value θ of blood alcohol given a measured value X of blood alcohol of 0.85 g/kg.

10.3.4 Propagation in mixed networks

When attempting to construct graphs to model a real-life system, it is often the case that both discrete and continuous variables will be involved within the system. Hence, it is important to consider the case of *mixed graphs*, in which both discrete and continuous nodes are of interest. Propagation through a mixed network is very similar in form to the propagation examples already covered. The *junction tree* is again formed and subsequent efficient propagation can be made through this junction tree. However, there is one very important restriction in the modelling conditions under which a mixed network can be used to propagate information. Exact mixed node propagation can only be implemented through the junction tree if the cliques of the DAG follow the strong running intersection property (Cowell et al. 1999). In order to understand this property, we must first consider the distinct difference in structure imposed by the mixed network. Unlike the cases of an all-continuous or an all-discrete network, the direction of propagation in the mixed network will clearly influence the form that this propagation takes. In other words, propagation passing information from a discrete node to a continuous node is clearly a different proposition to the reverse scenario. It is not possible to have, in HUGIN for example, a discrete child of a continuous parent. There is a natural specification of conditional distributions for the continuous node, given each of the discrete values the parent(s) could take, but the opposite scenario is not as clearly defined. The logistic distribution is a possible way of incorporating discrete nodes as children of continuous nodes into the network, but subsequent propagation is then considerably more complex.

10.3.5 Example of mixed network

Consider the following network which involves four nodes, three continuous and one discrete.

The network is a general one for trace evidence. The purpose of the network is to derive the distributions for the numerator and denominator in the likelihood ratio. Given control measurements x and recovered measurements y, prosecution and defence propositions H_p and H_d, and background information I, it can be shown that the likelihood ratio, V, is given by (Aitken and Taroni 2004):

$$V = \frac{f(y \mid x, H_p, I)}{f(y \mid H_d, I)}.$$

For the numerator, the instantiation of the nodes H (source) and X (the continuous measurement of some characteristic from a control item) gives the distribution for Y (the continuous measurement of some characteristic from a recovered item) by propagation through the network. This density function is $f(y \mid x, H_p, I)$. For the denominator, the instantiation of the node H to be H_d (alternative source) gives the distribution for Y (the continuous measurement of some characteristic from a recovered item) by propagation through the network, with density function $f(y \mid H_d, I)$. The general distributional results are given in Table 10.3.

The likelihood ratio, V, can then be determined by substituting the observed value for Y into the two density functions and determining the ratio.

An artificial example is given here to demonstrate the relationship between the distributional results and the output from HUGIN. Values of the parameters for this example are given in Table 10.4.

The likelihood ratio for the value $Y = 1.5$ is then given by the ratio of the Normal density functions, $N(1, 3.2)$ and $N(2, 5)$ evaluated at the point 1.5, from the results given in Table 10.5. For the numerator, $f(1.5 \mid 1, 3.2) = 0.2144$. For the denominator, $f(1.5 \mid 2, 5) = 0.1740$. The likelihood ratio is

$$0.2144/0.1740 = 1.2322.$$

Table 10.3 Definition of variables for a network to describe analysis of trace evidence. Node H is discrete and binary, representing the appropriate source-level proposition with states H_p and H_d (referring to the prosecution and defence propositions, respectively). Y, X and θ are continuous nodes representing, respectively, measurements made on recovered fragments, measurements made on control fragments and the background data.

Node	Child of	Parent of	Distribution
H :	–	Y	$Pr(H_p), Pr(H_d)$
θ :	–	X, Y	$N(\nu, \tau^2)$
X :	θ	Y	$(X \mid \theta, \sigma^2) \sim N(\theta, \sigma^2)$
Y :	H, X, θ	–	$(Y \mid X = x) \sim N(x, \sigma^2 + \frac{\sigma^2\tau^2}{\sigma^2+\tau^2})$ (H_p true)
			$(Y \mid \theta) \sim N(\theta, \sigma^2 + \tau^2)$ (H_d true)

Table 10.4 Illustrative values for the parameters in Table 10.3.

σ^2	τ^2	ν	X	Y	H_p	H_d
2	3	2	1	1.5	0.01	0.99

Table 10.5 Distributional results for the four variables from Table 10.3 using the parameter values in Table 10.4.

Node	Child of	Parent of	Distribution
H :	–	Y	$Pr(H_p) = 0.01$, $Pr(H_d) = 0.99$)
θ :	–	X, Y	$N(2, 3)$
X :	θ	Y	$(X \mid \theta, \sigma^2) \sim N(\theta, 2)$
Y :	H, X, θ	–	$(Y \mid X = 1) \sim N(1, 3.2)$ (H_p true) $(Y \mid \theta) \sim N(\theta, 5)$ (H_d true)

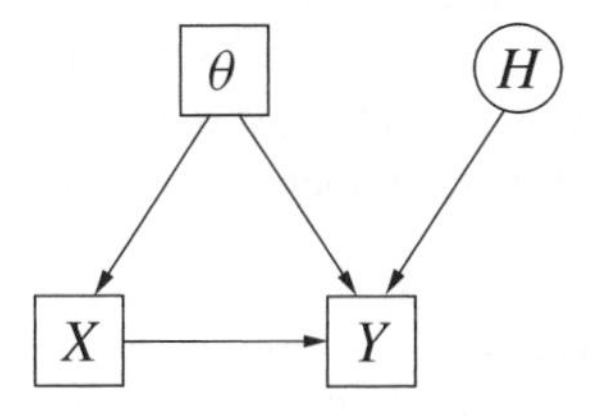

Figure 10.6 A mixed network of three continuous nodes X, Y and θ and one discrete node H. Node H represents the appropriate source-level proposition with states H_p and H_d (referring to the prosecution and defence propositions, respectively). Y, X and θ represent, respectively, measurements made on recovered fragments, measurements made on control fragments and the background data.

In Figure 10.6, the distributional result $Y \sim N(1, 3.2)$ for the numerator is obtained by instantiating nodes H_p to be true and X to equal 1 (*see* Figure 10.7(i)). The correct answer for the distribution of Y under H_d (*i.e.*, the denominator) is obtained by leaving X uninstantiated (*see* Figure 10.7(ii)).

10.4 Use of a continuous distribution which is not normal

The only continuous distribution which can be implemented directly in HUGIN is the Normal distribution. However, there are very many continuous distributions which are not Normal. HUGIN supports several of these. These are modelled in HUGIN through discretisation of the range of the continuous quantity into a finite set of intervals. An example is given here of the use of a beta distribution in conjunction with a binomial distribution. The derived model is used in a sample size problem, described in Aitken (1999).

The problem is to determine the size of a sample of items to be taken from a consignment of items in order to make an inference about the proportion of items in the consignment which have a particular characteristic. The application is to a limited case

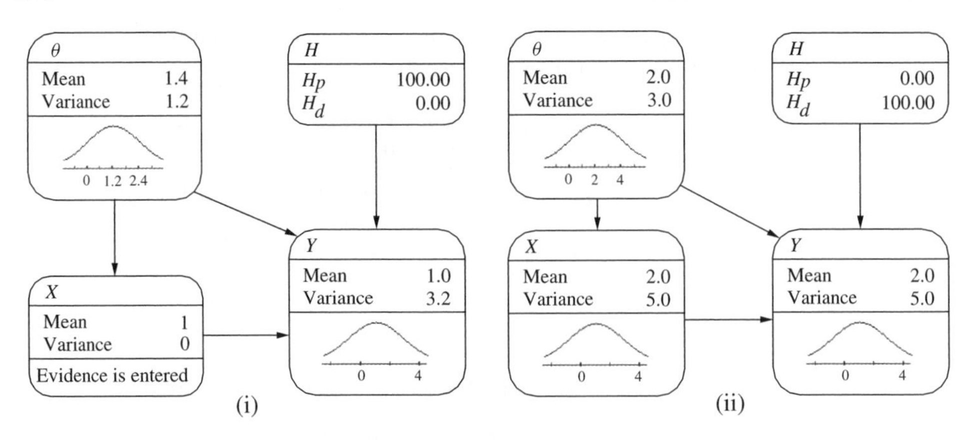

Figure 10.7 A mixed network of three continuous nodes X, Y and θ and one discrete node H. Node H represents the appropriate source-level proposition with states H_p and H_d (referring to the prosecution and defence propositions, respectively). Y, X and θ represent, respectively, measurements made on recovered fragments, measurements made on control fragments and the background data. Evaluation of the probability distribution for the variable Y given (i) a control measurement X of 1 and H_p, and (ii) given H_d only.

in which the items in the consignment are homogeneous in all respects except one. The characteristic in which the items are not homogeneous is a binary characteristic, that is one with only two responses, say positive and negative. It is of interest to know the proportion of items in the consignment which are positive. A sample is taken from the consignment and the numbers of positive and negative items in the sample are used to make an inference about the consignment. Three examples will help illustrate the problem. First, consider a consignment of white tablets, homogeneous with respect to colour, texture, shape and size; the requirement is an inference about the proportion of tablets which is illicit. Second, consider a consignment of CDs; the requirement is an inference about the proportion which is pirated. Third, consider a file of computer images; the requirement is an inference about the proportion which is pornographic.

A statistical model may be determined for these three examples. The structure is the same in all the examples as long as reasonable assumptions are made. First, each item in the consignment is assumed to be homogeneous in all respects except the characteristic of interest. The characteristic of interest can take one and only one of two possible values (*e.g.*, illicit or not, pirated or not, pornographic or not). The probabilities that an item, when inspected, may have each of the two characteristics are assumed constant over all items in the consignment; denote these as θ and $1-\theta$. These probabilities are independent from the inspection of one item to another.

Let X be the number of positive items selected in a sample of size n from a consignment sufficiently large that sampling may be considered to be with replacement. The model for this situation is a binomial model. The distribution of X is the binomial distribution with associated probability function

$$Pr(X = x \mid n, \theta) = \binom{n}{x} \theta^x (1-\theta)^{n-x}, \quad 0 < \theta < 1; \quad x = 0, \ldots, n.$$

The conjugate prior distribution for θ is a beta distribution which, with parameters α and β, has a probability density function

$$f(\theta \mid \alpha, \beta) = \frac{\Gamma(\alpha+\beta)}{\Gamma(\alpha)\Gamma(\beta)} \theta^{\alpha-1}(1-\theta)^{\beta-1}, \quad 0 < \theta < 1; \quad \alpha, \beta > 0,$$

where Γ is the gamma function, $\Gamma(z) = \int_0^\infty t^{z-1}e^{-t}dt$, with $\Gamma(z+1) = z\Gamma(z)$ for integer $z > 0$, $\Gamma(1) = 1$, $\Gamma(1/2) = \sqrt{\pi}$.

The posterior density function for θ is then

$$f(\theta \mid x+\alpha, n-x+\beta) = \frac{\Gamma(n+\alpha+\beta)}{\Gamma(x+\alpha)\Gamma(n-x+\beta)} \theta^{x+\alpha-1}(1-\theta)^{n-x+\beta-1}.$$

From this result it is possible to determine the probability that θ is greater than a certain proportion, say θ_0, from the equation

$$Pr(\theta > \theta_0 \mid n, x, \alpha, \beta) = \int_{\theta_0}^{1} f(\theta \mid x+\alpha, n-x+\beta)\, d\theta.$$

The application of this result to the determination of a sample size is described in detail in Aitken (1999) with a summary in Aitken and Taroni (2004).

A simple illustration, from the example of sampling from a consignment of white tablets, is given here of the use of the method. A criterion specified for sample size determination is that the size of the sample has to be chosen such that if all tablets inspected in the sample are illicit (positive), then one can be 95% certain that at least 50% of the consignment is illicit. Thus $x = n, n - x = 0$. It is also specified that the prior distribution is uniform, that is, $\alpha = \beta = 1$ and $f(\theta \mid 1, 1) = 1$. Then the sample size n is the smallest integer value of n such that

$$Pr(\theta > 0.5 \mid n+1, 1) = 0.95,$$

that is,

$$\int_{0.5}^{1} f(\theta \mid n+1, 1)\, d\theta = 0.95.$$

Use of a suitable statistical package (*e.g.*, R (2003)) shows that the solution of this equation is $n = 4$. Results for different levels of certainty associated with different proportions, different levels of the numbers of items that are negative and different values of α and β are available; see, for example, Aitken (1999) and Aitken and Taroni (2004). Guidelines have also been issued by the ENFSI (European Network of Forensic Science Institutes) Drugs Working Group (2004).

The sample size problem may also be approached using Bayesian networks and HUGIN. Figure 10.8 gives an example. There is one parent node 'Prop'. This represents the

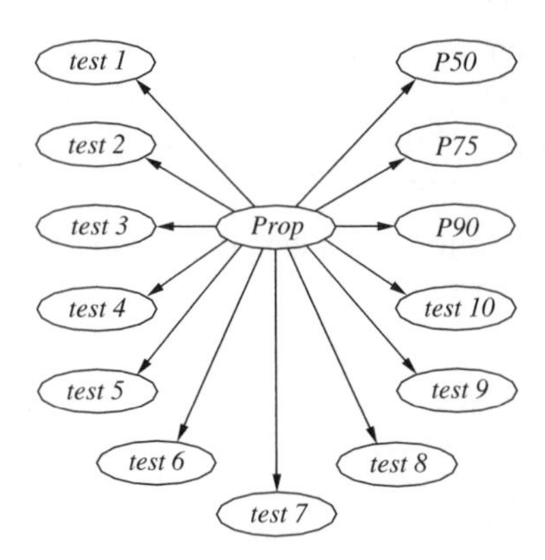

Figure 10.8 Bayesian network for the sampling problem. The node *Prop* represents the proportion θ of the consignment which is positive. *Prop* covers discrete intervals of size 0.05 over the range (0,1). The node 'test n' represents the outcomes of the inspection of the n-th item, which may either be 'positive' or 'negative'. The nodes P50, P75 and P90 with states 'yes' and 'no' are summary nodes and provide cumulative probabilities for the proportion (node *Prop*) being greater than 0.5, 0.75 and 0.9, respectively.

proportion θ of the consignment that is positive. The proportion is unknown and assigned a beta distribution. The beta distribution is not modelled exactly but interval probabilities can be determined by HUGIN using a special tool known as 'Expression builder' (*see* also Section 5.12.4). In the example given later, 20 intervals of equal width over the range (0,1) are specified and each is assigned a probability of 0.05, corresponding to a uniform prior distribution. In HUGIN language, this is obtained by specifying the expression `Beta(1,1)`.

In the construction of the network, the selection of the sample is broken down into the inspection of each individual item. Ten items are specified in Figure 10.8. The node 'test n' is a binary node with states 'positive' and 'negative', denoting the outcome of the inspection of the n-th item. In HUGIN, the conditional probability table of the nodes 'test n' can be completed, for example, through use of the expression `Distribution(Prop,1-Prop)`. This reflects the idea that the probability of a draw being positive is given by the proportion of positives in the consignment.

It is of interest to determine the probability that the true proportion is greater than a specified amount. In the example just described, the result was quoted that if four items were sampled and all found to be positive then one could be 95% certain that at least 50% of the consignment was positive. The three binary nodes 'P50', 'P75' and'P90' with states 'yes' and 'no' provide values for the probabilities that the true proportion of positive items in the consignment is greater than the three indicated percentages, 50%, 75% and 90%. Note that the conditional probability tables of the latter three nodes are specified – in HUGIN – by use of an expression of the following kind ($P = 0.5, 0.75, 0.9$):

```
Distribution(if(Prop>P,1,0),if(Prop>P,0,1)).
```

This expression assigns a probability of one to the state 'yes' if the current state of the node *Prop* is greater than P; otherwise, a zero probability is assigned. The probability assigned to 'no' is the complement of the probability assigned to 'yes'.

Agreement with the quoted result mentioned of 4 is obtained by instantiating the four nodes 'test 1', 'test 2', 'test 3' and 'test 4' to be all positive and leaving nodes 'test 5' to 'test 10' uninstantiated. The resultant probabilities for 'P50', 'P75' and 'P90' are 0.9689, 0.7631 and 0.4098, in agreement with data in Aitken (1999). Thus, one can be 96.89% certain that the true proportion is greater than 50%, 76.31% certain that the true proportion is greater than 75% and 40.98% certain that the true proportion is greater than 90%. Notice that the network also updates our expectation about the outcomes of future trials, by revising the probability that the next trial is positive to be the proportion of positive units observed so far, with the addition of the prior information of one positive unit out of two. The procedure used for updating is known as Laplace's law of succession (Cox and Hinkley 1974). The expectation of a beta distributed random variable with parameters α and β is $\alpha/(\alpha+\beta)$. The parameters for the posterior beta distribution, in which $\alpha = \beta = 1$ and the number of items selected equals n and the number of positives (x) is also n, are $n+\alpha = n+1$ and $\beta = 1$. The expectation is thus $(n+1)/(n+2)$. When $n = 4$, the probability that the next item selected is positive is then $(n+1)/(n+2) = 5/6 = 0.83$. This result may be observed in Figure 10.9 where the uninstantiated nodes 'test 5' to 'test 10' have updated probabilities of 0.83 for a positive outcome at the next selection (which may be from any one of the six nodes). The procedure can be considered analogous to a situation in which balls are drawn from a bag that contains only black or white balls. An inference is to be made about the proportion of black (and, by default, the proportion of white) balls in the bag. The total number of balls in the bag is assumed to be so large that the inference from sampling without replacement may be assumed to be the same as the inference from sampling with replacement.

10.5 Appendix

10.5.1 Conditional expectation and variance

As explained in the body of the text, it is possible to have a situation where one random variable, Y for example, is dependent on another random variable, X for example. This situation is different from the more familiar linear regression model in which there is one random variable which is dependent on a fixed variable whose values are known exactly and may be fixed by an experimenter.

Consider that Y is dependent on X. It is of interest to know the unconditional expectation and variance of Y; that is, unconditional on the value of the random variable X. The unconditional expectation and variance are given by the following formulae:

$$E(Y) = E(E(Y \mid X)) \tag{10.13}$$

and

$$Var(Y) = E(Var(Y \mid X)) + Var(E(Y \mid X)). \tag{10.14}$$

Examples will be given to illustrate these results. Proofs are not given here but may be found in books on probability theory, for example, O'Hagan (1988). Note that (10.13) has earlier been used in Section 7.4.4.3.

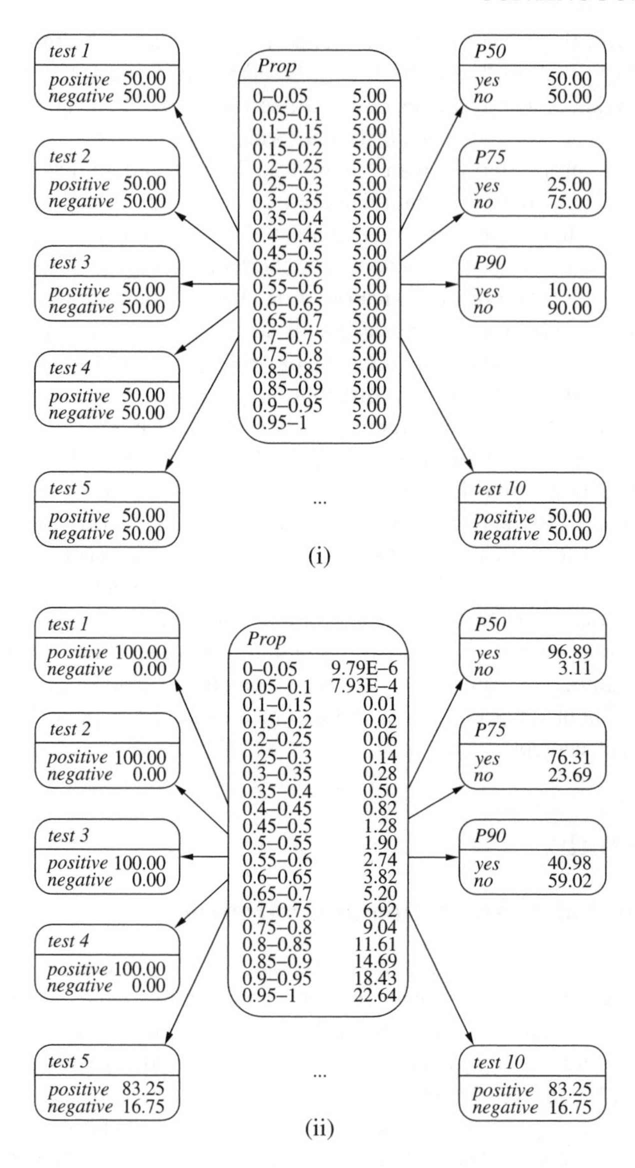

Figure 10.9 Bayesian network for the sampling problem as given by Figure 10.8: (i) initial state, and (ii) state after instantiation of the nodes ‘test 1’ to ‘test 4’.

10.5.2 Bayesian network for three serially connected continuous variables

Consider the Bayesian network from Figure 10.2. The marginal distributions of Y and Z may be calculated. For example,

$$f(y) = \int f(y \mid x) f(x)\, dx$$

$$= \frac{1}{2\pi} \int \exp\left\{-\frac{1}{2}(y-x)^2 - \frac{1}{2}x^2\right\} dx$$

$$= \frac{1}{2\pi} \exp\left(-\frac{1}{2}y^2\right) \int \exp\left\{-\frac{1}{2}(2xy + 2x^2)\right\} dx$$

$$= \frac{1}{2\pi} \exp\left(-\frac{1}{4}y^2\right) \int \exp\left\{-(x - \frac{1}{2}y^2)\right\} dx$$

$$= \frac{1}{\sqrt{2}\sqrt{2\pi}} \exp\left\{-\frac{1}{4}y^2\right\}.$$

This is the probability density function of a normally distributed random variable with expectation 0 and variance 2. This may be checked by replacing μ and σ^2 in (10.4) with 0 and 2, respectively.

These values of 0 and 2 may be obtained from (10.13) and (10.14). The distribution of $Y \mid X$ is $N(x, 1)$ and the distribution of X is $N(0, 1)$. Thus, $E(Y \mid X) = X$, and the quoted result that $E(Y) = E(E(Y \mid X))$ gives $E(Y) = E(X) = 0$. Also, $Var(Y \mid X) = 1$ so $E(Var(Y \mid X)) = 1$, and $E(Y \mid X) = X$ so $Var(E(Y \mid X)) = Var(X) = 1$. Thus, $Var(Y) = E(Var(Y \mid X)) + Var(E(Y \mid X)) = 2$. The distributional result is obtained from the result that if $(Y \mid X)$ and X are normally distributed then Y is normally distributed.

A generalisation of the this example is that if $X \sim N(\mu, \sigma^2)$ and $(Y \mid X = x) \sim N(x, \tau^2)$, then

$$Y \sim N(\mu, \sigma^2 + \tau^2).$$

A further generalisation is that if $X \sim N(\mu, \sigma^2)$ and $(Y \mid X = x) \sim N(\alpha x + \beta, \tau^2)$, then

$$Y \sim N(\alpha\mu + \beta, \alpha^2\sigma^2 + \tau^2).$$

These results may be checked using HUGIN.

Consider again the Bayesian network in Figure 10.2. Set

$$X \sim N(\mu, \sigma^2),$$

$$(Y \mid X = x) \sim N(\alpha x + \beta, \tau^2),$$

$$(Z \mid Y = y) \sim N(\gamma y + \delta, \eta^2).$$

Then, repeated use of these results gives the marginal distributions of Y and Z as

$$Y \sim N(\alpha\mu + \beta, \alpha^2\sigma^2 + \tau^2), \tag{10.15}$$

$$Z \sim N(\gamma(\alpha\mu + \beta) + \delta, \gamma^2(\alpha^2\sigma^2 + \tau^2) + \eta^2). \tag{10.16}$$

Some numbers may be inserted and the marginal results checked from these general results with those provided by an appropriate software tool (here, HUGIN is used).

Label the nodes in HUGIN, X, Y and Z as in Figure 10.2. Double click on the node labelled X and set the mean μ equal to 1 and the variance σ^2 equal to 2. Double click on the node labelled Y. Three cells to be filled in appear, labelled *Mean, Variance* and X, as the structure of the network is that Y is dependent on X. In the cell labelled *Mean* enter the value for β, 4 in this example. In the cell labelled *Variance* enter the value for τ^2, 5 in this example. In the cell labelled *X* enter the value for α, 3 in this example. Double click on the node labelled Z. Three cells to be filled in appear, labelled *Mean, Variance* and *Y*, as the structure of the network is that Z is dependent on Y. In the cell labelled *Mean* enter the value for δ, 2 in this example. In the cell labelled *Variance* enter the value for η^2, 7 in this example. In the cell labelled *Y* enter the value for γ, 8 in this example. Thus, the numerical relationships are

$$X \sim N(1, 2),$$
$$(Y \mid X = x) \sim N(3x + 4, 5),$$
$$(Z \mid Y = y) \sim N(8y + 2, 7).$$

Substitution of these values for $\alpha, \beta, \tau, \mu, \sigma, \gamma, \delta$ and η in (10.15) and (10.16) gives the following results

$$Y \sim N(7, 23),$$
$$Z \sim N(58, 1479).$$

10.5.3 Bayesian network for a continuous variable with a binary parent

Consider the Bayesian network given in Figure 10.10 where D is a binary variable which takes the value 0 with probability p_0 and the value 1 with probability p_1 such that $p_0 + p_1 = 1$, and

$$(Y \mid D = 0) \sim N(\mu_0, \sigma_0^2),$$
$$(Y \mid D = 1) \sim N(\mu_1, \sigma_1^2).$$

Then,

$$E(Y) = E(E(Y \mid D)) = p_0\mu_0 + p_1\mu_1.$$

Also,

$$Var(Y) = Var(E(Y \mid D)) + E(Var(Y \mid D)).$$

Now,

$$Var(Y \mid D = 0) = \sigma_0^2,$$
$$Var(Y \mid D = 1) = \sigma_1^2$$

so

$$E(Var(Y \mid D)) = p_0\sigma^2 + p_1\sigma^2.$$

Figure 10.10 A Bayesian network with one discrete (D) and one continuous (Y) node.

The variable $E(Y \mid D)$ may be considered as a binary variable, say T, which takes the value μ_0 with probability p_0 and the value μ_1 with probability $p_1 = (1 - p_0)$. Then, $E(T) = p_0\mu_0 + p_1\mu_1$ and

$$\begin{aligned} Var(T) &= E(T^2) - \{E(T)\}^2 \\ &= \mu_0^2 p_0 + \mu_1^2(1 - p_0) - (p_0\mu_0 + p_1\mu_1)^2 \\ &= p_0(1 - p_0)(\mu_0 - \mu_1)^2, \end{aligned}$$

and

$$\begin{aligned} Var(Y) &= Var(E(Y \mid D)) + E(Var(Y \mid D)) \\ &= p_0(1 - p_0)(\mu_0 - \mu_1)^2 + p_0\sigma^2 + p_1\sigma^2. \end{aligned}$$

10.5.4 Bayesian network for a continuous variable with a continuous parent and a binary parent, unmarried

Consider the Bayesian network in Figure 10.11 such that Y is a child of X and of D where X and Y are normally distributed random variables and D is a binary variable. Note that it is not permissible for a continuous node in a Bayesian network to have a discrete node as a child. The probabilistic structure is

$$X \sim N(\mu, \sigma^2),$$

$$\begin{aligned} Pr(D = 0) &= p_0, \\ Pr(D = 1) &= p_1, \\ (Y \mid X = x, D = 0) &\sim N(\alpha_0 x + \beta_0, \tau_0^2), \\ (Y \mid X = x, D = 1) &\sim N(\alpha_1 x + \beta_1, \tau_1^2), \end{aligned}$$

where $p_0 + p_1 = 1$.

From these distributional forms, the mean and the expectation of Y may be derived.

First, consider the expectation. When $D = 0$, $E(E(Y \mid X)) = \alpha_0\mu + \beta_0$ and when $D = 1$, $E(E(Y \mid X)) = \alpha_1\mu + \beta_1$. Thus,

$$E(E(E(Y \mid X, D))) = (\alpha_0\mu + \beta_0)p_0 + (\alpha_1\mu + \beta_1)p_1.$$

The results of Section 10.5.3 may be used, with μ_0 replaced by $\alpha_0 x + \beta_0$ and μ_1 replaced by $\alpha_1 x + \beta_1$, to derive the expression for the variance.

$$\begin{aligned} E(Y \mid X = x) &= p_0(\alpha_0 x + \beta_0) + p_1(\alpha_1 x + \beta_1) \\ &= (\alpha_0 p_0 + \alpha_1 p_1)x + (p_0\beta_0 + p_1\beta_1) \end{aligned}$$

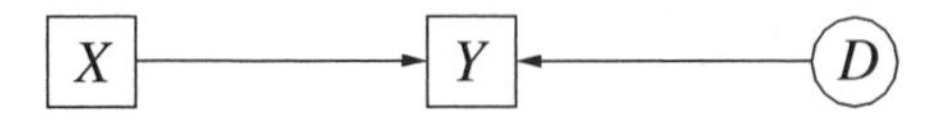

Figure 10.11 Bayesian network for a continuous variable, Y with a continuous parent, X, and a binary parent, D, unmarried.

$$Var(Y \mid X = x) = \tau^2 p_0 + \tau^2 p_1 + p_0 p_1 \{(\alpha_0 - \alpha_1)x + (\beta_0 - \beta_1)\}^2;$$
$$Var(E(Y \mid X)) = (\alpha_0 p_0 + \alpha_1 p_1)^2 \sigma^2,$$
$$E(Var(Y \mid X)) = \tau^2 p_0 + \tau^2 p_1 + p_0 p_1 E\{(\alpha_0 - \alpha_1)X + (\beta_0 - \beta_1)\}^2.$$

Then,

$$\begin{aligned} & E\{(\alpha_0 - \alpha_1)X + (\beta_0 - \beta_1)\}^2 \\ & = E\{(\alpha_0 - \alpha_1)^2 X^2 + (\beta_0 - \beta_1)^2 + 2(\alpha_0 - \alpha_1)(\beta_0 - \beta_1)X\} \\ & = (\alpha_0 - \alpha_1)^2(\sigma^2 + \mu^2) + (\beta_0 - \beta_1)^2 + 2(\alpha_0 - \alpha_1)(\beta_0 - \beta_1)\mu. \end{aligned}$$

Finally,

$$\begin{aligned} & Var(Y) \\ & = Var(E(Y \mid X)) + E(Var(Y \mid X)) \\ & = p_0 p_1[(\alpha_0 - \alpha_1)^2(\sigma^2 + \mu^2) + (\beta_0 - \beta_1)^2 + 2(\alpha_0 - \alpha_1)(\beta_0 - \beta_1)\mu] \\ & \quad + (\alpha_0 p_0 + \alpha_1 p_1)^2 \sigma^2 + \tau^2 p_0 + \tau^2 p_1. \end{aligned}$$

The following numerical example may be used to check the results given by these formulae with those given by HUGIN.

D:	$D = 0$	$D = 1$
Mean:	$\beta_0 = 2$	$\beta_1 = 5$
Variance:	$\tau_0^2 = 3$	$\tau_1^2 = 6$
X:	$\alpha_0 = 1$	$\alpha_1 = 4$

The other values used are $\mu = 8, \sigma^2 = 10, p_0 = 0.2, p_1 = 0.8$. Then

$$E(Y) = (8 + 2)0.2 + (32 + 5)0.8 = 31.6,$$

and

$$\begin{aligned} Var(Y) & = 0.16[3^2(10 + 64) + 3^2 + 2(-3)(-3)8] \\ & \quad + (0.2 + 3.2)^2 10 + (3 \times 0.2) + (6 \times 0.8) \\ & = 252.04. \end{aligned}$$

11

Further applications

Bayesian networks are a widely applicable formalism for supporting reasoning when information is incomplete. A particular instance of reasoning under uncertainty, not discussed so far in this book, is one where scientific evidence is used not to describe a link to a potential source but to infer characteristics of that source. Here, the primary aim is to assist criminal investigators in defining traits of, for example, the perpetrator(s) of a crime. Such analyses, also known as *offender profiling*, are presented in Section 11.1.

Evidence forms the basis from which probabilistic arguments are constructed to propositions of interest. Bayesian networks provide a coherent environment in which beliefs about target propositions can be re-evaluated in the light of newly acquired evidence. This constitutes a fundamental prerequisite for decision making under uncertainty. Bayesian networks can be extended to incorporate the basic ingredients necessary to perform Bayesian decision analysis, that is, decision and utility nodes, representing, broadly speaking, actions available to the scientist and values for possible consequences of these actions, respectively. The relevance of these extensions for forensic science is discussed in Section 11.2 in terms of an example involving DNA evidence.

11.1 Offender profiling

Offender profiling, or specific case analysis, is the name given to the procedure by which characteristics of a crime are used to predict characteristics of an offender. The prediction is uncertain; it is probabilistic. Conditional probabilities are associated with certain characteristics of the offender, conditional on certain characteristics of the crime and its scene. It is hoped that these probabilistic predictions will assist the investigators in their prioritisation of resources in that the predictions will suggest directions in which the investigators can guide the investigation so as to optimise the probability of identifying the offender.

The characteristics of the crime that can be considered may include characteristics of the victim and of the crime and its scene and observations by the victim of the offender. Characteristics of the victim that may be considered include age, sex and occupation. The victim may give an age estimate of the offender. Characteristics of the crime and its scene

Bayesian Networks and Probabilistic Inference in Forensic Science F. Taroni, C. Aitken, P. Garbolino and A. Biedermann

may include the cause of death, the location, the time of day, the place at which the victim was last seen, and the nature of the assault. The characteristics of the offender that may be predicted include age, marital status, the relationship to the victim, and whether there are previous convictions or not. It may not be the case that sex is a characteristic that is needed to be predicted, as many of the cases in which profiling is used are sexual in nature and for the vast majority of these the offender is male.

Provision of probabilistic predictions and the determination of the characteristics to be used require a data set of previous solved cases. The data set can then be used to construct probabilistic models. These can be tested on the original data set. A criterion is needed in order that the best model may be chosen, for example, the model that has the lowest number of incorrect answers. Care has to be taken with the bias that is introduced through testing the model on the same data as that used to derive the model. One approach, known as *cross-validation*, by which this bias is reduced is to omit each member of the database in turn, fit the model on the remainder of the database and test on the omitted member. This process is repeated for each member of the database. The model chosen is the one that performs best under the chosen criterion. The values of the parameters of the model are then determined by fitting the model to the full database. The model is then available for use in future unsolved cases. It may, of course, also be used on the previous unsolved cases in the hope of shedding new light on them.

The use of Bayesian networks in offender profiling is illustrated here with a single example, but is described in greater detail in Aitken et al. (1996a) and Aitken et al. (1996b). The example is that of child murders with a sexual connotation. The database was compiled for all such murders in Great Britain from 1960 to 1991. Cases continue to be added to the database but the details here relate to that subset. The data under consideration were restricted to those cases in which there was a single offender and a single victim.

A network consisting of seven nodes was constructed in three stages in discussion with a senior detective. Such staged approaches for eliciting knowledge from domain experts have also been reported, for instance, in medical disciplines (*e.g.*, Helsper and van der Gaag (2003)). First, the choice of the characteristics (nodes) to be included was made. The network was designed for predicting the characteristics of the offender from the characteristics of the victim and those of the crime scene. The description of the seven characteristics and their possible states are given in Table 11.1.

The second stage determined which nodes were to be joined by edges, as shown in Figure 11.1. These edges can be determined subjectively using expert opinion. In certain cases, an objective determination is possible. An example for continuous data is given in Aitken et al. (2005), where the existence of an edge is determined by the corresponding value of the partial correlation coefficient measuring the strength of the partial linear association between two variables.

The third stage is the specification of the conditional probabilities. Again, this can be done subjectively using expert opinion. In some examples, objective evaluation may be made from the database with the use of appropriate cross-tabulations for discrete data or appropriate conditional probability density functions for continuous data.

Once the network has been constructed, with associated probability tables it is possible to use it to provide probabilities for characteristics of the offender given characteristics of the victim and crime scene. Two examples are given in Table 11.2.

Table 11.1 Description of nodes and possible states for a seven-node network to model child murders with sexual connotations. [Reproduced by permission of The Forensic Science Society.]

Node description	States
Age of victim	0 – 7, 8 – 12, 13+
Sex of victim	Male, female
Location of last sighting	Home, others
Method of killing	Strangulation, others
Marital status of offender	Living with partner, others
Relationship of offender to victim	Known, stranger
Pre-conviction status of offender	Yes, no

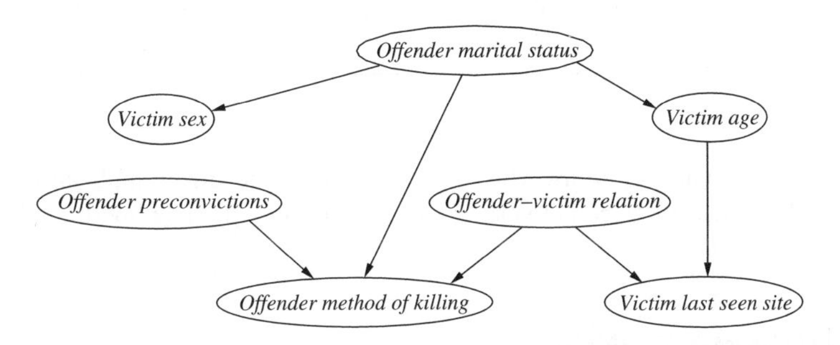

Figure 11.1 A seven-node network showing the relationships among three characteristics of the offender (previous convictions, relationship to victim and marital status) and four characteristics of the victim and the crime scene (age, sex, method of killing and site last seen). Six of the nodes are binary nodes with the following response: offender marital status (living with partner or not), victim's sex (male or female), victim last seen site (home or elsewhere), offender pre-convictions (yes or no), offender method of killing (strangulation or other), and offender–victim relation (known or a stranger). One node has three categories: victim age (0–7, 8–12, 13+). [Reproduced by permission of The Forensic Science Society.]

The main change in the revised probabilities compared to the initial probabilities arises with the relationship between the offender and the victim. If the victim is murdered in her own home, the probability that the offender is known to the victim is 0.66, that is, the odds are 0.66/0.34 or approximately 2 to 1, on that the offender is to be found among the people known to the victim. However, if the victim is murdered outwith her own home, the probability that the offender is known to the victim is 0.11, that is, the odds are 0.11/0.89, or approximately 8 to 1 against, that the offender is to be found among the people known to the victim.

Note that in this context of offender profiling, the network is an aid to an investigation and is not evidence in itself. A Bayesian network can provide assistance to investigators to help them identify suspects. However, it is not evidence and the output of their use should

Table 11.2 Probabilities for the characteristics of the offender for a female victim, aged 0–7 years, found strangled (i) in her own home and (ii) outwith her own home. [Reproduced by permission of The Forensic Science Society.]

Characteristic	Outcome	Probability		
		Initial	Revised	
			In own home	Outwith own home
Living with partner	Yes	0.24	0.33	0.36
	No	0.76	0.67	0.64
Relationship	Known to victim	0.57	0.66	0.11
	Unknown to victim	0.43	0.34	0.89
Previous convictions	Yes	0.73	0.73	0.70
	No	0.27	0.27	0.30

not be produced as such in court. This is in contrast to the many applications described earlier in the book in which the network can be used to provide likelihood ratios and hence provide an evaluation of evidence.

11.2 Decision making

Forensic scientists are routinely faced with the problems of making decisions under circumstances of uncertainty (*e.g.*, to perform or not perform a test). In Chapter 8, two questions have been presented: (1) what are the chances of obtaining a likelihood ratio that will support H_p or H_d in this scenario? and (2) how can the laboratory or the customer take a rational decision on the necessity to perform tests after an estimate of possible values of the likelihood ratio is obtained? The first question refers to the process of *case pre-assessment* and has been discussed in Chapter 8, the second refers to *decision making* and will be presented here briefly.

In the early 1950s, a debate was initiated on how people should make decisions involving money that were in some sense rational, and on how people in fact made monetary decisions and whether these could be regarded as rational (Lindley 1985; Luce and Raiffa 1957; Raiffa 1968; Savage 1951, 1972; Smith 1988). This debate – through the use of the notion of a so-called *utility function* – was extended in the context of financial decision making. It has been proposed to develop these ideas in a forensic context involving scientific evidence. In forensic science, decision theory can be used to develop general approaches for the determination of optimal choices given a certain body of evidence and values. The perspective will be that of an individual decision maker, who is either the customer or acts on behalf of the customer (*i.e.*, the forensic scientist) and who is interested in the determination of an optimal course of action using formal modelling.

Graphical models can be extended to deal with the kind of decision problems that are of interest here.

11.2.1 Decision analysis

The process of making a decision consists in the choice, given personal objectives, from two or more possible outcomes of the one that is considered as the most suitable when the consequences of the choice are uncertain.

To be a rational decision maker, according to Bayesian decision theory, it is required to choose the decision offering the highest expected utility. This rule is known as the *rule of the maximisation of the expected utility*. Thus, decision theory (a theory for making decisions) provides a unified framework for integrating all aspects of a decision problem (Lindley 1965, 1985).

A first task in any decision problem is to draw up an exhaustive list of actions that are available: $d_1, d_2, \ldots, d_m \in \Delta$. The space of decisions Δ is provided with a partial pre-ordering, denoted by $\succeq$, where $d_i \succeq d_j$ denotes that d_i is a better decision, in some sense, than d_j. This means that it is all the time possible to detect which decision is suitable or whether two decisions are equivalent. Secondly, a list of n exclusive and exhaustive uncertain events (also called *states of nature*) is needed: $\theta_1, \theta_2, \ldots, \theta_n \in \Omega$, where Ω denotes the entirety of these states of nature.

It is possible to measure uncertainty on the events using a suitable probability distribution P over Ω. Therefore, each alternative is associated with a probability distribution and a choice among probability distributions. In such a case, the scientist is faced with making a decision under risk.

The decision d and the event θ are related to one another. The main problem is that of choosing a member of the first list (decision) without knowing which member of the second list (state of nature) would happen.

The combination of decision d_i with state of nature θ_j will result in a foreseeable consequence. This consequence will be written as $C_{ij} = C_{d_i}(\theta_j)$. Varying d_i, $i = 1, \ldots, m$, and θ_j, $j = 1, \ldots, n$, a space of consequences is obtained. Consequences are defined in such a way that it is possible for them to be ranked, with the first in the ranking called the 'best' and the last the 'worst'. It follows that the next task is to provide something more than just a ranking. In order to do this, a standard is introduced, and a coherent comparison (*i.e.*, a comparison that does not produce inconsistencies) with this standard provides a numerical assessment.

Assume that C is the happiest consequence and c is the worst of them. C and c is a reference pair of highly desirable and highly undesirable consequences, respectively. It follows that any consequence C_{ij} may be compared unfavourably with C and favourably with c. Associated with any consequence C_{ij} is a unique number $u \in (0, 1)$ such that C_{ij} is just as desirable as a probability u of C and $1 - u$ of c. The number associated with C_{ij} will be denoted as $u(C_{ij})$ or $u(d_i, \theta_j)$ and will be called as the *utility* of C_{ij}.

Utility is a measure of the desirability of consequences of a course of action that applies to decision making under risk. The measure of the utility is assumed to reflect the preferences of the decision maker. The numerical order of utilities for consequences has to preserve the decision maker's preference order among the consequences.

So, for example, for decision (action) d_1 and three mutually exclusive states of nature, θ_1, θ_2 and θ_3, if the decision maker prefers consequence C_{13} to C_{12} and C_{12} to C_{11}, the utilities assigned must be such that $u(C_{13}) > u(C_{12}) > u(C_{11})$. In general, it can be said that decision d_i weakly dominates decision d_k if and only if $u(d_i, \theta) \geq u(d_k, \theta)$ for every

$\theta \in \Omega$:

$$d_i \succeq d_k \Leftrightarrow u(d_i, \theta) \geq u(d_k, \theta) \quad \forall \theta \in \Omega.$$

This utility is a probability: u is by definition the probability of obtaining the best consequence. Numbers are associated with decisions in such a way that the best decision is that with the highest number. If decision d_i is taken and if state θ_j occurs, the probability of obtaining the consequence C is $u(C_{ij})$:

$$Pr(C \mid d_i, \theta_j) = u(C_{ij}).$$

The *expected utility* of decision d_i is

$$E(U \mid d_i) = \sum_{j=1}^{n} Pr(C \mid d_i, \theta_j) Pr(\theta_j)$$

$$= \sum_{j=1}^{n} u(C_{ij}) Pr(\theta_j).$$

The expected utility $E(U \mid d_i)$ gives a numerical value to the probability of obtaining the best consequence C if decision d_i is taken. A decision problem is solved by maximising the expected utility[1]. The numerical order of expected utilities of actions preserves the decision maker's preference order among these actions.

Utility is not just a number describing the attractiveness of a consequence but is a number measured on a probability scale and obeys the laws of probability. It is a measure of the value of the decision d_i: the greater the expected utility, the greater the desirability of the decision because it offers a greater probability of obtaining the best consequence. Note that the same result would have been obtained had other standards been used. A property of the utility is that it is unaffected by a linear change. The process is not influenced by the reference points C and c: the suitable decision does not change with a varying origin or scale of utility.

Decision theory provides a useful framework for the exploration of alternatives. It forces the decision maker (*i.e.*, the scientist) to recognise that deciding not to take an action is just as much a decision as deciding which action to take. The theory forces the decision maker to recognise that he may err either by taking an unnecessary action or by failing to take a necessary action. It helps to formalise and categorise the thinking to make sure that all relevant possibilities have been considered.

11.2.2 Bayesian networks and decision networks

An extension of the ideas of Bayesian networks allows the scientist to obtain an aid to support decision making. *Bayesian decision networks* (BDNs) incorporate an explicit representation of the decisions under consideration and the values (*utilities*) of the resulting

[1] It is possible to formulate decision problems in an alternative way in terms of loss or regret associated with each pair (θ, d) by defining a loss function. The loss function $L(\theta, d)$ is the difference between the utility of the outcome of action d for state θ and the utility of the outcome of the best action for that state. Therefore, the action minimising the expected loss is the same as the action maximising the expected utility.

outcomes, that is, the states that may result from a decision. Sometimes, a decision is also called an *action*. BDNs combine probabilistic reasoning with utilities to make decisions that maximise the expected utility.

A BDN consists of three types of nodes:

- *Chance nodes*: These nodes represent random variables (as in Bayesian networks).
- *Decision nodes*: These are used to represent the decision being made at a particular time. The decision node covers the actions from among which the decision maker must choose.
- *Utility nodes*: These nodes take the task of representing the decision maker's utility function. Utility tables are used for this purpose. These specify a utility value for each configuration of the utility node's parents, which can be discrete chance and decision nodes.

Figure 11.3 represents a simple BDN for a genetical problem involving the determination of kinship to assist with problems of inheritance. Two individuals, say A and B, wish to know if they are full sibs or unrelated (Taroni et al. 2005a). The two questions of interest (before performing the DNA profile test) in this scenario are:

1. Can a value of the likelihood ratio be obtained that will be such as to support H_p or H_d? Notice that H_p and H_d are defined, respectively, as 'The pair of individuals A and B are full sibs' and 'The pair of individuals A and B are unrelated'.
2. How can the laboratory take a rational decision on the need to perform a DNA test?

Case pre-assessment is an effective tool to enable scientists to answer the former question (*see* Chapter 8). Decision analysis deals with the latter. Fundamental to the pre-assessment is the estimation of useful distributions for evidence under the two competing propositions. For this purpose, simulation techniques can be useful. For example, given allele frequencies (at different *loci*) from a selected population database, two databases are to be generated: one database of a large number of pairs of siblings and one of a large number of pairs of unrelated individuals. Then, for a given couple of individuals in the first database (siblings), a likelihood ratio, V, is estimated:

$$V = \frac{Pr(gt_A, gt_B \mid H_p)}{Pr(gt_A, gt_B \mid H_d)},$$

where gt_A and gt_B represent the genotypes of individuals A and B, respectively. The same procedure is performed for couples of individuals from the database of unrelated individuals. Two distributions of the possible values of likelihood ratios are obtained. The first assesses the support for full sibship when the individuals are related (brothers, H_p), and the second assesses the support for full sibship when the individuals are unrelated (H_d). So they provide the answer to the question 'could we obtain a value supporting the hypothesis H_p or H_d in this scenario?' Informative results can be obtained from such simulations (*see* Figure 11.2).

In the context of paternity and kinship, it may be acceptable to make a digression from considering solely the likelihood ratio V and to consider the probability that individuals

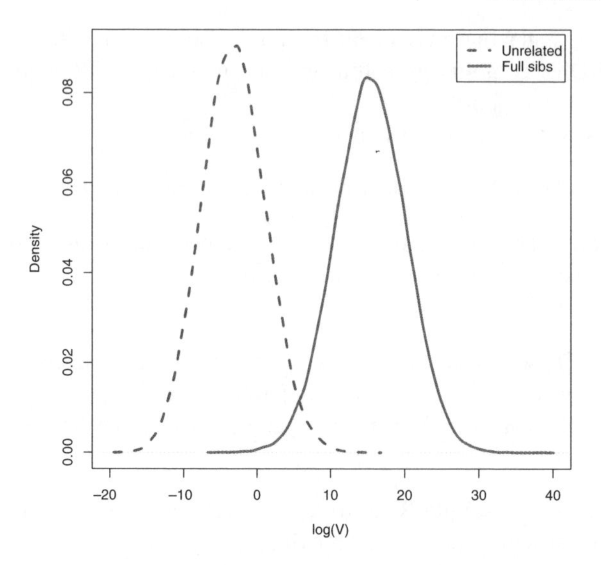

Figure 11.2 Distribution of the logarithm of the likelihood ratio, $\log(V)$, for full siblings *versus* unrelated individuals. [Reproduced with permission of ASTM International.]

A and B are full sibs; that is, the probability that H_p is true. This probability is known as the *probability of sibship*. Hummel (1971, 1983) provides a verbal scale, given here in Table 11.3, Columns 1 and 2. Hummel's scale can be used to characterise states of nature (θ_j) in the decision-making approach, because legal decision in kinship cases is closely related to this scale by jurisprudence.

Note that states of nature are specified from a particular point of view, for example, the perspective of full siblings, where an individual is interested in proving the sibling relationship. On the contrary, if an individual is interested in proving unrelatedness, then the order of Column 3 in Table 11.3 is inverted: θ_1 represents the state *Not useful*, and so on.

Table 11.3 Hummel's scale and states of nature. [Reproduced with permission of ASTM International.]

Probability of sibship	Likelihood of sibship	States of nature
Greater than 0.9979	Practically proved	θ_1
0.9910–0.9979	Extremely likely	θ_2
0.9500–0.9909	Very likely	θ_3
0.9000–0.9499	Likely	θ_4
0.8000–0.8999	Undecided	θ_5
Less than 0.8000	Not useful	θ_6

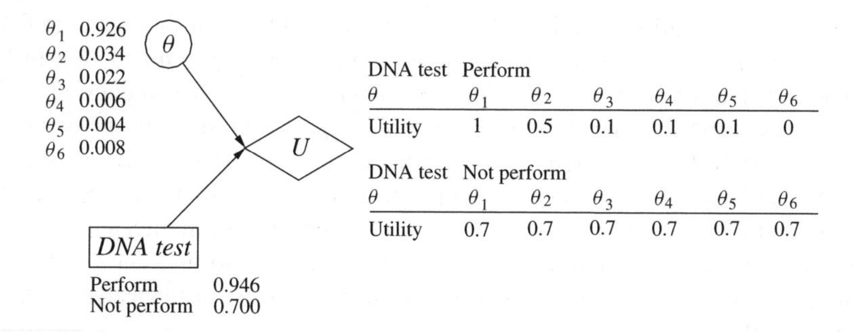

Figure 11.3 Kinship problem represented by a decision network. The decision node *DNA test* has two states, perform or do not perform the test. The chance node θ has six states, ranging from 'practically proved' (θ_1) to 'not useful' (θ_6). The utility node has 12 states corresponding to the 12 combinations of the decision and chance node states.

Decisions of interest are 'Perform the DNA profile test' and 'Not perform the DNA profile test,' denoted as d_1 and d_2, respectively.

Figure 11.3 illustrates that the decision (d_1 or d_2) and the chance node θ influence the utility node as expressed by the arcs from those nodes to the utility node U. The utility node describes the value of the consequence $C_{ij}, i = 1, 2, j = 1, \ldots, 6$. The number associated with it is denoted as $u(d_i, \theta_j)$. Utilities equal to 1 ($u(d_1, \theta_1)$) and 0 ($u(d_1, \theta_6)$) are assigned to states 'Practically proven' and 'Not useful', respectively, which are considered to be the best and the worst consequences.

Utilities u assigned to intermediate consequences C_{1j}, $j = 2, \ldots, 5$ are specified by answering the following question:

> Does the decision maker prefer the intermediate consequence or does he prefer the best consequence ('Practically proven') with probability set equal to u?

For example, consider $C_{ij} = C_{13}$. Does the decision maker prefer to perform the test and learn that the kinship is very likely (plausibility between 0.9500 and 0.9909) or to accept that if he performs the test the result that kinship is practically proved (plausibility greater than 0.9979) will be obtained with probability 0.1 (the value of u assigned to θ_3 for d_1)?

The utility of the consequence under d_2 is fixed, for the sake of illustration from Figure 11.3, as 0.7: this means that it is assumed the decision maker is indifferent between not performing the test and performing the test and obtaining the best consequence ('Practically proven') with probability 0.7. The utility of not performing the test is a direct consequence of the cost of performing it. This cost may be purely monetary. More generally, it may combine monetary costs with other non-financial burdens, such as inconvenience, intrusiveness, and so on. Different decision makers may state very different values for $u(C_{1j})$, $j = 1, \ldots, 6$ and $u(C_{2\cdot})$, depending on the kind of interest there is in determining the level of parentage, and on their adversity to risk.

The probability table associated with the chance node presents probabilities for $\theta_1, \ldots, \theta_6$ obtained through simulation techniques as previously presented for two individuals under the proposition H_p of full sibship for a fixed prior probability of sibship equal to 0.1. Note that there is no parent node to the decision node (no arc pointing to this node). In fact, no state of nature is known at the time the decision is made, so no link is requested.

The network allows the decision maker to check expected utilities for the two decisions to be able to choose the more suitable decision. So, if a couple of individuals are really full siblings, the DNA test will generally confirm it and d_1 should always be used ($E(U \mid d_1) = 0.946$ and $E(U \mid d_2) = 0.7$). The situation becomes more complicated depending on the chosen prior probabilities and if the DNA test is requested to find out a different level of parentage, such as half sibship versus unrelatedness or full sibship versus half sibship (Taroni et al. 2005a).

11.2.3 Forensic decision analyses

Decision analyses help the scientist to understand the problem to be faced better and make better and more consistent decisions, for example, as to why a chosen action is appropriate. Practical implementations of decision theory can be achieved through the use of graphical models that represent, in an economic, simple and intuitive way, the probabilistic relations existing among the ingredients of situations in which a decision must be taken. They allow the scientists to approach a problem, structure and solve it without entering too deeply into mathematical backgrounds whilst performing calculations automatically.

Many situations can be approached using BDNs. For example, forensic scientists are frequently concerned with situations in which more than one item of evidence is found. Two or more traces may be recovered, for example, at different locations where, at temporally distinct instances, crimes have been committed. At some stage of an investigation, it may well be that a potential source is not available for comparison. However, analyses may be performed on crime samples with the aim of evaluating possible linkages between the cases. Decision theory can assist investigators in a rational approach to the question of whether cases should be considered as linked. In fact, scientists may face difficult situations when being asked to provide a decision on whether, based on the scientific evidence considered, there is a 'link' between two cases. It is in this context that a useful distinction can be drawn between, on the one hand, scientific evidence, which forms a basis for reasoning and the construction of arguments relative to propositions of interest (the items of evidence come from the same source, or the two offences have been committed by the same individual) and, on the other hand, the fact or state of an actual belief in the truth or otherwise of such propositions given the evidence. The former can be seen as a problem in inductive inference whereas the latter is a problem in decision making. A decision network dealing with such an aspect is presented in Taroni et al. (2005b), where it is emphasised that graphical models provided by forensic scientists can be integrated in decision networks covering issues that are in the competence of other actors of the legal system.

There are yet others questions of importance for forensic scientists where BDNs could potentially be useful. A frequently encountered question is, for example, which method is appropriate for the analysis of a particular item of evidence. Once a method or technique has been chosen, how many analyses or tests need to be performed? In the context of DNA profiling analyses, it may be a question, for example, of how many STR loci need to be analysed in order to approach a comparison between a (biological) crime sample and a sample provided by a suspect (Taroni et al. 2005c). The question, in this context, appears to be simple. For practical reasons in DNA profiling, the scientists use commercially available kits that offer results for 16 markers, for example, at the same time.

The questions of choice of method and of sample size, however, arise in many other forensic scientific disciplines. For these disciplines, the questions are fundamental and, as yet, have no simple answers.

Bibliography

Aitken CGG 1995 *Statistics and the Evaluation of Evidence for Forensic Scientists.* John Wiley & Sons, Chichester.

Aitken CGG 1999 Sampling – How big a sample? *Journal of Forensic Sciences* **44**(4), 750–760.

Aitken CGG, Connolly T, Gammerman A, Zhang G, Bailey D, Gordon R and Oldfield R 1996a Statistical modelling in specific case analysis. *Science & Justice* **36**, 245–255.

Aitken CGG and Gammerman A 1989 Probabilistic reasoning in evidential assessment. *Journal of the Forensic Science Society* **29**, 303–316.

Aitken CGG, Gammerman A, Zhang G, Connolly T, Bailey D, Gordon R and Oldfield R 1996b Bayesian belief networks with an application in specific case analysis In *Computational Learning and Probabilistic Reasoning* (ed. Gammerman A), pp. 169–184. John Wiley & Sons, Chichester.

Aitken CGG, Lucy D, Zadora G and Curran JM 2005 Evaluation of trace evidence for three-level multivariate data with the use of graphical models. *Computational Statistics and Data Analysis.* In press.

Aitken CGG and Taroni F 1997 A contribution to the discussion on "Bayesian analysis of the deoxyribonucleic acid profiling data in forensic identification applications", Foreman et al. *Journal of the Royal Statistical Society, Serie A* **160**, 463.

Aitken CGG and Taroni F 2004 *Statistics and the Evaluation of Evidence for Forensic Scientists*, 2nd edn. John Wiley & Sons, New York.

Aitken CGG, Taroni F and Garbolino P 2003 A graphical model for the evaluation of cross-transfer evidence in DNA profiles. *Theoretical Population Biology* **63**, 179–190.

Anderson T and Twining W 1998 *Analysis of Evidence: How to do Things with Facts Based on Wigmore's Science of Judicial Proof*, 2nd edn. Northwestern University Press, Evanston, Ill.

Balding DJ 1997 Errors and misunderstandings in the second NRC report. *Jurimetrics Journal* **37**, 469–476.

Balding DJ 2000 Interpreting DNA evidence: Can probability theory help? In *Statistical Science in the Courtroom* (ed. Gastwirth JL), pp. 51–70. Statistics for Social Science and Public Policy Springer.

Balding DJ 2003 Likelihood-based inference for genetic correlation coefficients. *Theoretical Population Biology* **63**, 221–230.

Balding DJ 2005 *Weight-of-Evidence for Forensic DNA Profiles.* John Wiley & Sons, Chichester.

Balding DJ and Donnelly P 1996 Evaluating DNA profile evidence when the suspect is identified through a database search. *Journal of Forensic Sciences* **41**(4), 603–607.

Balding DJ, Greenhalgh M and Nichols RA 1996 Population genetics of STR loci in Caucasians. *International Journal of Legal Medicine* **108**, 300–305.

Bayesian Networks and Probabilistic Inference in Forensic Science F. Taroni, C. Aitken, P. Garbolino and A. Biedermann

Balding DJ and Nichols RA 1994 DNA profile match probability calculation: how to allow for population stratification, relatedness, database selection and single bands. *Forensic Science International* **64**, 125–140.

Balding DJ and Nichols RA 1995 A method for quantifying differentiation between populations and multi-allelic loci and its implications for investigating identity and paternity In *Human Identification: the Use of DNA Markers* (ed. Weir BS), pp. 3–12. Kluwer Academic Publishers, Netherlands.

Balding DJ and Nichols RA 1997 Significant genetic correlations among Caucasians at forensic DNA loci. *Heredity* **78**, 583–589.

Bayes T 1763 An essay towards solving a problem in the doctrine of chances In *Studies in the History of Statistics and Probability* (1970) (eds. Pearson ES and Kendall MG), vol. 1, pp. 134–153. Griffin, London.

Bernardo JM and Smith AFM 1994 *Bayesian Theory*. John Wiley & Sons, Chichester.

Biasotti AA 1959 A statistical study of the individual characteristics of fired bullets. *Journal of Forensic Sciences* **4**, 34–50.

Biasotti AA and Murdock J 1997 Firearms and toolmark identification: the scientific basis of firearms and toolmark identification In *Modern Scientific Evidence: The Law and Science of Expert Testimony* (eds. Faigman L, Kaye DH, Saks MJ and Sanders J), pp. 144–150. West Publishing, St. Paul, MN.

Biedermann A and Taroni F 2004 Bayesian networks and the assessment of scientific evidence: the use of qualitative probabilistic networks and sensitivity analyses. *14th Interpol Forensic Science Symposium Lyon, France*.

Biedermann A and Taroni F 2005 A probabilistic approach to the joint evaluation of firearms and GSR evidence. *Forensic Science International*. In press.

Biedermann A, Taroni F, Delemont O, Semadeni C and Davison AC 2005a The evaluation of evidence in the forensic investigation of fire incidents (part I): an approach using Bayesian networks. *Forensic Science International* **147**, 49–57.

Biedermann A, Taroni F, Delemont O, Semadeni C and Davison AC 2005b The evaluation of evidence in the forensic investigation of fire incidents (part II): practical examples of the use of Bayesian networks. *Forensic Science International* **147**, 56–69.

Booth G, Johnston F and Jackson G 2002 Case assessment and interpretation – application to a drugs supply case. *Science & Justice* **42**, 123–125.

Buckleton JS, Evett IW and Weir BS 1998 Setting bounds for the likelihood ratio when multiple hypotheses are postulated. *Science & Justice* **38**, 23–26.

Buckleton JS and Triggs CM 2005 Relatedness and DNA: are we taking it seriously enough? *Forensic Science International* **152**, 115–119.

Buckleton JS, Triggs CM and Walsh SJ 2004 *Forensic DNA Evidence Interpretation*. CRC Press, Boca Raton, FL.

Bunch SG 2000 Consecutive matching striation criteria: a general critique. *Journal of Forensic Sciences* **45**(5), 955–962.

Castillo E, Gutierrez JM and Hadi S 1997 *Expert Systems and Probabilistic Network Models*. Springer, New York.

Cavallini D, Corradi F and Guagnano G 2004 OOBN for forensic identification via a search in a database of DNA profiles. Technical report 2004-4, Università degli Studi di Firenze, Firenze.

Champod C and Evett IW 2001 A probabilistic approach to fingerprint evidence. *Journal of Forensic Identification* **51**, 101–122.

Champod C and Jackson G 2000 European fibres group workshop: case assessment and Bayesian interpretation of fibres evidence. *Proceedings of the 8th Meeting of European Fibres Group*, pp. 33–45, Krakow.

Champod C, Lennard C, Margot P and Stoilovic M 2004 *Fingerprints and Other Ridge Skin Impressions*. CRC Press, Boca Raton, FL.

Champod C and Taroni F 1999 The Bayesian approach In *Forensic Examination of Fibres* (eds. Robertson J and Grieve M), pp. 379–398. Taylor and Francis, London.

Charniak E 1991 Bayesian networks without tears. *AI Magazine* **12**(4), 50–63.

Cohen JL 1977 *The Probable and the Provable*. Clarendon Press, Oxford.

Cohen JL 1988 The difficulty about conjunction in forensic proof. *The Statistician* **37**, 415–416.

Conan Doyle A 1953 *The Complete Sherlock Holmes*. Doubleday & Company, Garden City.

Cook R, Evett IW, Jackson G, Jones PJ and Lambert JA 1998a A hierarchy of propositions: deciding which level to address in casework. *Science & Justice* **38**, 231–239.

Cook R, Evett IW, Jackson G, Jones PJ and Lambert JA 1998b A model for case assessment and interpretation. *Science & Justice* **38**, 151–156.

Cook R, Evett IW, Jackson G, Jones PJ and Lambert JA 1999 Case pre-assessment and review in a two-way transfer case. *Science & Justice* **39**(2), 103–111.

Cooper GF 1990 The computational complexity of probabilistic inference using Bayesian belief networks. *Artificial Intelligence* **42**, 393–405.

Cooper GF 1999 An overview of the representation and discovery of causal relationships using Bayesian networks In *Computation, Causation, and Discovery* (eds. Glymour C and Cooper GF), pp. 3–62. The MIT Press, Cambridge, MA.

Coupé VMH and Van der Gaag LC 1998 Practicable sensitivity analysis of Bayesian belief networks In *Prague Stochastics '98 – Proceedings of the Joint Session of the 6th Prague Symposium of Asymptotic Statistics and the 13th Prague Conference on Information Theory, Statistical Decision Functions and Random Processes* (eds. Hušková M, Lachout P and Víšek JA), pp. 81–86. Union of Czech Mathematicians and Physicists.

Coupé VMH, Van der Gaag LC and Habbema JDF 2000 Sensitivity analysis: an aid for belief-network quantification. *The Knowledge Engineering Review* **15**(3), 215–232.

Cowell RG 2003 FINEX: a probabilistic expert system for forensic identification. *Forensic Science International* **134**, 196–206.

Cowell RG, Dawid AP, Lauritzen S and Spiegelhalter DJ 1999 *Probabilistic Networks and Expert Systems*. Springer, New York.

Cowell RG, Lauritzen SL and Mortera J 2004 Identification and separation of DNA mixtures using peak area information. Technical Report 25, Cass Business School, London.

Cox DR and Hinkley DV 1974 *Theoretical Statistics*. Chapman & Hall, London.

Curran JM 2005 An introduction to Bayesian credible intervals for sampling error in DNA profiles. *Law, Probability and Risk* **4**, 115–126.

Curran JM, Buckleton JS, Triggs CM and Weir BS 2002 Assessing uncertainty in DNA evidence caused by sampling effects. *Science & Justice* **42**, 29–37.

Curran JM, Hicks TN and Buckleton JS 2000 *Forensic Interpretation of Glass Evidence*. CRC Press, Boca Raton, FL.

D'Agostini GD 2003 *Bayesian Reasoning in Data Analysis: A Critical Introduction*. World Scientific, New York.

Darboux JG, Appell PE and Poincaré JH 1908 *Examen critique des divers systèmes ou études graphologiques auxquels a donné lieu le bordereau* In *L'affaire Dreyfus – La révision du Procès de Rennes – Enquête de la Chambre Criminelle de la Cour de Cassation* Ligue Française des Droits de L'homme et du Citoyen Paris.

Daston L 1988 *Classical Probability in the Enlightment*. Princeton University Press, Princeton, NJ.

Dawid AP 1987 The difficulty about conjunction. *The Statistician* **36**, 91–97.

Dawid AP 2000 Causal inference without counterfactuals (with discussion). *Journal of the American Statistical Association* **95**, 407–448.

Dawid AP 2002 Influence diagrams for causal modelling and inference. *International Statistical Review* **70**, 161–189.

Dawid AP 2003 An object-oriented Bayesian network for estimating mutation rates. *Proceedings of the Ninth International Workshop on Artificial Intelligence and Statistics* Key West, Florida, USA.

Dawid AP and Evett IW 1997 Using a graphical method to assist the evaluation of complicated patterns of evidence. *Journal of Forensic Sciences* **42**, 226–231.

Dawid AP and Vovk V 1999 Prequential probability: principles and properties. *Bernoulli* **5**, 125–162.

Dawid AP, Mortera J, Pascali VL and van Boxel D 2002 Probabilistic expert systems for forensic inference from genetic markers. *Scandinavian Journal of Statistics* **29**, 577–595.

de Finetti B 1930a Fondamenti logici del ragionamento probabilistico. *Bollettino Della Unione Matematica Italiana* **9**, 258–261.

de Finetti B 1930b Funzione caratteristica di un fenomeno aleatorio. *Memorie Della Regia Accademia dei Lincei* **4**, 86–133.

de Finetti B 1937 La prévision. ses lois logiques, ses sources subjectives. *Annales de l'Institut Henri Poincaré* **7**, 1–68. English translation in etc.

de Finetti B 1975 *Theory of Probability*. John Wiley & Sons, New York.

de Finetti B 1989 La probabilità: guardarsi dalle contraffazioni! In *La logica Dell'incerto* (ed. Mondadori M), pp. 149–188, Il Saggiatore, Milan.

Diaconis P and Freedman D 1981 The persistence of cognitive illusions. *Behavioural and Brain Sciences* **4**, 333–334.

Diaconis P and Zabell S 1982 Updating subjective probability. *Journal of the American Statistical Association* **77**, 882–830.

Dillon JH 1990 The Modified Griess Test: a chemically specific chromophoric test for nitrite compounds in gunshot residues. *AFTE Journal* **22**(3), 243–250.

Domotor Z, Zanotti M and Graves H 1980 Probability kinematics. *Synthese* **44**, 421–442.

Donkin WF 1851 On certain questions relating to the theory of probability. *The London, Edinburgh and Dublin Philosophical Magazine and Journal of Science, Series 4* **1**, 353–368.

Druzdzel MJ 1996 Explanation in probabilistic systems: is it feasible? will it work? *Proceedings of the Workshop Intelligent Information Systems V*, pp. 1–13, Deblin.

Druzdzel MJ 1997 Five useful properties of probabilistic knowledge representations from the point of view of intelligent systems. *Fundamenta Informaticae* **30**, 241–254. Special Issue on Knowledge Representation and Machine Learning.

Druzdzel MJ and Henrion M 1993a Belief propagation in qualitative probabilistic networks In *Qualitative Reasoning and Decision Technologies* (eds. Carrete NP and Singh MG), pp. 451–460. CIMNE, Barcelona.

Druzdzel MJ and Henrion M 1993b Efficient reasoning in qualitative probabilistic networks. *Proceedings of the 11th Annual Conference on Artificial Intelligence (AAAI-93)*, pp. 548–553, Washington, DC.

Druzdzel MJ and Henrion M 1993c Intercausal reasoning with uninstantiated ancestor nodes. *Proceedings of the Ninth Annual Conference on Uncertainty in Artificial Intelligence (UAI-93)*, pp. 317–325, Washington, DC.

Earman J 1992 *Bayes or Bust?* The MIT Press, Cambridge, MA.

Edwards AWF 1987 *Pascal's Arithmetical Triangle*. Griffin & Oxford University Press, London, New York.

Edwards W 1991 Influence diagrams, Bayesian imperialism, and the *Collins* case: an appeal to reason. *Cardozo Law Review* **13**, 1025–1074.

ENFSI (European Network of Forensic Science Institutes) Drugs Working Group 2004 Guidelines on representative drug sampling Den Haag.

Evett IW 1984 A quantitative theory for interpreting transfer evidence in criminal cases. *Applied Statistics* **33**, 25–32.

Evett IW 1987 On meaningful questions: a two-trace transfer problem. *Journal of the Forensic Science Society* **27**, 375–381.

Evett IW 1992 Evaluating DNA profiles in a case where the defence is "It was my brother". *Journal of the Forensic Science Society* **32**(1), 5–14.

Evett IW 1993 Establishing the evidential value of a small quantity of material found at a crime scene. *Journal of the Forensic Science Society* **33**, 83–86.

Evett IW, Gill P and Lambert JA 1998a Taking accout of peak areas when interpreting mixed DNA profiles. *Journal of Forensic Sciences* **43**, 62–69.

Evett IW, Lambert JA and Buckleton JS 1998b A Bayesian approach to interpreting footwear marks in forensic casework. *Science & Justice* **38**(4), 241–247.

Evett IW, Jackson G and Lambert JA 2000 More in the hierarchy of propositions: exploring the distinction between explanations and propositions. *Science & Justice* **40**, 3–10.

Evett IW, Gill PD, Jackson G, Whitaker J and Champod C 2002a Interpreting small quantities of DNA: the hierarchy of propositions and the use of Bayesian networks. *Journal of Forensic Sciences* **47**, 520–530.

Evett IW, Jackson G, Lambert JA and McCrossan S 2002b The impact of the principles of evidence interpretation and the structure and content of statements. *Science & Justice* **40**, 233–239.

Evett IW and Weir BS 1998 *Interpreting DNA Evidence*. Sinauer Associates Incorporated, Sunderland.

Fenton N and Neil M 2000 The 'jury observation fallacy' and the use of Bayesian networks to present probabilistic legal arguments. *Mathematics Today – Bulletin of the IMA* **36**, 180–187.

Finkelstein MO and Fairley WB 1970 A Bayesian approach to identification evidence. *Harvard Law Review* **83**, 489–517.

Foreman LA, Lambert JA and Evett IW 1998 Regional genetic variation in Caucasians. *Forensic Science International* **95**, 27–37.

Foreman LA, Smith AFM and Evett IW 1997 A Bayesian approach to validating STR multiplex databases for use in forensic casework. *International Journal of Legal Medicine* **110**, 244–250.

Friedman RD 1986a A close look at probative value. *Boston University Law Review* **66**, 733–759.

Friedman RD 1986b A diagrammatic approach to evidence. *Boston University Law Review* **66**, 571–622.

Friedman RD 1996 Assessing evidence. *Michigan Law Review* **94**, 1810–1838.

Friedman N and Goldszmidt M 1996 Learning Bayesian networks with local structure In *Uncertainty in Artificial Intelligence: Proceedings of the Twelfth Conference* (eds. Horvitz E and Jensen FV), pp. 252–262. Morgan Kaufmann Publishers, San Mateo, CA.

Galavotti MC 2005 *Philosophical Introduction to Probability*. CSLI Publications, Stanford, CA.

Garbolino P 2001 Explaining relevance. *Cardozo Law Review* **22**, 1503–1521.

Garbolino P and Taroni F 2002 Evaluation of scientific evidence using Bayesian networks. *Forensic Science International* **125**, 149–155.

Geiger D and Heckerman D 1991 Advances in probabilistic reasoning In *Uncertainty in Artificial Intelligence: Proceedings of the Seventh Conference* (eds. Bonissone P, D'Ambrosio B and Smets P), pp. 118–126. Morgan Kaufmann Publishers, San Mateo, CA.

Good IJ 1985 Weight of evidence: a brief survey (with discussion) In *Bayesian Statistics 2* (eds. Bernardo JM, DeGroot MH, Lindley DV and Smith AFM), pp. 249–270. North Holland, Amsterdam.

Graham J, Curran J and Weir BS 2000 Conditional genotypic probabilities for microsatellite loci. *Genetics* **155**, 1973–1980.

Hacking I 1975 *The Emergence of Probability*. Cambridge University Press, Cambridge.

Halliwell J, Keppens J and Shen Q 2003 Linguistic Bayesian networks for reasoning with subjective probabilities in forensic statistics. *9th International Conference on Artificial Intelligence and Law (ICAIL 2003)*, pp. 42–50, Edinburgh, Scotland.

Harbison SA and Buckleton JS 1998 Applications and extensions of subpopulation theory: a caseworkers guide. *Science & Justice* **38**, 249–254.

Harman G 1965 The inference to the best explanation. *Philosophical Review* **77**, 88–95.

Helsper EM and van der Gaag LC 2003 Ontologies for probabilistic networks. Technical Report UU-CS-2003-042, Institute of Information and Computing Science, Utrecht University, Netherlands.

Hempel CG 1965 *Aspects of Scientific Explanation and Other Essays in the Philosophy of Science*. The Free Press, New York.

Henrion M 1986 Uncertainty in artificial intelligence: is probability epistemically and heuristically adequate? In *Expert Judgement and Expert Systems (NATO ISI Series F)* (eds. Mumpower J, Phillips LD, Renn O and Uppuluri VRR), pp. 105–130, vol. 35. Springer, Berlin .

Henrion M and Druzdzel MJ 1991 Qualitative propagation and scenario-based approaches to explanation of probabilistic reasoning In *Uncertainty in Artificial Intelligence 6* (eds. Bonissone PP, Henrion M, Kanal LN and Lemmer JF), pp. 17–32. Elsevier, North Holland.

Hitchcock C 2002 Probabilistic causation In *Stanford Encyclopedia of Philosophy* (ed. Zalta EN) http://plato.stanford.edu/archives/fall2002/entries/causation-probabilistic/ (last accessed April 2005).

Horwich P 1982 *Probability and Evidence*. Cambridge University Press, Cambridge.

Howson C and Urbach P 1993 *Scientific Reasoning: The Bayesian Approach*, 2nd edn. Open Court, La Salle.

Hummel K 1971 Biostatistical opinion of parentage based upon the results of blood group tests In *Biostatistische Abstammungsbegutachtung mit Blutgruppenbefunden* (ed. Schmidt P). Gustav Fisher, Stuttgart.

Hummel K 1983 Selection of gene frequency tables In *Inclusion Probabilities in Parentage Testing* (ed. Walker R). American Association of Blook Banks, Arlington, VA.

Jackson G 2000 The scientist and the scales of justice. *Science & Justice* **40**, 81–85.

Jaynes ET 1983 Where do we stand on maximum entropy? In *Papers on Probability, Statistics and Statistical Physics* (ed. Rosenkrantz RD), pp. 211–314. Reidel Publishing, Dordrecht.

Jaynes ET 2003 *Probability Theory: The Logic of Science*. Cambridge University Press, Cambridge.

Jeffrey RC 1983 *The Logic of Decision*, 2nd edn. University of Chicago Press, Chicago, Ill.

Jeffrey RC 1992 *Probability and the Art of Judgment*. Cambridge University Press, Cambridge.

Jeffrey RC 2004 *Subjective Probability (The Real Thing)*. Cambridge University Press, Cambridge.

Jensen FV 2001 *Bayesian Networks and Decision Graphs*. Springer, New York.

Jensen FV, Lauritzen SL and Olesen KG 1990 Bayesian updating in causal probabilistic networks by local computation. *Computational Statistical Quarterly* **4**, 269–282.

Jowett C 2001a Lies, damned lies, and DNA statistics: DNA match testing, Bayes' theorem, and the criminal courts. *Medicine, Science & Law* **41**, 194–205.

Jowett C 2001b Sittin' in the dock with the Bayes. *New Law Journal Expert Witness Supplement* **151**, 201–212.

Kadane JB and Schum DA 1996 *A Probabilistic Analysis of the Sacco and Vanzetti Evidence*. John Wiley & Sons, New York.

Katterwe H 2003 True or false. *Information Bulletin for Shoeprint/Toolmark Examiners* **9**(2), 18–25.

Kim JH and Pearl J 1983 A computational model for combined causal and diagnostic reasoning in inference systems. *Proceedings of IJCAI-83*, pp. 190–193, Karlsruhe.

Kind SS 1994 Crime investigation and the criminal trial: a three chapter paradigm of evidence. *Journal of the Forensic Science Society* **35**, 155–164.

Kirk PL and Kingston CR 1964 Evidence evaluation and problems in general criminalistics. *16th Annual Meeting of the American Academy of Forensic Sciences*, Chicago, Ill.

Kjærulff U and Van der Gaag LC 2000 Making sensitivity analysis computationally efficient In *Proceedings of the Eleventh Conference on Uncertainty in Artificial Intelligence* (eds. Boutilier C and Goldszmidt M), pp. 362–367. Morgan Kaufmann Publishers, San Francisco, CA, Quebec.

Köller N, Niessen K, Riess M and Sadorf E 2004 *Probability Conclusions in Expert Opinions on Handwriting. Substantiation and Standardization of Probability Statements in Expert Opinions.* Luchterhand, München.

Koller D and Pfeffer A 1997 Object-Oriented Bayesian networks. *Proceedings of the Thirteenth Annual Conference on Uncertainty in Artificial Intelligence (UAI-97)*, pp. 302–313, Providence, Rhode Island, USA.

Kullback S 1959 *Information Theory and Statistics.* John Wiley & Sons, New York.

Lauritzen SL and Mortera J 2002 Bounding the number of contributors to mixed DNA stains. *Forensic Science International* **130**, 125–126.

Lauritzen SL and Spiegelhalter DJ 1988 Local computations with probabilities on graphical structures and their application to expert systems. *Journal of the Royal Statistical Society, Series B* **50**(2), 157–224.

Lee PM 2004 *Bayesian Statistics, an Introduction*, 3rd edn. Arnold, London.

Lee JW, Lee HS and Hwang JJ 2002 Statistical analysis for estimating heterogenity of korean population in dna typing using STR loci. *International Journal of Legal Medicine* **116**, 153–160.

Leibniz GW 1930 *Sämtliche Schriften und Briefe*, vol. 6. Otto Reichl, Darmstadt.

Lemmer JF and Barth SW 1982 Efficient minimum information updating for Bayesian inferencing in expert systems. *Proceedings of the 2nd American National Conference on AI*, pp. 424–427, Pittsburgh, PA.

Lempert RO 1977 Modeling relevance. *Michigan Law Review* **75**, 1021–1057.

Levitt TS and Blackmond Laskey K 2001 Computational inference for evidential reasoning in support of judicial proof. *Cardozo Law Review* **22**, 1691–1731.

Lichtenberg W 1990 Methods for the determination of shooting distance. *Forensic Science Review* **2**, 37–62.

Lindley DV 1965 *Introduction to Probability and Statistics from a Bayesian Viewpoint.* Cambridge University Press, Cambridge.

Lindley DV 1985 *Making Decisions*, 2nd edn. John Wiley & Sons, Chichester.

Lindley DV 1991 Probability In *The Use of Statistics in Forensic Science* (eds. Aitken CGG and Stoney DA), pp. 27–50. Ellis Horwood, New York.

Lindley DV 2004 Foreword In *Statistics and the Evaluation of Evidence for Forensic Scientists*, 2nd edn, (eds. Aitken CGG and Taroni F). John Wiley & Sons, New York.

Lindley DV and Eggleston R 1983 The problem of missing evidence. *The Law Quarterly Review* **99**, 86–99.

Luce RD and Raiffa H 1957 *Games and Decisions. Introduction and Critical Survey.* John Wiley & Sons, Chichester.

Mackie JL 1974 *The Cement of the Universe.* Clarendon Press, Oxford.

Madsen AL and Jensen FV 1999 Lazy propagation: a junction tree inference algorithm based on lazy evaluation. *Artificial Intelligence* **113**, 203–245.

Mahoney SM and Laskey KB 1999 Representing and combining partially specified CPTs In *Uncertainty in Artificial Intelligence: Proceedings of the Fifteenth Conference* (eds. Laskey KB and Prade H), pp. 391–400. Morgan Kaufmann Publishers, San Francisco, CA.

Morgan MG and Henrion M 1990 *Uncertainty. A Guide to Dealing with Uncertainty in Quantitative Risk and Policy Analysis*. Cambridge Unviersity Press.

Mortera J 2003 Analysis of DNA mixtures using probabilistic expert systems In *Highly Structured Stochastic Systems* (eds. Green PL, Hjort NL and Richardson S). Oxford University Press, Oxford.

Mortera J, Dawid AP and Lauritzen SL 2003 Probabilistic expert systems for DNA mixture profiling. *Theoretical Population Biology* **63**, 191–205.

Mueller C and Kirkpatrick L 1988 *Federal Rules of Evidence*. Little, Brown and Company, Boston, MA.

National Research Council 1992 *DNA Technology in Forensic Science*. National Academy Press, Washington, DC.

Neil M, Fenton N and Nielsen L 2000 Building large-scale Bayesian networks. *The Knowledge Engineering Review* **15**, 257–284.

Nichols RG 1997 Firearms and toolmark identification criteria: a review of the literature. *Journal of Forensic Sciences* **42**(3), 466–474.

Nichols RG 2003 Firearms and toolmark identification criteria: a review of the literature, Part II. *Journal of Forensic Sciences* **48**(2), 318–327.

O'Hagan A 1988 *Probability: Methods and Measurements*. Chapman & Hall, London.

Pearl J 1988 *Probabilistic Reasoning in Intelligent Systems: Networks of Plausible Inference*. Morgan Kaufmann Publishers, San Mateo, CA.

Pearl J 1999 Graphs, structural models, and causality In *Computation, Causation, and Discovery* (eds. Glymour C and Cooper GF), pp. 95–138. The MIT Press, Cambridge, MA.

Pearl J 2000 *Causality: Models, Reasoning, and Inference*. Cambridge University Press, Cambridge.

Pearl J and Verma T 1987 The logic of representing dependencies by directed graphs *Proceedings of the National Conference on AI (AAAI-87)*, pp. 374–379. AAAI Press/MIT Press, Menlo Park, CA.

R Development Core Team 2003 *R: A Language and Environment for Statistical Computing*. R Foundation for Statistical Computing, Vienna, ISBN 3-900051-00-3.

Raiffa H 1968 *Decision Analysis. Introductory Lectures on Choices under Uncertainty*. Addison-Wesley, Reading, MA.

Ramsey FP 1931 Truth and probability In *The Foundations of Mathematics and Other Logical Essays* (ed. Braithwaite RB), pp. 156–198. Routledge & Kegan Paul Ltd. Reprinted in *Studies in Subjective Probability* (1980) 2nd edn, (eds. Kyburg HE and Smokler HE), pp. 61–92. Dover Publications, New York .

Ribaux O and Margot P 2003 Case based reasoning in criminal intelligence using forensic case data. *Science & Justice* **43**(3), 135–143.

Robertson B and Vignaux GA 1993 Taking fact analysis seriously. *Michigan Law Review* **91**, 1442–1464.

Robertson B and Vignaux GA 1995 *Interpreting Evidence. Evaluating Forensic Science in the Courtroom*. John Wiley & Sons, Chichester.

Rowe WF 2000 Range In *Encyclopedia of Forensic Science* (eds. Siegel JH, Saukko PJ and Knupfer GC), vol. 2 pp. 949–953. Academic Press, San Diego, CA.

Salmon WC 1990 Rationality and objectivity in science or Tom Kuhn meets Tom Bayes In *Scientific Theories, Minnesota Studies in the Philosophy of Science* (ed. Savage CW), vol. 14, pp. 175–204. University of Minnesota Press, Minneapolis, MN.

Salmon WC 1998 *Causality and Explanation*. Oxford University Press, New York, Oxford.

Salmon WC 1999 Scientific explanation In *Introduction to the Philosophy of Science* (eds. Salmon MH, Earman J, Glymour C, Lennox JG, Machamer P, McGuire JE, Norton JD, Salmon WC and Schaffner KF), 2nd edn, pp. 7–41. Hackett Publishing, Indianapolis, IN.

Salmon W, Jeffrey RC and Greeno JG 1971 *Statistical Explanation and Statistical Relevance*. University of Pittsburgh Press, Pittsburgh, PA.

Savage LJ 1951 Theory of statistical decision. *Journal of the American Statistical Association*, vol. 46, pp. 55–67.

Savage LJ 1972 *The Foundations of Statistics*, 2nd edn. Dover Publications, New York.

Scheines R 1997 An introduction to causal inference In *Causality in Crisis?* (eds. McKim VR and Turner S), pp. 185–199. University of Notre Dame Press, Notre Dame.

Schervish MJ 1995 *Theory of Statistics*. Springer-Verlag, Berlin, New York.

Schum DA 1994 *Evidential Foundations of Probabilistic Reasoning*. John Wiley & Sons, New York.

Shafer G and Vovk V 2001 *Probability and Finance. It's Only a Game!* John Wiley & Sons, New York.

Shenoy PP and Shafer G 1990 Axioms for probability and belief-function propagation In *Uncertainty in Artificial Intelligence, Proceedings of the Fifth Conference* (eds. Shachter RD, Levitt TS, Kanal LN and Lemmer JF), pp. 169–198. North Holland, Amsterdam.

Sinha SK 2003 Y-chromosome: genetics, analysis, and application in forensic science. *Forensic Science Review* **15**(2), 77–201.

Sjerps M and Kloosterman AD 1999 On the consequences of DNA profile mismatches for close relatives of an excluded suspect. *International Journal of Legal Medicine* **112**, 176–180.

Smith JQ 1988 *Decision Analysis, a Bayesian Approach*. Chapman & Hall, London.

Spirtes P, Glymour C and Scheines R 2001 *Causation, Prediction and Search*, 2nd edn. The MIT Press, Cambridge, MA.

Stockton A and Day S 2001 Bayes, handwriting and science. *Proceedings of the 59th Annual ASQDE Meeting – Handwriting & Technology: At the Crossroads*, pp. 1–10, Des Moines, IA.

Stoney DA 1991 Transfer evidence In *The Use of Statistics in Forensic Science* (eds. Aitken CGG and Stoney DA), pp. 107–138. Ellis Horwood, New York.

Stoney DA 1994 Relaxation of the assumption of relevance and an application to one-trace and two-trace problems. *Journal of the Forensic Science Society* **34**(1), 17–21.

Taroni F and Aitken CGG 1998 Probabilités et preuve par ADN dans les affaires civiles et criminelles. Questions de la cour et réponses fallacieuses des experts. *Revue Pénale Suisse* **116**, 291–313.

Taroni F, Aitken CGG and Garbolino P 2001 De Finetti's subjectivism, the assessment of probabilities and the evaluation of evidence: a commentary for forensic scientists. *Science & Justice* **41**, 145–150.

Taroni F and Biedermann A 2005 Inadequacies of posterior probabilities for the assessment of scientific evidence. *Law, Probability and Risk* **4**, 89–114.

Taroni F, Bozza S and Aitken CGG 2005a Decision analysis in forensic science. *Journal of Forensic Sciences* **50**, 894–905.

Taroni F, Bozza S, Bernard M and Champod C 2005c Value of information for DNA tests: a decision perspective. *Journal of Forensic Sciences*. Accepted for publication.

Taroni F, Bozza S and Biedermann A 2005b Two items of evidence, no putative source: an inference problem in forensic intelligence. *Journal of Forensic Sciences*. Accepted for publication.

Taroni F, Champod C and Margot P 1998 Forerunners of Bayesianism in early forensic science. *Jurimetrics Journal* **38**, 183–200.

Thagart P 1988 *Computational Philosophy of Science*. The MIT Press, Cambridge, MA.

Thagart P 2003 Why wasn't O.J. convicted? emotional coherence and legal inference. *Cognition and Emotion* **17**, 361–383.

Thompson WC 1995 Subjective interpretation, laboratory error and the value of DNA evidence: three case studies. *Genetica* **96**, 153–168.

Thompson WC and Ford S 1991 The meaning of a match: sources of ambiguity in the interpretation of DNA prints In *Forensic DNA Technology* (eds. Farley M and Harrington J), pp. 93–152. CRC Press, New York.

Thompson WC, Taroni F and Aitken CGG 2003 How the probability of a false positive affects the value of DNA evidence. *Journal of Forensic Sciences* **48**, 47–54.

Tikochinsky Y, Tishby NZ and Levine RD 1984 Consistent inference of probabilities for reproducible experiments. *Physical Review Letters* **52**, 1357–1360.

van Fraassen BC 1986 A demonstration of Jeffrey conditionalization rule. *Erkenntnis* **24**, 17–24.

van Fraassen BC 1989 *Laws and Symmetry*. Clarendon Press, Oxford.

Vicard P and Dawid AP 2003 Estimating mutation rates from paternity data *Atti del Convegno Modelli Complessi e Metodi Computazionali Intensivi per la Stima e la Previsione*, pp. 415–418. Università Cà Foscari, Venezia.

Weir BS 1996 *Genetic Data Analysis II*. Sinauer Associates, Sunderland, MA.

Weir BS 2000 Statistical analysis In *Encyclopedia of Forensic Sciences* (eds. Siegel JA, Saukko PJ and Knupfer GC), pp. 86–97. Academic Press, San Diego, CA.

Weir BS 2001 Forensics In *Handbook of Statistical Genetics* (eds. Balding DJ, Bishop M and Cannings C), pp. 721–739. John Wiley &Sons, Chichester.

Wellman MP 1990a Fundamental concepts of qualitative probabilistic networks. *Artificial Intelligence* **44**(3), 257–303.

Wellman MP 1990b Qualitative probabilistic networks for planning under uncertainty In *Readings in Uncertain Reasoning* (eds. Shafer G and Pearl J), pp. 711–722. Morgan Kaufmann Publishers, San Meteo, CA.

Wellman MP and Henrion M 1993 Explaining "explaining away". *IEEE Transactions on Pattern Analysis and Machine Intelligence* **15**(3), 287–292.

Whittaker J 1990 *Graphical Models in Applied Multivariate Statistics*. John Wiley & Sons, Chichester.

Wigmore JH 1913 The problem of proof. *Illinois Law Review* **8**(2), 77–103.

Wigmore JH 1937 *The Science of Judicial Proof: As Given by Logic, Psychology, and General Experience and Illustrated in Judicial Trials*, 3rd edn. Little, Brown and Company, Boston, MA.

Williams PM 1980 Bayesian conditionalisation and the principle of Minimum information. *British Journal for the Philosophy of Science* **31**, 131–144.

Williamson J 2004 *Bayesian Nets and Causality: Philosophical and Computational Foundations*. Clarendon Press, Oxford.

Wright S 1922 Coefficients of inbreeding and relationship. *American Naturalist* **56**, 330–338.

Author Index

Subject Index

Statistics in Practice

Human and Biological Sciences

Berger – Selection Bias and Covariate Imbalances in Randomized Clinical Trials
Brown and Prescott – Applied Mixed Models in Medicine
Chevret (Ed) – Statistical Methods for Dose-Finding Experiments
Ellenberg, Fleming and DeMets – Data Monitoring Committees in Clinical Trials: A Practical Perspective
Lawson, Browne and Vidal Rodeiro – Disease Mapping with WinBUGS and MLwiN
Lui – Statistical Estimation of Epidemiological Risk
*Marubini and Valsecchi – Analysing Survival Data from Clinical Trials and Observation Studies
Parmigiani – Modeling in Medical Decision Making: A Bayesian Approach
Senn – Cross-over Trials in Clinical Research, Second Edition
Senn – Statistical Issues in Drug Development
Spiegelhalter, Abrams and Myles – Bayesian Approaches to Clinical Trials and Health-Care Evaluation
Whitehead – Design and Analysis of Sequential Clinical Trials, Revised Second Edition
Whitehead – Meta-Analysis of Controlled Clinical Trials

Earth and Environmental Sciences

Buck, Cavanagh and Litton – Bayesian Approach to Interpreting Archaeological Data
Glasbey and Horgan – Image Analysis in the Biological Sciences
Helsel, Nondetects and Data Analysis: Statistics for Censored Environmental Data
McBride – Using Statistical Methods for Water Quality Management
Webster and Oliver – Geostatistics for Environmental Scientists

Industry, Commerce and Finance

Aitken and Taroni – Statistics and the Evaluation of Evidence for Forensic Scientists, Second Edition
Balding – Weight-of-evidence for Forensic DNA Profiles
Lehtonen and Pahkinen – Practical Methods for Design and Analysis of Complex Surveys, Second Edition

Ohser and Mücklich – Statistical Analysis of Microstructures in Materials Science
Taroni, Aitken, Garbolino and Biedermann – Bayesian Networks and Probabilistic Inference in Forensic Science

*Now available in paperback